Hypergeometric Functions on Domains of Positivity, Jack Polynomials, and Applications

Recent Titles in This Series

(Continued in the back of this publication)

CONTEMPORARY MATHEMATICS

138

Hypergeometric Functions on Domains of Positivity, Jack Polynomials, and Applications

Proceedings of an AMS Special Session
held March 22–23, 1991
in Tampa, Florida

Donald St. P. Richards
Editor

American Mathematical Society
Providence, Rhode Island

EDITORIAL BOARD

The AMS Special Session on Hypergeometric Functions on Domains of Positivity, Jack Polynomials, and Applications was held at the 865th Meeting of the American Mathematical Society in Tampa, Florida, on March 22–23, 1991.

1991 *Mathematics Subject Classification*. Primary 33C80.

Library of Congress Cataloging-in-Publication Data

Hypergeometric functions on domains of positivity, Jack polynomials, and applications: proceedings of an AMS special session held March 22–23, 1991 in Florida/Donald St. P. Richards, editor.
 p. cm.—(Contemporary mathematics, ISSN 0271-4132; 138)
 Contributions in English and French.
 "AMS Special Session on Hypergeometric Functions on Domains of Positivity, Jack Polynomials, and Applications held at the 865th Meeting of the American Mathematical Society..."—T. P. verso.
 ISBN 0-8218-5159-4 (alk. paper)
 1. Hypergeometric functions—Congresses. I. Richards, Donald St. P. II. AMS Special Session on Hypergeometric Functions on Domains of Positivity, Jack Polynomials, and Applications (1991: Tampa, Fla.) III. Series: Contemporary mathematics (American Mathematical Society); v. 138.
QA353.H9H97   1992
515′.55—dc20
                   92-26610
                   CIP

Contents

Preface

This volume is largely based on the lectures presented during a special session of the 865th meeting of the American Mathematical Society, convened in Tampa, Florida, during the period March 22 - 23, 1991. This special session, entitled "Hypergeometric functions on domains of positivity, Jack polynomials and applications," was centered on a branch of research initiated some forty years ago by Bochner. The initial impetus for Bochner's work came from questions in analytic number theory. It is remarkable that, since then, these hypergeometric functions have been found to be important in areas as diverse as combinatorics, harmonic analysis, molecular chemistry, multivariate statistics, partial differential equations, probability theory, representation theory and mathematical physics. In addition, the scope of these functions has been broadened considerably. While the initial investigations of these hypergeometric functions were carried out within the context of matrix spaces, the articles within this volume relate these functions to the study of domains of positivity and root systems.

At this stage, some brief, historical remarks are in order. Given a real, positive definite (symmetric) $n \times n$ matrix, Λ, with *integer* entries, let $r(\Lambda)$ denote the number of $k \times n$ matrices T, with integer matrices, and where $k \geq n$, such that $T'T = \Lambda$. A problem considered by Bochner (and others) is to investigate the asymptotic behavior of $r(\Lambda)$ as "$\Lambda \to \infty$." A natural approach to studying the asymptotic behavior of $r(\Lambda)$ is to study a generating function for $r(\Lambda)$. Thus if Z is also positive definite and $n \times n$, and $\text{etr}(Z) \equiv \exp(\text{tr } Z)$, define the theta function

$$\Theta(Z) = \sum_T \text{etr}(-\pi T Z T').$$

It is not difficult to derive Jacobi's formula

$$\Theta(Z) = (\det Z)^{k/2} \, \Theta(Z^{-1}),$$

hence

(1) $$\sum_\Lambda r(\Lambda) \, \text{etr}(-\pi \Lambda Z) = (\det Z)^{k/2} \sum_\Lambda r(\Lambda) \, \text{etr}(-\pi \Lambda Z^{-1}).$$

The formula (1) is, of course, a *modular relation*:

$$(2) \qquad \sum_M a_M \, \mathrm{etr}(-MZ) = (\det Z)^{-\delta} \sum_\Lambda b_\Lambda \, \mathrm{etr}(-\Lambda Z^{-1}),$$

where $\delta \in \mathbb{C}$ with $\mathrm{Re}(\delta)$ sufficiently large. To extend (2) to more general modular relations of the form

$$(3) \qquad \sum_M a_M f(M) = (\det Z)^{-\delta} \sum_\Lambda b_\Lambda g(\Lambda),$$

for functions f and g, Bochner defined the Bessel (or $_0F_1$ hypergeometric) function of matrix argument, by the integral equation

$$(4) \quad (\det Z)^{-\delta}\mathrm{etr}(-\Lambda Z^{-1}) = \int_{M>0} (\det M)^{\delta-(n+1)/2}\mathrm{etr}(-MZ)\,_0F_1(\delta;\Lambda M) \, dM.$$

Here $\{M > 0\}$ denotes the space of positive definite matrices, and dM is the corresponding Lebesgue measure. Of course, it must be verified that the function $_0F_1$ is well-defined by (4). On doing so, Bochner then proved that the modular relation (2) extends to (3), for certain classes of functions f, whenever g is the *Hankel transform* of f:

$$(5) \qquad g(\Lambda) = \int_{M>0} f(M) \,_0F_1(\delta;\Lambda M) \, (\det M)^{\delta-(n+1)/2} \, dM.$$

If we write $g = \mathcal{H}_\delta f$ whenever (5) holds, then a more general theorem of Bochner (*Ann. Math.*, 1951) is the following.

THEOREM. (Bochner, 1951) *Suppose that R and S are positive Borel measures on $\{\Lambda > 0\}$ such that*

$$\int_{\Lambda>0} \mathrm{etr}(-\Lambda Z) \, R(d\lambda) = (\det Z)^{-\delta} \int_{M>0} \mathrm{etr}(-MZ^{-1}) \, S(dM),$$

where $Z > 0$. If $g = \mathcal{H}_\delta f$ and f is completely monotone (the Laplace of a positive Borel measure on the cone of positive definite matrices) then

$$\int_{\Lambda>0} f(\Lambda)R(d\lambda) = (\det Z)^{-\delta} \int_{M>0} g(M)S(dM).$$

The proof of this result consists of writing the function f in the form

$$f(\Lambda) = \int_{\xi>0} \mathrm{etr}(-\Lambda\xi) \, \nu(d\xi),$$

where ν is a positive measure, and computing the Hankel transform of f by repeated interchanges of integration.

After Bochner's initial investigations, there was rapid development of the theory of hypergeometric functions of matrix argument and their more general relatives. In particular, Herz used the Laplace transform to define the general family of matrix argument functions, $_pF_q$; James and Constantine developed the zonal (spherical) polynomial series expansions for the class of $_pF_q$ hypergeometric functions, and defined the hypergeometric functions of two matrix arguments. The names and results of Gindikin, Jack, Macdonald, Muirhead, and others are now well-known in this area. The theory has found its most extensive applications in multivariate statistical analysis, notably in the area of "noncentral distribution theory."

As for the papers appearing here, I am pleased to note that they cover a broad range of applications of these hypergeometric functions. A common thread running through all the articles is the use of spherical functions, in the form of zonal or Jack polynomials or even as matrix entries for irreducible representations. The reader will find connections with mathematical physics, p-adic analysis, harmonic analysis, random walks, combinatorics, root systems, q-hypergeometric series, representation theory, random matrices, and operator theory. As important is the fact that they provide many references to the literature, and point towards open problems. Additional directions for future work can be found in the abstracts of talks which were presented at the special session but not included here because of the exigencies of time. In particular, I take all blame for the fact that applications to multivariate statistics, total positivity, and related topics will have to appear elsewhere.

Before closing, I do wish to draw the reader's attention to one type of open problem, with important implications for multivariate statistics and probability theory. Suppose that $_pF_q(X, Y)$ is a generalized hypergeometric function of two "matrix" arguments. In the classical matrix argument context, X and Y are real symmetric matrices; in the general Jack-polynomial context, $X = (x_1, \dots, x_n)$ and $Y = (y_1, \dots, y_n)$ are n-dimensional vectors. It is of interest to statisticians to determine the nonnegativity of the functions

$$\frac{\partial^2}{\partial x_1 \partial x_2} \log[V(X)_p F_q(X, Y)]$$

and

$$\frac{\partial^2}{\partial x_1 \partial y_1} \log[V(X)V(Y)\, _pF_q(X, Y)]$$

where $V(X)$ is an anti-symmetric function of $x_1, \dots, x_n$. In the case where we begin with hypergeometric functions defined on the cone of *Hermitian* positive definite matrices, in which case the Jack polynomials are the familiar Schur functions, much is known; in this case, it is no surprise that the appropriate choice for V is $V(X) = \prod_{1 \le i < j \le n}(x_i - x_j)$. An extensive study in the Hermitian case is currently in preparation (Chang, Peddada and Richards, 1992). But for no other situation do any results appear to be available. The reason for this paucity of results in the non-Hermitian cases is that only in the Hermitian case are explicit formulas for the general matrix argument $_pF_q$ functions available in terms of their classical counterparts (Gross and Richards, *Bull (N.S.) Amer. Math. Soc.*, 1991).

It is a great pleasure to acknowledge the support I received from the speakers at the special session. Bearing in mind that all speakers traveled at their own expense, the international flavor of the special session is more greatly appreciated. In particular, I note that the American, Canadian, Dutch, French, Jamaican and Russian schools were represented at the meeting. I thank the contributors for their patience during the processing of the volume. I also thank the referees for their prompt and accurate replies.

Finally, I thank Barbara Palmore who persevered, during difficult times, to convert all the contributions into a uniform format; John Stembridge who provided expert advice on $\mathcal{AMS}$-TEX and a template for other articles; Donna Harmon and her production staff for their extreme patience; the American Mathematical Society for providing me with the opportunity to convene the special session and produce this volume; Ken Gross and Ray Kunze who encouraged me to organize the session; and last, but by no means, least, Mercedes, Chandra and Suzanne Richards who allowed me to neglect them during the final stages of the production process.

Donald St. P. Richards

Charlottesville

May, 1992

Contemporary Mathematics
Volume **138**, 1992

Special Functions on Finite Upper Half Planes

JEFF ANGEL, NANCY CELNIKER, STEVE POULOS, AUDREY
TERRAS, CINDY TRIMBLE AND ELINOR VELASQUEZ

1. Introduction

The plan is to define some graphs which are finite analogs of the familiar Poincaré upper half plane. Then we will study some special functions living on these graphs – mostly eigenfunctions of the adjacency operator. Here we give a brief summary of our motivation. More details will be given later in the paper. See also the earlier papers [7] and [32].

1) There are many applications of the finite circle $\mathbb{Z}/n\mathbb{Z}$; that is, the quotient ring of integers modulo n. In particular, Fourier analysis on the finite circle is necessary for swift computation of approximations to functions on the usual circle $\mathbb{R}/\mathbb{Z}$. So why not study a finite analog H_q of the Poincaré upper half plane H with hopes of casting some light on the darker corners of H or H/Γ? Here q is normally a power of an odd prime p. In fact, we will find that there are finite counterparts for all the special functions on H which we considered in Terras [33]; e.g., spherical, Bessel functions and modular forms. Much of our work concerns eigenfunctions of combinatorial analogs of the Laplacian. Equivalently we study eigenfunctions of adjacency operators of upper half plane graphs. These operators may also be viewed as finite versions of the Radon transform (see Velasquez [34]) as well as finite convolution operators. The K-Bessel function analogs were considered by Velasquez [34]. The spherical functions theory presented here was given by Poulos [26]. The analogs of modular forms and other examples in Sections 2.3.1 and 2.3.2 below were developed by Trimble. Many authors have considered finite analogs of classical special functions. See for example Evans [10].

1991 *Mathematics Subject Classification.* Primary 33C80, 43A90, Secondary 05E30, 11F37.

Key words and phrases. Modular form, spherical function, finite field, highly regular graph, Ramanujan graph, Bessel function, Eisenstein series, Kloosterman sum.

This paper is in final form and no version of it will be submitted for publication elsewhere .

2) A congruence subgroup Γ of $\text{SL}(2, \mathbb{Z})$ acts on H by fractional linear transformations. A fundamental domain for Γ is a subset of D of H which can be identified with $\Gamma \backslash H$. Such a fundamental domain can be formed of a finite union of r copies of the fundamental domain D_1 for $\text{SL}(2, \mathbb{Z})$. One can produce a graph with r vertices by connecting those corresponding to neighboring copies of D_1. Such graphs can be obtained by the construction we give here. See Figures 1, 2 and 3. In the simplest cases we can also obtain graphs which lie on the regular solids. It is easy to see that our H_3 graph is the octahedron. H. Stark found that the 3 different graphs corresponding to the finite upper half plane H_5 can all be put on the dodecahedron. Angel found that one of the H_4 graphs is the icosahedron.

3) It is of interest in graph theory to obtain graphs with large expansion constants. This led Lubotsky, Phillips and Sarnak [22] to define Ramanujan graphs. See (1.8) below for the definition. Such graphs are good expanders and their Ihara zeta functions satisfy the Riemann hypothesis (see Sunada [31]). We have shown that many of our graphs are Ramanujan, but the general question is still open. There is another question from graph theory that our graphs solve. One wants to know (see Lubotsky [21]) if there are k-regular graphs X such that the girth g of X satisfies:

$$\frac{4}{3} \log_k(|X|) < g \le 2 \log_k(|X|).$$

We will see that an infinite number of our graphs have this property. Winnie Li [20] has also found graphs that satisfy this inequality.

4) Our work involves finite matrix groups such as $\text{SL}(2, \mathbb{F}_q)$ and Fourier analysis on these groups. Some references are Diaconis [9], Gelfand, Graev, and Piatetski-Shapiro [11], Piatetski-Shapiro [25], Harish-Chandra [12], and Velasquez [34]. In [32], we used the representation theory of the affine group over the field with p elements. We saw in [7] that this could be somewhat eliminated by studying the explicit eigenfunctions of the adjacency operator given by analogs of K-Bessel functions. Poulos [26] shows that we do not need to use representation theory to get our hands on the rest of the eigenfunctions. Instead we can use the collapsed adjacency operators defined by Bollobás [5, p. 158]. So this paper will not make use of any representation theory. However, we do not mean to say that we are not using the subject surreptitiously.

5) In asking whether the spectra of the adjacency operators of our graphs for the H_q satisfy the Ramanujan bound, we are asking to solve a problem with surprising connections with some old problems in number theory; e.g., the estimation of exponential sums, the number of points on curves over finite fields, the Weil conjectures on zeta functions of such curves. See Ireland and Rosen [15] and Schmidt [27]. R. Evans and H. Stark showed (see [32]) that one can use elementary results from Schmidt [27] to bound "most" of the eigenvalues of the adjacency operators of our graphs. And various computers have spewed out

reams of paper verifying that the rest of the eigenvalues, for H_q with $q < 130$ or so, satisfy the Ramanujan bound. But as someone just wrote in an email message: "the ULTIMATE AIM in pure mathematics is PROVING theorems." And so far the proof has eluded us for the conjecture that all the finite upper half plane graphs are Ramanujan. But the formula for the missing eigenvalues or spherical functions given by Soto-Andrade [35] looks promising in that it is an exponential sum that may be amenable to estimation.[1]

What is new about this paper? How does it differ from [7] and [32]? Well, the big difference is that here we consider arbitrary finite fields, not just those with an odd prime number of elements. When the finite field is $\mathbb{F}_q$, for $q = p^r$, for $r > 1$, the Galois group of $\mathbb{F}_q$ over $\mathbb{F}_p$ acts on the graphs, meaning that there are fewer nonisomorphic graphs. And some of the proofs of results in [7] require some modification when $r > 1$ or $p = 2$; e.g., the proof that our graphs are connected. And more computer studies had to be carried out. We sketch connections with graphs built from fundamental domains for congruence subgroups of the modular group (Figures 1-3). Section 2.1 is fairly easy to get from [7]. Section 2.2 has little intersection with [7] or [32]. We did not consider the finite analogs of spherical functions in the earlier papers – or collapsed adjacency matrices. Section 2.3 is also new – containing finite analogs of modular forms motivated by Harish-Chandra [2], a way to define the missing eigenfunction of the Laplacian when $p = 3$, and an extension of the theory to fields of characteristic 2.

The Poincaré upper half plane H is perhaps the simplest example of a continuous non-Euclidean geometry. It can also be considered as the simplest non-compact symmetric space. See Terras [33, Vol. I] for more information on this subject. Here we give only a brief summary to motivate what follows.

DEFINITION. The *Poincaré upper half plane* H is defined by

$$H = \{z = x + iy : x, y \in \mathbb{R}, y > 0\},$$

with arc length element ds defined by

$$ds^2 = y^{-2}(dx^2 + dy^2)$$

and Laplacian

$$\Delta = y^2 \left\{ \frac{\partial^2}{\partial x^2} + \frac{\partial^2}{\partial y^2} \right\}.$$

The special linear group $G = \mathrm{SL}(2, \mathbb{R})$ of 2×2 real matrices g with $\det g = 1$ acts on $z \in H$ by fractional linear transformation; i.e., if $g = \begin{pmatrix} a & b \\ c & d \end{pmatrix}$, then $gz = (az + b)/(cz + d)$. Moreover this group action leaves invariant both the

---

[1]N. Katz has recently proved the desired estimate to show that *all* the graphs are indeed Ramanujan.

arc length and Laplacian above. We can identify H with G/K, where K is the special orthogonal group

$$K = SO(2) = \{g \in G \mid {}^t g g = I\}.$$

We are interested in eigenfunctions of the Laplacian on H. This allows one to generalize Fourier analysis to H. There are 4 *basic types of eigenfunctions* of Δ on H.

1) The simplest eigenfunctions of Δ are the power functions defined for $s \in \mathbb{C}$, $z = x + iy \in H$ by $p_s(z) = y^s$.

2) The K-invariant eigenfunctions of Δ, or *spherical functions*, are defined by

$$h_s(z) = \int_{k \in K} p_s(kz)dk,$$

where dk is Haar measure on K normalized to give K measure 1. It can be shown that if $z = ke^{-r}i$, for $r > 0$, $k \in K$, then $h_s(z) = P_{-s}$ (cosh r), where P_s denotes the Legendre function. See Terras [33, Vol. I]. A good reference for the classical special functions such as Legendre's is Lebedev [19].

3) Let N be the subgroup of G consisting of matrices of the form $\begin{pmatrix} 1 & b \\ 0 & 1 \end{pmatrix}$. The N-invariant eigenfunctions of Δ are K-*Bessel functions*. They can be defined for $s \in \mathbb{C}$ with Re $s > 0$, $z \in H$, $a \in \mathbb{R}$ by:

$$k(s|z, a) = \int_{u \in \mathbb{R}} p_s\left(\frac{1}{z + u}\right) e^{2\pi i a u} du.$$

It is an exercise in Terras [33, Vol. I, pp. 136-7] to show that

$$k(s|z, \frac{a}{\pi}) = 2e^{2iax}\pi^{1/2}\Gamma(s)^{-1}|a|^{s-1/2}y^{1/2}K_{s-1/2}(2|ay|),$$

when $a \neq 0$, Re $s > 0$. Here $K_s(y)$ denotes the classical K-Bessel function and $\Gamma(s)$ is the gamma function (see Lebedev [19]).

4) Suppose that Γ is some discrete subgroup of G such as the modular group $\mathrm{SL}(2, \mathbb{Z})$. The Γ-invariant eigenfunctions of Δ are called *modular forms*, or, more precisely, Maass wave forms. They were first studied by H. Maass as analogs of the holomorphic Γ-invariant modular forms which have been much used in number theory and applied mathematics. One simple example of a Maass wave form is the *non-holomorphic Eisenstein series*, defined for $z \in H$, $s \in \mathbb{C}$, with Re $s > 1$, when $\Gamma = \mathrm{SL}(2, \mathbb{Z})$ by:

$$E_s(z) = \sum_{\gamma \in \Gamma_\infty \backslash \Gamma} p_s(\gamma z) = \frac{1}{2}p_s(z) \sum_{\mathrm{g.c.d}(c,d)=1} N(cz + d)^{-s}.$$

Here $Nz = z\bar{z}$, and Γ_∞ is the subgroup fixing the cusp at ∞; i.e., the upper triangular matrices. There are also cusp forms (which vanish at the cusp). The cusp forms are much more mysterious. See Hejhal [13] for computations of eigenvalues of Δ for some cusp forms on congruence subgroups. Gelfand, Graev, and Piatetski–Shapiro [11] show that dimensions of spaces of Maass wave forms are multiplicities of certain continuous series of representations of $G = \mathrm{SL}(2, \mathbb{R})$. They also show that the dimensions of spaces of holomorphic modular forms are multiplicities of discrete series of representations of G. An example of a holomorphic modular form for $\mathrm{SL}(2, \mathbb{Z})$ is the *holomorphic Eisenstein series* defined for $z \in H$ and $k = 4, 6, 8, \ldots$, by:

$$G_k(z) = \sum_{(m,n) \in \mathbb{Z}^2 - 0} (mz + n)^{-k}.$$

All the holomorphic modular forms for $\mathrm{SL}(2, \mathbb{Z})$ can be built from the Eisenstein series. See Terras [33, Vol. I, pp. 184-187]. Let us consider some examples of discrete subgroups of G.

DEFINITION. The *principal congruence subgroup* is defined for a positive integer N by

$$\Gamma(N) = \left\{ \gamma = \begin{pmatrix} a & b \\ c & d \end{pmatrix} \;\middle|\; a, b, c, d \in \mathbb{Z}, \gamma \equiv I (\mathrm{mod}\ N) \right\}.$$

Figures 1 and 2 give a few fundamental domains for these congruence groups. For more examples see Klein and Fricke [18].

FIGURE 1. FUNDAMENTAL DOMAIN D₁ FOR SL(2,Z).

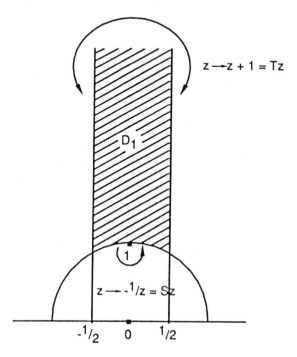

FIGURE 2. FUNDAMENTAL DOMAIN D$_3$ FOR Γ(3).

Each copy γD$_1$ is labelled with γ.

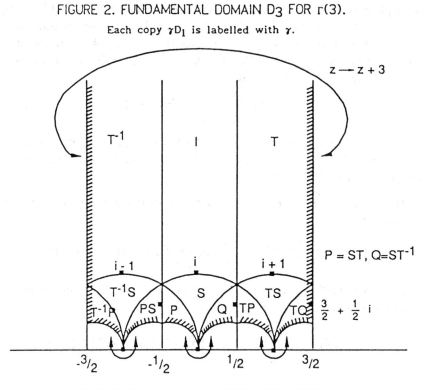

FIGURE 3. H$_3$ ≅ X$_3$(-1,1).   THE OCTAHEDRON.

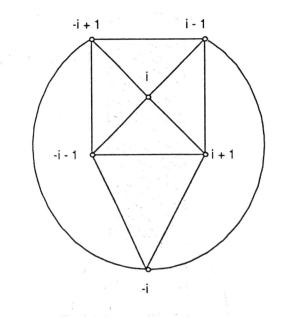

After this brief introduction to special functions on the Poincaré upper half plane, let us consider a finite analog in which the field of real numbers $\mathbb{R}$ is replaced by the finite field $\mathbb{F}_q$ with $q = p^r$ elements (p a prime). For most of this paper we will assume that p is an *odd* prime. Section 2.3.3 says a little about the case $p = 2$. Our construction of a finite upper half plane was motivated by Stark's p-adic upper half plane (see Stark [30]) which was designed to study modular forms for arithmetic groups coming from algebraic number fields. More details on the constructions can be found in Terras [32], Celniker et al. [7], Celniker [6], Poulos [26], and Velasquez [34].

Suppose that $\delta \in \mathbb{F}_q$ is a non-square; e.g., δ generates the multiplicative group $\mathbb{F}_q^* = \mathbb{F}_q - 0$.

DEFINITION. The *finite upper half plane* H_q is defined by

$$(1.1) \qquad H_q = \left\{ z = x + y\sqrt{\delta} \,\middle|\, x, y \in \mathbb{F}_q, y \neq 0 \right\}.$$

Note that here $\sqrt{\delta}$ plays the role of $i = \sqrt{-1}$ in the Poincaré upper half plane. Note also that H_q is contained in $\mathbb{F}_q(\sqrt{\delta})$, the field with q^2 elements, just as H is contained in $\mathbb{C} \cong \mathbb{R}(i)$.

NOTATION. For $z = x + y\sqrt{\delta} \in H_q$, let

$$(1.2) \qquad \begin{cases} y = \operatorname{Im} z = \text{imaginary part of } z, \\ x = \operatorname{Re} z = \text{real part of } z, \\ \bar{z} = x - y\sqrt{\delta} = \text{conjugate of } z, \\ \bar{z} = z^q = \text{the image of } z \text{ under the nontrivial element} \\ \qquad\qquad \text{of the Galois group of } \mathbb{F}_q(\sqrt{\delta})/\mathbb{F}_q, \\ Nz = z\bar{z} = \text{norm of } z, \\ \operatorname{Tr} z = z + \bar{z} = \text{trace of } z. \end{cases}$$

Define $\mathrm{GL}(2, \mathbb{F}_q)$ to be the group of all nonsingular 2×2 matrices with entries in $\mathbb{F}_q$. If $g = \begin{pmatrix} a & b \\ c & d \end{pmatrix} \in \mathrm{GL}(2, \mathbb{F}_q) = G$, define the *action* of g on $z \in H_q$ by

$$(1.3) \qquad gz = \frac{az + b}{cz + d} \in H_q.$$

Then it is not hard to see that we can identify H_q with $\mathrm{GL}(2, \mathbb{F}_q)/K$, where

$$(1.4) \qquad K = \{ g \in G \mid g\sqrt{\delta} = \sqrt{\delta} \}.$$

And one finds that

$$K = \left\{ \begin{pmatrix} a & b\delta \\ b & a \end{pmatrix} \,\middle|\, a, b \in \mathbb{F}_q \right\}.$$

The subgroup K of G is easily seen to be isomorphic to the (cyclic) multiplicative group of the field $\mathbb{F}_q(\sqrt{\delta}) \cong \mathbb{F}_{q^2}$, via the map

$$\begin{pmatrix} a & b\delta \\ b & a \end{pmatrix} \longrightarrow a + b\sqrt{\delta}.$$

Note that the "determinant-one" subgroup $\mathrm{SL}(2, \mathbb{F}_q)$ also acts transitively on H_q, as can be seen by a counting argument. Here we find $\mathrm{GL}(2, \mathbb{F}_q)$ more convenient.

Next we want to define a G-invariant graph associated to H_q. To do this, we use a *point-pair invariant;* i.e., a map $d : H_q \times H_q \to \mathbb{F}_q$ such that $d(gz, gw) = d(z, w)$, for all $z, w \in H_q$ and all $g \in G$.

DEFINITION. We define our point-pair invariant by

$$(1.5) \qquad\qquad d(z, w) = \frac{N(z - w)}{(\mathrm{Im}\ z)(\mathrm{Im}\ w)} \in \mathbb{F}_q.$$

This is not a distance as it has values in a finite field, but it is a good substitute for the Poincaré distance. Its definition was motivated by the p-adic point-pair invariant given by Stark in [30]. It is a straightforward computation to verify the following Lemma.

LEMMA 1. *The function $d(z, w)$ defined by (1.5) is a point-pair invariant.*

DEFINITION. The *graph* of H_q has vertices given by the elements $z \in H_q$. Fix an element $a \in \mathbb{F}_q$ with $a \neq 0$ and $a \neq 4\delta$. Connect two elements $z, w \in H_q$ by an edge if $d(z, w) = a$. We call this graph $X_q(\delta, a)$.

LEMMA 2. *The graph $X_q(\delta, a)$ is a $(q+1)$-regular graph, assuming that $a \neq 0$ or 4δ. The points $z = x + y\sqrt{\delta} \in H_q$ which are adjacent to $\sqrt{\delta}$ are those for which x, y solve the equation $x^2 = ay + \delta(y - 1)^2$.*

The proof of Lemma 2 is the same as that given in Celniker et al. [7] for $q = p$ prime. See Celniker [6] for the prime power case. Furthermore it was noticed by Stark that the graphs $X_q(\delta, a)$ and $X_q(\delta c^2, ac^2)$ are isomorphic for any nonzero $c \in \mathbb{F}_q$. For each q, we fix δ to be a generator of the multiplicative group $\mathbb{F}_q^*$. Then we get $q - 2$ graphs $X_q(\delta, a)$, for $a \in \mathbb{F}_q, a \neq 0, 4\delta$.

EXAMPLES.

1) $q = p = 3$.

There is only one graph for $p = 3$. It is $X_3(-1, 1)$. It is easily checked that this graph is the octahedron. See Figure 3. Note that you can get this graph from the fundamental domain D_3 for $\Gamma(3) \setminus H$ drawn in Figure 2. Let the vertices be the various copies of the fundamental domain D_1 for $\Gamma(1) \setminus H$ which make up D_3 (or just the images of i in these copies). Connect two vertices iff they are adjacent images of D_1. Recall that if $Tz = z + 1$, then TD_1 and $T^{-1}D_1$ are adjacent via the boundary identifications in D_3. It may be enlightening to note that $\Gamma/\Gamma(3) \cong \mathrm{SL}(2, \mathbb{F}_3)$.

2) $q = p = 5$.

It was noticed by H. Stark that the 3 graphs $X_5(3, a)$, for $a = -1, 1, 3$, lie on a dodecahedron. See Celniker et al. [7] for the pictures.

3) $q = 4$.

See the end of this paper where it is shown that one such graph is the icosahedron.

We are interested in the spectra of adjacency matrices of graphs. If X is a graph with vertices $v_1, \dots, v_n$ then the *adjacency matrix* $A = A(X)$ is defined to be the $n \times n$ matrix with (i, j) entry 1 if v_i is adjacent to v_j, and 0 otherwise. We will also need the *adjacency operator* A acting on functions $f : X \to \mathbb{C}$ by

$$Af(z) = \sum_{w \text{ adjacent to } z} f(w).$$

There are various standard facts about A. See Biggs [4] or Bollobás [5]. For example, the largest eigenvalue of the adjacency operator of our graph $X_q(\delta, a)$ is $q + 1$. The corresponding eigenfunction is constant. There are several ways to view this adjacency operator. In Section 2.2, we view it as a convolution operator on the finite affine group defined by (1.6). In Velasquez [34] it is viewed as a finite Radon transform.

Clearly we can identify our finite upper half plane with the *affine group* defined by

$$(1.6) \qquad \mathrm{Aff}(q) = \{ \begin{pmatrix} y & x \\ 0 & 1 \end{pmatrix} \mid x, y \in \mathbb{F}_q, y \neq 0 \}.$$

Poulos [26] shows that the set

$$(1.7) \qquad S_q(\delta, a) = \{ \begin{pmatrix} v & u \\ 0 & 1 \end{pmatrix} \mid u^2 = av + \delta(v - 1)^2 \}$$

is a symmetric set of generators for $\mathrm{Aff}(q)$, when $q = p$, an odd prime, and a is not 0 or 4δ. The proof requires some modification for $q = p^r$, with $r > 1$. Here a set S is "symmetric" if $g \in S$ implies $g^{-1} \in S$. Thus our graph $X = X_q(\delta, a)$ can be viewed as the *Cayley graph* attached to $G = \mathrm{Aff}(q)$ with generating set $S_q(\delta, a)$. That is, we connect vertices $z, w \in G$ iff $z = ws$ for some $s \in S$. Thus we can use the group structure to help in the analysis of the spectrum of the adjacency operator. And the sets $S_q(\delta, a)$ will be essential to the study of spherical functions on the finite upper half plane in Section 2. For they represent the K-double cosets of $G = \mathrm{GL}(2, \mathbb{F}_q)$, as was proved by Poulos [26]. See Proposition 3 below.

One interesting question about our graphs concerns whether they satisfy the following definition which was made in Lubotsky, Phillips and Sarnak [22].

DEFINITION. A k-regular graph X is *Ramanujan* if any eigenvalues λ of the adjacency operator such that $|\lambda| \neq k$ satisfies

$$(1.8) \qquad |\lambda| \leq 2\sqrt{k - 1}.$$

A Ramanujan graph has many nice properties. It is a good expander graph, meaning that it is good for rapid transmission of information at minimal cost. Moreover, its Ihara zeta function satisfies the Riemann hypothesis. See Bien [3], Fan Chung [8], Maria Klawe [17], Lubotsky [21], Sunada [31], and Terras [32] for more information. Ramanujan graphs are constructed by Lubotsky, Phillips and Sarnak [22], making use of the full solution of the Ramanujan conjecture on Fourier coefficients of modular forms of weight 2 in order to prove the graphs are indeed Ramanujan.

Celniker [6] shows that our graphs $X_q(\delta, a)$ are Ramanujan for $q \leq 121$ using various computers at UCSD. Moreover, it is shown in Celniker et al. [7] that all of what we call the 1-dimensional eigenvalues of the adjacency operator of $X_q(\delta, a)$ satisfy the Ramanujan bound (see the definition of 1-dimensional eigenvalues after (2.2) below). This result was found by R. Evans and H. Stark using different methods to reduce it to a result in Schmidt [27].

Celniker [6] found that equality can occur in (1.8) for our graphs $X_q(\delta, a)$ when q is an even power of a prime.

EXAMPLE. H_9 is viewed as the graph $X_9(1 + \sqrt{2}, 2 + 2\sqrt{2})$.

Consider $\mathbb{F}_9 = \mathbb{F}_3(\sqrt{2})$. A generator of the multiplicative group of $\mathbb{F}_9$ is $\xi = 1 + \sqrt{2}$. Let $\alpha = 2 + 2\sqrt{2}$. Consider the graph $X_9(\xi, \alpha)$. It is the Cayley graph of the affine group Aff(9) associated to the set $S_9(\xi, \alpha)$ of solutions of the equation

$$x^2 = \alpha y + \xi(y - 1)^2.$$

The 1-dimensional eigenvalues of the adjacency operator are:

$$0, -2, 0, -6, 0, -2, 0, 10.$$

The rest of the eigenvalues are from the following list, each repeated 8 times:

$$0, 0, 0, 0, -2, 4, 4, -6.$$

This graph is Ramanujan, but the boundary eigenvalue $-6 = -2\sqrt{q}$ does occur. And it occurs in the 1-dimensional part of the spectrum. The rest of the graphs for H_9 do not have this property that a boundary eigenvalue occurs.

The other graphs associated to H_9 have eigenvalues listed in Table 1 below which were computed by Celniker. In this table the 1-dimensional eigenvalues are in the first column and those in parentheses repeat twice. The second column has the rest of the eigenvalues and they repeat 8 times. The second column is computed as in [7] and [32]. Clearly, the same sorts of symmetries that we noted in [6, Table 1] are occurring here. See Proposition 4,5,6 of [7], which said that the eigenvalues for $X_q(\delta, \alpha)$ are $\pm$ those for $X_q(\delta, 4\delta - \alpha)$. Here we find that there is also a better symmetry. The eigenvalues of $X_9(1 + \sqrt{2}, a + b\sqrt{2})$ are the same as those of $X_9(1 + \sqrt{2}, b + a\sqrt{2})$. In fact, the graph $X_9(1 + \sqrt{2}, a + b\sqrt{2})$ is isomorphic to $X_9(1 + \sqrt{2}, b + a\sqrt{2})$, where a and b are switched. This is a result of the non-trivial Galois automorphism of $\mathbb{F}_9 = \mathbb{F}_3(\sqrt{2})$ over $\mathbb{F}_3$. Thus we should omit the graphs for $\alpha = 2 + \sqrt{2}, \sqrt{2}$, and $2\sqrt{2}$. See Celniker [6] for more extensive computations.

Another interesting quantity associated to a graph is the girth.

DEFINITION. The *girth* g of a graph X is the length of a minimal circuit or cycle.

Celniker [6] and Poulos [26] prove that the girth of $X_q(\delta, a)$ is either 3 or 4. When the girth is 3, we have examples of graphs of a type sought by graph theorists (see Bollobás [5] or Lubotsky [21]). Li [20] also produces examples of such graphs using abelian groups. The question was to give examples of k-regular graphs X such that

$$(1.9) \qquad \frac{4}{3}\log_k(|X|) < g \leq 2\log_k(|X|)$$

When $X_q(\delta, a)$ has girth 3, it satisfies (1.9). Celniker [6] proves that for any prime p the girth of $X_p(\delta, a)$ is 3 if a and $a - 3\delta$ are both squares. And she shows that the girth of $X_p(\delta, 2\delta)$ is 3 if $p \equiv 3 \pmod 4$. Poulos [26] gives another proof. However, it was pointed out to us by Lubotsky and Sarnak that our examples are not quite what was wanted since the degrees of our graphs grow with $|X|$.

Table 1. Eigenvalues of adjacency operators for $X_9(1 + \sqrt{2}, \alpha)$

α	1-dimensional eigenvalues	rest of spectrum
1	10, (-1.4142,1.4142,4), -2	-2 ,-2, -2, 4 ±4.4721, ±1.4142
2	10, (-2,-.8284,4.8284), 2	-3.8541, -3.6180, -2, -1.3820 -.8284, 2, 2.8541, 4.8284
$1+2\sqrt{2}$	10, (-4.8284,-2,.8284), 2	-4.8284, -3.8541, -2, .8284 1.3820, 2, 2.8541, 3.6180
$2+\sqrt{2}$	10, (-4.8284,-2,.8284), 2	-4.8284, -3.8541, -2, .8284 1.3820, 2, 2.8541, 3.6180
$2+2\sqrt{2}$	10, (-2,0,0), -6	-6, -2, 0, 0, 0, 0, 4, 4
$\sqrt{2}$	10, (-1.4142,1.4142,4), -2	-2, -2, -2, 4 ±4.4721, ±1.4142
$2\sqrt{2}$	10, (-2,-.8284,4.8284), 2	-3.8541, -3.6180, -2, -1.3820 -.8284, 2, 2.8541, 4.8284

2. Special Functions on H_q

In this section we consider analogs of some of the well-known special functions for the Poincaré upper half plane mentioned in the introduction. In Section 2.1, we consider counterparts of the power function $p_s(z)$ as well as the K-Bessel function. In Section 2.2, spherical functions are discussed. Finally Section 2.3 concerns analogs of modular forms and some miscellaneous other functions, as well as a sketch of the theory for fields of characteristic 2.

2.1. Power Functions and K-Bessel Functions.

When δ is a fixed non-square in $\mathbb{F}_q$ and a is any element of $\mathbb{F}_q$ except 0 and 4δ, we defined graphs $X_q(\delta, a)$ representing H_q before Lemma 2 of Section 1. In this section we study certain eigenfunctions of the adjacency operators A_a of $X_q(\delta, a)$ for all $a \neq 0$ or 4δ. Equivalently we seek eigenfunctions of all the Laplacians $\Delta = A_a - (q+1)I$ for H_q. Note that, as for the non-Euclidean Laplacian on the Poincaré upper half plane, the adjacency operators are invariant under left translation by elements of the general linear group. This means that for each $g \in G = \mathrm{GL}(2, \mathbb{F}_q)$ and for each $f : H_q \longrightarrow \mathbb{C}$, defining $f^g(z) = f(gz)$, for $z \in H_q$, we have

$$A(f^g) = (Af)^g.$$

The simplest eigenfunctions of the adjacency operators A_a for the graphs $X_q(\delta, a)$ are the *power functions* defined for any multiplicative character χ of $\mathbb{F}_q^*$ by:

$$(2.1) \qquad\qquad p_\chi(z) = \chi(\mathrm{Im}\ z).$$

It is easy to see that p_χ is an eigenfunction of the adjacency operators for all of our graphs; i.e., for all $a \neq 0, 4\delta$, $A p_\chi = R_\chi p_\chi$, where

$$(2.2) \qquad\qquad R_\chi = \sum_{w \in S_q(\delta, a)} \chi(\mathrm{Im}\ w).$$

Here the set $S_q(\delta, a)$ is defined in (1.7). It is a symmetric set of generators of the affine group $\mathrm{Aff}(q)$, for $a \neq 0$ or 4δ. We call the eigenvalues R_χ in (2.2) *one-dimensional* because they come from one-dimensional representations of $\mathrm{Aff}(q)$. See Terras [32]. Note that if χ is the trivial representation, i.e. $\chi(y) = 1$ for all $y \in \mathbb{F}_q^*$, then $R_\chi = q + 1$ and p_χ is constant.

PROPOSITION 1. *The one-dimensional eigenvalues R_χ, satisfy the Ramanujan bound:*

$$(2.3) \qquad\qquad |R_\chi| \leq 2\sqrt{q},$$

for χ non-trivial.

The proof is the same as in Terras [32] using arguments of R. Evans and H. Stark to reduce the problem to a theorem in Schmidt [27].

We can use the power functions to build up other eigenfunctions of the adjacency operator, just as we did for the Poincaré upper half plane in Terras [33]. In this section we consider analogs of K-Bessel functions. We use the uppercase K to denote Kloosterman sums given in (2.7). These are related by Proposition 2 to the lower case animal defined by (2.6). These k-Bessel functions are invariant with respect to the unipotent (or nilpotent) subgroup N of G defined by

$$(2.4) \qquad N = \left\{ \begin{pmatrix} 1 & x \\ 0 & 1 \end{pmatrix} \,\Big|\, x \in \mathbb{F}_q \right\}$$

Note that the group N is isomorphic to the additive group of $\mathbb{F}_q$. Most of this section comes from Celniker et al. [7].

The real K-Bessel functions are well known to number theorists, since they appear in Fourier expansions of modular forms such as the Eisenstein series although they are not always named (see Selberg [28, pp. 367-370, 521-545] and Terras [33, Vol. I]). There are applications to the Kronecker limit formula and the theory of complex multiplication, for example. The finite analog or Kloosterman sum appears in Fourier coefficients of modular forms as well. See Selberg [28, pp. 506-520].

Now we can define what we mean by a k-Bessel function, imitating the definition in Terras [33, p. 136].

DEFINITION. A k-*Bessel function* $f : H_q \to \mathbb{C}$ is an eigenfunction of all the adjacency operators A of $X_q(\delta, a)$ for $a \neq 0, 4\delta$, such that $f(z)$ transforms by N according to the additive character ψ of $\mathbb{F}_q$; i.e.,

$$(2.5) \qquad Af = \lambda f, \quad f(z + u) = \psi(u) f(z) \ \forall \ z \in H_q, \ u \in \mathbb{F}_q \ .$$

It is easy to construct such a k-Bessel function making use of the power function. Define the k-Bessel function $k(z|\chi, \psi)$, for $z \in H_q, \chi$ a multiplicative character and ψ an additive character of $\mathbb{F}_q$, by

$$(2.6) \qquad k(z|\chi, \psi) = \sum_{u \in \mathbb{F}_q} \chi(\mathrm{Im}(\frac{-1}{z+u})) \psi(u).$$

It is not hard to see that as a function of $z, k(z|\chi, \psi)$ satisfies (2.5) with the eigenvalue $\lambda = R_\chi$, the same eigenvalue as the power function p_χ. Here you just need to use the fact that A is a left G-invariant operator. For this implies that

$$Ak(g\sqrt{\delta}|\chi, \psi) = \sum_{u \in \mathbb{F}_q, s \in S} p_\chi \left\{ \begin{pmatrix} 0 & -1 \\ 1 & 0 \end{pmatrix} \begin{pmatrix} 1 & u \\ 0 & 1 \end{pmatrix} gs\sqrt{\delta} \right\} \psi(u),$$

with the sum over $u \in \mathbb{F}_q$ and $s \in S = S_q(\delta, a)$. Then recall that p_χ is an eigenfunction of A. When χ is a nontrivial multiplicative character, it is true that $k(z|\chi, \psi)$ is not the 0-function. See Celniker et al. [7, Lemma 3].

The functions $k(z|\chi, \psi)$ and $k(z|\chi, \psi')$ are orthogonal if ψ and ψ' are distinct additive characters of $\mathbb{F}_q$. Moreover $k(z|\chi, \psi)$ is orthogonal to the power function $p_{\chi'}$ if ψ is any nontrivial additive character and if χ, χ' are any two multiplicative characters of $\mathbb{F}_q^*$. See Celniker et al. [7, Prop. 3]. As a corollary, we see that for χ nontrivial, the one-dimensional eigenvalues R_χ of A must occur with multiplicity $\geq q$. All these eigenvalues satisfy the Ramanujan bound. The rest of the eigenvalues except $q+1$ have multiplicity $\geq q-1$. See Celniker et al. [7], Terras [32], and Velasquez [34] for the details.

Just as on the real Poincaré upper half plane, there is another formula for a Bessel function. In this case, it is also called a Kloosterman sum.

DEFINITION. Given an additive character ψ of $\mathbb{F}_q$ and a multiplicative character χ of $\mathbb{F}_q^*$, define the *Kloosterman sum* for $a, b \in \mathbb{F}_q$ by:

$$(2.7) \qquad K_\psi(\chi|a,b) = \sum_{t \in \mathbb{F}_q^*} \chi(t)\psi(at + bt^{-1}).$$

This should perhaps be called a generalized Kloosterman sum. Such sums have been much studied by number theorists, since they occur in Fourier coefficients of modular forms. Thus bounds for them were wanted in connection with the Ramanujan conjectures. See Selberg [28, pp. 506-520]. Recently there has been a certain amount of Kloostermania associated to them. See Huxley [14] and Iwaniec [16]. Gelfand, Graev, and Piatetski-Shapiro [11, p. 160] construct p-adic analogs of this function as matrix entries of continuous series representations of $\mathrm{SL}(2, \mathbb{Q}_p)$. In a footnote, they also construct discrete series representations of $\mathrm{SL}(2, \mathbb{F}_q)$ with a similar sort of function, which is a sum over $t \in \mathbb{F}_{q^2}$ such that $Nt = 1$ (loc. cit., p. 185).

One obtains the relation between the k- and K-Bessel functions in the same way as in the continuous case (see Terras [33, Vol. I, pp. 136-71]). For this we need an analog of the Γ-function known as a Gauss sum, a sum which is even more familiar to number theorists.

DEFINITION. The *Gauss sum* for ψ an additive character and χ a multiplicative character of $\mathbb{F}_q$ is defined by

$$(2.8) \qquad \Gamma(\chi, \psi) = \sum_{x \in \mathbb{F}_q^*} \chi(x)\chi(x).$$

The Gauss sums were used by Gauss to give one of his many proofs of the quadratic reciprocity law. See Berndt and Evans [2] and Ireland and Rosen [15, pp. 146, 162] for more information about them. For example,

$$(2.9) \qquad |\Gamma(\chi, \psi)| = \sqrt{q}.$$

It is not news that many classical special functions have counterparts over finite fields. See Evans [10] for some other examples.

The formula relating the k and K-Bessel functions for H_q is given in the following Proposition.

PROPOSITION 2. *Suppose that ψ is an additive character of $\mathbb{F}_q$ while χ is a multiplicative character. Given the definitions (2.6)-(2.8), we have*

$$(2.10) \qquad \Gamma(\chi, \psi)k(z|\chi, \psi) = g\chi(y)\psi(-x)K(\chi'| - \delta y^2, -1/4),$$

where

$$g = \sum_{w \in \mathbb{F}_q} \psi(w^2)$$

is a Gauss sum, and $\chi'(y) = \lambda(y)\chi(y)$ *where*

$$\lambda(y) = \begin{cases} 1, & \text{if } y \text{ is a square in } \mathbb{F}_q, \\ -1, & \text{otherwise.} \end{cases}$$

When $q = p$ is a prime, $\lambda(y)$ is the Legendre symbol. The proof goes through as in Celniker et al. [7, Prop. 7].

2.2. Spherical Functions.

Laplace and Legendre introduced spherical harmonics, that is, eigenfunctions of the Laplacian on the sphere, in the 1780's. See any undergraduate Fourier Series and Boundary Value Problems course for a discussion or Terras [33, Vol. I, Chapter 2]. The analog for the Poincaré upper half plane is the Legendre function mentioned in Section 1. The theory of spherical functions for continuous Lie groups has been carried out by many people in this century; e.g., Cartan, Weyl, Gelfand, and Harish-Chandra. Cartan noted the various characterizations of spherical functions on compact symmetric spaces G/K in 1929. He saw these functions as matrix entries of representations π of G with $\pi(K)$ having a fixed vector. Gelfand studied the convolution algebra of K bi-invariant functions on G. The main point is that this is a commutative algebra if G/K is a symmetric space (i.e., a Riemannian manifold with a geodesic-reversing isometry at each point). See Terras [33, Vol. II, Section 4.2.3]. Gelfand theory emphasizes the convolution integral operators rather than the Laplacian. The analog for p-adic groups exists also (see Macdonald [23]). Selberg made use of the theory of spherical functions in deriving the Selberg trace formula.

The translation of this theory to finite groups does not always go as expected. In the finite abelian case, one obtains Krawtchouk polynomials which are of interest in coding theory (see Jessie MacWilliams and Sloane [24]). See the discussion of $\mathbb{F}_2^3$ in Example 1 later in this section. Some references for the finite spherical functions are: Bannai and Ito [1], Diaconis [9], Lubotsky [21], Poulos [26], Stanton [29], and Velasquez [34]. We could view spherical functions in the language of representation theory or that of association schemes. Instead we will stick to a graph theoretic approach, using collapsed adjacency matrices discussed below and in Bollobás [5, p. 158]. This approach is due to Poulos [26].

DEFINITION. A (zonal) *spherical function* $f : H_q \to \mathbb{C}$ is a K-invariant eigenfunction of all the Laplacians for the graphs we have attached to H_q such that $f(I) = 1$, where K is defined by (1.4).

We can identify the finite upper half plane H_q with G/K, where $G = \text{GL}(2, \mathbb{F}_q)$ and K is, as in (1.4), the subgroup fixing $\sqrt{\delta}$. It is standard to note that there are many equivalent definitions of spherical functions. One can view them as eigenfunctions of K-invariant differential operators or K-invariant integral operators. See Terras [33, Vol. I, p. 69]. In the finite case, however, the differential and integral operators approaches coincide. In the discussion that follows we shall identify functions on H_q with functions on G by identifying $f(g)$ with $f(g\sqrt{\delta})$ for $g \in G = \text{GL}(2, \mathbb{F}_q)$. We will also often identify functions on G with functions on G/K or $K \setminus G/K$.

In particular, suppose that $f : G/K \to \mathbb{C}$ is a K-invariant eigenfunction of the adjacency operator, then it is also an eigenfunction of the *mean value operator*

$$(2.11) \qquad M_y f(x) = |K|^{-1} \sum_{k \in K} f(xky),$$

for each $x, y \in G$. For M_y is a constant multiple of the adjacency operator for the graph $X_q(\delta, a)$ if y is in the generating set $S_q(\delta, a)$ defined by (1.7). Suppose then that $M_y f = \lambda_y f$ for some $\lambda_y \in \mathbb{C}$. Then $f(I) = 1$ implies that the eigenvalue $\lambda_y = f(y)$. To see this, write out what it means to say $M_y f(x) = \lambda_y f(x)$ at $x =$ the identity, assuming that f is K-invariant. Thus we see that the eigenvalues of the mean value operators are the eigenfunctions. Another way to say this is to note that a K-bi-invariant function f on G such that $f(I) = 1$ is spherical iff

$$(2.12) \qquad \sum_{k \in K} f(xky) = |K| f(x) f(y),$$

for all $x, y \in G$.

One important thing about G and K is that (G, K) is a *Gelfand pair*. This means that the convolution algebra of all K-bi-invariant functions $f : G \to \mathbb{C}$ is commutative. There are other equivalent definitions of Gelfand pairs. It can be shown that it is equivalent to say that the decomposition of the regular representation of G on $L(G/K)$ is multiplicity free. See Diaconis [9, p. 54], Bannai and Ito [1], or Stanton [29]. You can use this result to derive a formula for a spherical function as an average over K of left translates of a character of an irreducible representation of G which appears in the regular representation of G on $L^2(G/K)$; i.e., the induced representation $\text{Ind}_K^G 1$. See formula (2.13) below for a special case.

We write A_a for the adjacency operator associated to the generating set $S_a = S_q(\delta, a)$, for $a \in \mathbb{F}_q$, defined in (1.7). For what follows, we need to know that the S_a represent the K-orbits in H_q or the K-double cosets of G. This is given in the follow proposition which is proved in Poulos [26, Section 2.1].

PROPOSITION 3.

1) *Every set $S_a = S_q(\delta, a) = K g_a$, for some $g_a \in S_q(\delta, a)$. Here $S_q(\delta, a)$ is defined by (1.7).*

2) *The K-orbits in H_q are the sets S_a, for $a \in \mathbb{F}_q$.*

You also need to recall that the S_a are invariant under inversion; i.e., that $g \in S_a$ implies that $g^{-1} \in S_a$.

DEFINITION. The *adjacency operator* A_a is defined by

$$A_a f(x) = \sum_{d(z,w)=a} f(w).$$

Note that $A_0 = I$ and $A_{4\delta} f(z) = f(\bar{z})$. The other A_a are adjacency operators of $(q+1)$-regular graphs. Identify a function $f(z)$ on H_q with a function on the group $G = \mathrm{GL}(2, \mathbb{F}_q)$ by writing $f(g) = f(g\sqrt{\delta})$, for $g \in G$. Then we can define *convolution* of functions f, g, on H_q by:

$$(f * g)(u) = \sum_{v \in G} f(v) g(uv^{-1}).$$

It follows that the adjacency operator A_a is actually convolution with the delta function for the set S_a; i.e., $A_a f = \delta_{S_a} * f$, where $\delta_S(u) = 1$ for $u \in S$ and 0 otherwise.

PROPOSITION 4. *The adjacency operators A_i, $i \in \mathbb{F}_q$, form a commutative set of operators.*

PROOF. We need to show that $S_a S_b = S_b S_a$, where

$$S_a S_b = \{xy \mid x \in S_a, y \in S_b\}.$$

Pick $s_a \in S_a$, $s_b \in S_b$. The product $s_a s_b = s_c \in S_c$, for some $c \in \mathbb{F}_q$. We know that s_c^{-1} is also in S_c. Thus $s_c^{-1} = k s_c$, for some $k \in K$. Then $(s_a s_b)^{-1} = k s_c$. It follows that the element $(k^{-1} s_b^{-1})$, of S_b, times the element s_a^{-1}, of S_a, equals the original element, s_c, of S_c. Thus summing over $S_a S_b$ is the same as summing over $S_b S_a$, proving the theorem.

COROLLARY. *There is a common basis of orthogonal eigenfunctions for the adjacency operators $A_i, i \in \mathbb{F}_q$. They are orthogonal with respect to the standard inner product*

$$(f, g) = \sum_{z \in H_q} f(z)\overline{g(z)}.$$

This means that the spherical functions do indeed exist. For they were defined to be common eigenfunctions of the adjacency operators for all of the graphs attached to H_q. The adjacency operators could be studied using algebraic combinatorics as in Bannai and Ito [1]. The adjacency relations form a symmetric association scheme. But we do not have a $P-$ or $Q-$ polynomial association scheme. Our space is not 2-point homogeneous. Nevertheless the space has many of the properties obtained by Bannai and Ito [1] and Stanton [29].

We want to imitate the theory of the Poincaré upper half plane as described, for example, in Terras [33, Vol. I]. In that case one has $G = \mathrm{SL}(2, \mathbb{R}) = KAK$, where $K = SO(2)$ and $A = $ diagonal matrices in G. This does not happen for $\mathrm{GL}(2, \mathbb{F}_q)$, since the sets S_a need not be represented by diagonal matrices. And in the Poincaré upper half plane, one has Harish-Chandra's integral formula for spherical functions, which writes *any* spherical function as an integral over K of a power function (the formula we gave in Section 1). See Terras [33, Vol. II, p. 70]. This result does produce some spherical functions in the finite case - but *only* those corresponding to 1-dimensional eigenvalues and not arbitrary spherical functions. See (2.13) below.

Poulos [26] shows that our graphs $X_q(\delta, a)$, for $a \neq 0, 4\delta$, are not only regular, but also what Bollobás [5, p. 158] calls "highly regular." This allows us to collapse the adjacency matrix to a $q \times q$ matrix with the same minimal polynomial.

DEFINITION. A connected graph $X(V, E)$ is *highly regular* with collapsed adjacency matrix $C = (c_{ij})$ iff for every vertex $v \in V$, there is a partition of V into sets V_i, $i = 1, \ldots, n$ with $V_1 = \{v\}$, such that each vertex $y \in V_i$ is adjacent to exactly c_{ij} vertices in V_j.

THEOREM. (Poulos [26])

1) *For $a \neq 0, 4\delta$, the graph $X_q(\delta, a)$ is highly regular.*

2) *We can partition H_q as a union of sets $S_i = S_q(\delta, i)$, $i \in \mathbb{F}_q$. Then the collapsed adjacency operator $C = (c_{ij})$ associated to the graph $X_q(\delta, a)$ has (i, j) entry c_{ij} equal to the number of vertices $y \in S_j$ to which each vertex $v \in S_i$ is adjacent in $X_q(\delta, a)$. This number is independent of the choice of v in S_i.*

3) *When $(i, j) \neq (0, a)$ or $(4\delta, 4\delta - a)$, we have $c_{ij} \leq 2$.*

See Poulos [26] for the proof of this result.

Before giving an example of Theorem 1 from our theory of finite upper half planes, let us consider an example which is more well-known. See Stanton [29] or Velasquez [34] for more general examples of this sort. There are applications in coding theory (see MacWilliams and Sloane [24]).

EXAMPLE 1. The cube or $\mathbb{F}_2^3$.

More details on this example as well as the general case of $(\mathbb{Z}/n\mathbb{Z})^k$ can be found in Velasquez [34].

We can view $\mathbb{F}_2^3$ as G/K, with G the semi-direct product of $\mathbb{F}_2^3$ with the symmetric group S_3. Then $K = S_3$. Define the Hamming distance between $x, y \in \mathbb{F}_2^3$ by $d(x, y) =$ the number of coordinates for which $x_i \neq y_i$. This is G-invariant. Partition the vertices using the sets $S_i = \{x \mid d(x, 0) = i\}$, $i = 0, 1, 2, 3$. Define the adjacency operator by

$$Af(x) = \sum_{d(x,y)=1} f(y).$$

Then the collapsed adjacency matrix is

$$C = \begin{pmatrix} 0 & 3 & 0 & 0 \\ 1 & 0 & 2 & 0 \\ 0 & 2 & 0 & 1 \\ 0 & 0 & 3 & 0 \end{pmatrix}.$$

The K-invariant eigenfunctions of A are the eigenfunctions of C. They satisfy a 3-term recursion:

$$(3 - r)f(r + 1) + rf(r - 1) = \lambda f(r), \quad r = 0, 1, 2, 3.$$

One finds that the solutions are expressed in terms of the *Krawtchouk polynomials*

$$K_k(x; n) = \sum_{j=0}^{k} (-1)^j \binom{x}{j} \binom{n - x}{k - j}$$

where $\binom{a}{b}$ stands for the binomial coefficient. In particular, the eigenvalue is $\lambda = 3 - 2k = K_1(k; 3)$, and the corresponding eigenfunction is $f_k(r) = K_k(r; 3)$.

EXAMPLE 2. H_3.

In this case, $\delta = -1$ and we write $i = \sqrt{-1}$. There is only one 4-regular graph, the octahedron of Figure 3. The K orbits in H_3 are:

$$S_0 = \{i\}, \quad S_1 = \{\pm i \pm 1\}, \quad S_{-1} = \{-i\}.$$

The spherical functions are determined by their values on the S_i. And the collapsed adjacency matrix is easily seen to be

$$\begin{pmatrix} 0 & 4 & 0 \\ 1 & 2 & 1 \\ 0 & 4 & 0 \end{pmatrix}.$$

The eigenvalues are $0, 4, -2$.
The eigenvalues 0 and 4 are 1-dimensional, but -2 is not.
The eigenvectors are $\{1, 0, -1\}$, $\{1, 1, 1\}$, and $\{1, -1/2, 1\}$.
These eigenfunctions give the spherical functions for the octahedron H_3.

The collapsed adjacency matrix can also be interpreted as the adjacency matrix of a multigraph; i.e., a collection of vertices connected by $0, 1$, or more directed edges.

Since the spherical function f is constant on S_i, we write $f(S_i) = f(i)$. Then $Af = \lambda f$ iff

$$4f(1) = \lambda f(0), \quad f(0) + 2f(1) + f(-1) = \lambda f(1), \quad 4f(1) = \lambda f(-1).$$

In this case we are looking at a 3-term recursion for the spherical functions. However, as soon as q is larger than 3, this fails (except possibly for the $a = 2\delta$ graph), since Poulos [26] proves that $z \in H_q$ with $z \neq \pm\sqrt{\delta}$ is adjacent to at most 2 elements of any S_i. This means that if you arrange the collapsed adjacency matrix so that the first and last rows correspond to $\sqrt{\delta}$ and $-\sqrt{\delta}$, respectively, then the middle rows have entries that are ≤ 2.

EXAMPLE 3.. H_5.

In what follows we take $\delta = 2$ and list the sets S_i in the following order to take advantage of symmetry:

$$S_0, S_{-4\delta=2}, \quad S_{-8\delta=4}, \quad S_{-12\delta=1}, \quad S_{-16\delta=3=4\delta} \ .$$

The collapsed adjacency matrix for $X_5(2,1)$ is

$$\begin{pmatrix} 0 & 0 & 0 & 6 & 0 \\ 0 & 1 & 2 & 2 & 1 \\ 0 & 2 & 2 & 2 & 0 \\ 1 & 2 & 2 & 1 & 0 \\ 0 & 6 & 0 & 0 & 0 \end{pmatrix} \ .$$

The eigenvalues are: $6, -2, -3, 2, 1$.

The collapsed adjacency matrix for $X_5(2,2)$ is

$$\begin{pmatrix} 0 & 6 & 0 & 0 & 0 \\ 1 & 2 & 2 & 1 & 0 \\ 0 & 2 & 2 & 2 & 0 \\ 0 & 1 & 2 & 2 & 1 \\ 0 & 0 & 0 & 6 & 0 \end{pmatrix} \ .$$

The eigenvalues are: $6, 3, 1, -2, -2$.

The collapsed adjacency matrix for $X_5(2,4)$ is

$$\begin{pmatrix} 0 & 0 & 6 & 0 & 0 \\ 0 & 2 & 2 & 2 & 0 \\ 1 & 2 & 0 & 2 & 1 \\ 0 & 2 & 2 & 2 & 0 \\ 0 & 0 & 6 & 0 & 0 \end{pmatrix} \ .$$

The eigenvalues are: $6, -4, 2, 0, 0$.

The common eigenfunctions for these 3 collapsed adjacency matrices are:

$$\begin{pmatrix} 1 \\ 1 \\ 1 \\ 1 \\ 1 \end{pmatrix}, \ \begin{pmatrix} 6 \\ 1 \\ -4 \\ 1 \\ 6 \end{pmatrix}, \ \begin{pmatrix} 6 \\ -2 \\ 2 \\ -2 \\ 6 \end{pmatrix}, \ \begin{pmatrix} 6 \\ 3 \\ 0 \\ -3 \\ -6 \end{pmatrix}, \ \begin{pmatrix} 1 \\ -2 \\ 0 \\ 2 \\ -1 \end{pmatrix} \ .$$

The middle 3 vectors should be multiplied by $\frac{1}{6}$ to make them correspond to spherical functions.

Harish-Chandra's integral formula for spherical functions on the real Poincaré upper half plane has an analog in the finite case for χ a character of the multiplicative group $\mathbb{F}_q^*$:

$$(2.13) \qquad\qquad h_\chi(z) = |K|^{-1} \sum_{k \in K} p_\chi(kz).$$

Unfortunately this gives only the spherical functions corresponding to the 1-dimensional eigenvalues. These can also be viewed as the spherical functions corresponding to what in the real case would be called continuous series representations of G. Since we already know that the 1-dimensional eigenvalues satisfy the Ramanujan bound, we are not too excited by (2.13).

What are the non-1-dimensional spherical functions and how can we bound them or equivalently their eigenvalues? We have a similar result in general with p_χ replaced by the character of a higher dimensional representation of G found by decomposing $\mathrm{Ind}_K^G 1$. See Diaconis [9], Poulos [26], Velasquez [34]. Moreover Soto-Andrade [35] has rewritten this formula for the character of a discrete series representation of $\mathrm{GL}(2, \mathbb{F}_q)$ corresponding to a non-trivial multiplicative character ω of the group U of elements z of $\mathbb{F}_q(\sqrt{\delta})^*$ such that $Nz = z\bar{z} = 1$. Define ϵ to be the character of $\mathbb{F}_q^*$ which is 1 on squares, -1 on non-squares, and 0 on 0. Then Soto-Andrade's formula says that the spherical function h_ω at $z \in S_q(\delta, a)$, for $a \neq 0$ or 4δ, is:

$$(q + 1)h_\omega(z) = \sum_{Nz=1} \epsilon(c + 2 \ \mathrm{Re} \ z)\omega(z),$$

where $c = (a/\delta) - 2$. One would hope that there is some way to estimate this.[2] The formula is valid when it is not identically 0.

2.3. Miscellaneous Remarks.

2.3.1. Modular Forms.

The only reference that we know about for finite analogs of modular forms is Harish-Chandra [12]. This section is a collaboration of Trimble and Terras.

We know that if χ is a multiplicative character of $\mathbb{F}_q^*$, then according to (2.13) we have a spherical function given by:

$$h_\chi(z) = |K|^{-1} \sum_{k \in K} \chi(\mathrm{Im}(kz)).$$

Now $k = \begin{pmatrix} a & b\delta \\ b & a \end{pmatrix}$, $\det k = N(a + b\sqrt{\delta})$, and

$$\mathrm{Im}(kz) = \frac{\det k \ \mathrm{Im} \ z}{N(bz + a)} = \mathrm{Im} \ z \ \ N\left\{\frac{a + b\sqrt{\delta}}{a + bz}\right\}.$$

Thus we can rewrite our spherical function as:

$$h_\chi(z) = |K|^{-1}\chi(\mathrm{Im} \ z) \sum_{a+b\sqrt{\delta}} \chi \circ N\left\{\frac{a + b\sqrt{\delta}}{a + bz}\right\},$$

where the sum is over $a + b\sqrt{\delta} \in \mathbb{F}_q(\sqrt{\delta})^*$. This is reminiscent of the formula for the Eisenstein series on the real Poincaré upper half plane, both holomorphic

[2]N. Katz has found a way.

and non-holomorphic, defined in Section 1. Of course, here we are replacing the discrete subgroup $\mathrm{SL}(2,\mathbb{Z})$ with K, the analog of the compact group $SO(2)$. However, in the finite case, all subgroups are compact and discrete. Choosing other subgroups such as N or $\mathrm{GL}(2,\mathbb{F}_p)$ can be tried. See Harish-Chandra [12] for the case N, which doesn't seem to be so good.

Next let us define an analog of holomorphic Eisenstein series for the finite upper half plane.

DEFINITION. *Given a multiplicative character π of $\mathbb{F}_q(\sqrt{\delta})^*$, define the Eisenstein sum associated to the subgroup K by*

$$(2.14) \qquad G_\pi(z) = |K|^{-1} \sum_{k \in K} \pi \left\{ \frac{a + b\sqrt{\delta}}{a + bz} \right\}.$$

The function G_π is not constant on K-orbits in H_q. Instead it transforms by a multiplier.

DEFINITION. *Given a multiplicative character π of $\mathbb{F}_q(\sqrt{\delta})^*$, define the multiplier J_π by*

$$(2.15) \qquad J_\pi(\gamma, z) = \pi(cz + d) \quad \text{if} \quad \gamma = \begin{pmatrix} a & b \\ c & d \end{pmatrix} \in \mathrm{GL}(2, \mathbb{F}_q).$$

The following Lemma says that the multiplier has the usual property. It is an exercise to do the proof.

LEMMA 3. *The multiplier has the property that*

$$J_\pi(\gamma\eta, z) = J_\pi(\gamma, \eta z) J_\pi(\eta, z),$$

for all $\gamma, \eta \in GL(2, \mathbb{F}_q)$ and all $z \in H_q$.

Now we can define a vector space of modular forms with a given multiplier for any subgroup Γ of $G = \mathrm{GL}(2, \mathbb{F}_q)$; e.g., $\Gamma = K$. Here we assume that the multiplier $m : \Gamma \times H_q \to \mathbb{C}$ has the property of Lemma 3.

DEFINITION. *The vector space of modular forms with multiplier m for Γ is*

$$\mathcal{M}_q(\Gamma, m) = \{ f : H_q \to \mathbb{C} \mid f(\gamma z) = m(\gamma, z) f(z), \forall \gamma \in \Gamma, \forall z \in H_q \}.$$

LEMMA 4. *The Eisenstein sum G_π defined by (2.14) is in $\mathcal{M}_q(K, m)$, where the multiplier m is*

$$m(k, z) = \frac{J_\pi(k, z)}{J_\pi(k, \sqrt{\delta})},$$

for $k \in K$. Here J_π is defined by (2.15).

PROOF.

$$|K| \, G_\pi(k, k'z) = \sum_{k \in K} \frac{J_\pi(k, \sqrt{\delta})}{J_\pi(k, k'z)} = \sum_{k \in K} \frac{J_\pi(k, k'\sqrt{\delta})}{J_\pi(kk', z)} J_\pi(k', z)$$

$$= \frac{J_\pi(k', z)}{J_\pi(k', \sqrt{\delta})} \sum_{k \in K} \frac{J_\pi(kk', \sqrt{\delta})}{J_\pi(kk', z)}.$$

REMARKS. To obtain functions with multiplier 1, we could take quotients f/g for $f, g \in \mathcal{M}_q(K, m)$, for g nonvanishing. But powers of f seem even better since $m^{q-1} = 1$. For if the function f corresponds to multiplier m, then f^e corresponds to multiplier m^e.

EXAMPLE. Let $q = p = 3$. Then $\zeta = 1 + i$ generates $\mathbb{F}_3(i)^*$. Define the multiplicative character π by $\pi(1 + i) = \exp(2\pi i/8)$. The Eisenstein sum is determined by its values on elements of the sets Sa. Recall that $S_0 = \{i\}$, $S_1 = \{\pm i \pm 1\}$, $S_{-1} = \{-i\}$. It is easy to see that

$$G_\pi \begin{pmatrix} i \\ 1+i \\ -i \end{pmatrix} = \begin{pmatrix} 1 \\ (1+i)/4 \\ 0 \end{pmatrix}, \ G_{\pi^4} \begin{pmatrix} i \\ 1+i \\ -i \end{pmatrix} = \begin{pmatrix} 1 \\ 0 \\ 1 \end{pmatrix}, \ G_{\pi^8} \begin{pmatrix} i \\ 1+i \\ -i \end{pmatrix} = \begin{pmatrix} 1 \\ 1 \\ 1 \end{pmatrix}$$

and therefore

$$(G_\pi)^8 \begin{pmatrix} i \\ 1+i \\ -i \end{pmatrix} = \begin{pmatrix} 1 \\ 4^{-12} \\ 1 \end{pmatrix}, \ (G_{\pi^4})^2 \begin{pmatrix} i \\ 1+i \\ -i \end{pmatrix} = \begin{pmatrix} 1 \\ 0 \\ 1 \end{pmatrix}.$$

So the last 3 functions give 3 linearly independent elements of $\mathcal{M}_3(K, 1)$. We already knew this space was 3-dimensional, since there is a basis of spherical functions or eigenfunctions of the collapsed adjacency matrix.

It might also be interesting to consider finite analogs of Poincaré series.

2.3.2 An Intriguing Example.

We are left with our quest for simple formulas for the missing eigenfunctions of the adjacency operator corresponding to non-1-dimensional eigenvalues. Here we mention only the case $p = q = 3$, for which C. Trimble has an intriguing example based on a formula in Gelfand-Graev and Piatetski-Shapiro [11, p. 185]. See Velasquez [34] for some other ideas about getting our missing eigenfunctions from representations of $\mathrm{GL}(2, \mathbb{F}_q)$.

EXAMPLE. (Trimble) Again we assume that $p = q = 3$. Suppose that ψ is an additive character of $\mathbb{F}_q$. Define the function

(2.16) $$f(z) = \psi(\mathrm{Re}\ z).$$

Then

$$Af(z) = \sum_{u^2=av+\delta(v-1)^2} f\left\{ \begin{pmatrix} y & x \\ 0 & 1 \end{pmatrix} \begin{pmatrix} v & u \\ 0 & 1 \end{pmatrix} \sqrt{\delta} \right\} = \psi(x) \sum_{u^2=av+\delta(v-1)^2} \psi(yu).$$

So we find that $Af(z) = \lambda(y)f(z)$, where

$$\lambda(y) = \sum_{u^2=av+\delta(v-1)^2} \psi(yu).$$

In this case where $p = 3$, $\lambda(y)$ does not really depend on y, and we have an eigenfunction of the adjacency operator, which corresponds to the missing non 1-dimensional eigenvalue, -2.

2.3.3. Characteristic 2.

As usual, fields of characteristic 2, require some different treatment. This is of interest, not just for completeness, but because it actually gives some insight into methods we will need for the group $\mathrm{GL}(3, \mathbb{F}_q)$. This section is the work of Angel.

For the remainder of this section, we consider the case of finite fields $\mathbb{F}_q$, where $q = 2^r$. We must make modifications in our definitions, since everything is a square in this field. So we pick an irreducible polynomial $p(x) = x^2 + bx + c$, with $b, c \in \mathbb{F}_2$. Let $p(\theta) = 0$. Then $\bar{\theta} = \theta^q$ is the other root of $p(x)$. We can assume that θ generates the multiplicative group of $\mathbb{F}_q(\theta) = \mathbb{F}_{q^2} = \mathbb{F}_{2^{2r}}$.

DEFINITION. The finite upper half plane H_{2^r} is defined by

(2.17) $$H_{2^r} = \{z = x + y\theta \mid x, y \in \mathbb{F}_{2^r}, y \neq 0\}.$$

As before, we can think of H_{2^r} as a subset of $\mathbb{F}_{2^r}(\theta)$. We define, for $z = x+y\theta$, $x = \mathrm{Re}\ z$, $y = \mathrm{Im}\ z$, $\bar{z} = x + y\bar{\theta} = z^q$, $Nz = z\bar{z}$, just as for odd primes.

The *action* of $g = \begin{pmatrix} a & b \\ c & d \end{pmatrix}$ on $z \in H_{2^r}$ is given by fractional linear transformation

$$gz = \frac{az + b}{cz + d} = u + v\theta,$$

where

$$u = \mathrm{Re}\ (gz) = \frac{ac\ Nz + bd + (ad + bc)x}{N(cz + d)},$$

and

$$v = \mathrm{Im}\ (gz) = \frac{\det g\ \mathrm{Im}\ x}{N(cz + d)}.$$

The *point-pair invariant* is defined, as before, by:

$$d(z, w) = \frac{N(z - w)}{(\mathrm{Im}\ z)\,(\mathrm{Im}\ w)}.$$

The graph of H_{2^r} is obtained again by connecting z to w iff $d(z, w) = a$, for fixed a not equal to 0 or $(\theta - \bar{\theta})^2$. Note that in both even and odd characteristics we must not allow a to equal either 0 or $b^2 - 4c$. Angel shows that then we get a $(q + 1)$-regular graph. Note that our equation for points z adjacent to θ is $N(z - \theta) = ay$, which is equivalent to

$$(2.18) \qquad (z - \theta)^{q+1} = ay = a\frac{z - z^q}{\theta - \theta^q}.$$

This is a polynomial equation in z of degree $q + 1$. To find out what values of a give $q + 1$ solutions, you need to take the derivative. One can obtain the icosahedron graph in this way. The collapsed adjacency matrix is

$$\begin{pmatrix} 0 & 5 & 0 & 0 \\ 1 & 2 & 2 & 0 \\ 0 & 2 & 2 & 1 \\ 0 & 0 & 5 & 0 \end{pmatrix}.$$

Its eigenvalues are: 5, $\pm\sqrt{5} \cong \pm 2.2361$, -1.

REFERENCES

1. E. Bannai and T. Ito, *Algebraic Combinatorics I, Association Schemes*, Benjamin/Cummings, Menlo Park, CA, 1984.
2. B. Berndt and R. Evans, *The determination of Gauss sums*, Bull. (N.S.) Amer. Math. Soc. **5** (1981), 107-129.
3. F. Bien, *Construction of telephone networks by group representations*, Notices of the Amer. Math. Soc. **36** (1989), 187-196.
4. N. Biggs, *Algebraic Graph Theory*, Cambridge U. Press, London, 1974.
5. B. Bollobás, *Graph Theory: An Introductory Course*, Springer-Verlag, New York, 1979.
6. N. Celniker, Ph.D. Thesis, U.C.S.D., 1991.
7. N. Celniker, S. Poulos, A. Terras, C. Trimble, and E. Velasquez, *Is there life on finite upper half planes?* Grosswald Memorial Volume, Contemporary Math., (to appear).
8. F. R. K. Chung, *Diameters and eigenvalues*, Journal of the American Math. Soc **2** (1989), 187-196.
9. P. Diaconis, *Group Representations in Probability and Statistics*, Institute of Math. Statistics, Hayward, CA, 1988.
10. R. Evans, *Hermite character sums*, Pacific J. of Math. **122** (1986), 357-390.
11. I. M. Gelfand, M. I. Graev, and I. I. Piatetski-Shapiro, *Representation Theory and Automorphic Forms*, Academic Press, Boston, 1990.
12. Harish-Chandra, *Eisenstein series over finite fields*, in: "Collected Papers," Vol. 4, pp. 8-21, Springer-Verlag, New York, 1984.
13. D. Hejhal, *Eigenvalues of the Laplacian for Hecke triangle groups*, preprint.
14. M. N. Huxley, *Introduction to Kloostermania*, in "Elementary and Analytic Theory of Numbers" (H. Iwaniec, ed.), pp. 217-306, Polish Scientific Publ., Warsaw, 1985.
15. K. Ireland and M. Rosen, *A Classical Introduction to Modern Number Theory*, Springer-Verlag, New York, 1982.
16. H. Iwaniec, *Selberg's lower bound for the first eigenvalue for congruence groups*, in: "Number Theory, Trace Formulas, and Discrete Groups," (K. Aubert, E. Bombieri, D. Goldfeld, eds.), pp. 371-375, Academic Press, New York, 1989.
17. M. Klawe, *Limitations on explicit constructions of expanding graphs*, S.I.A.M. Journal of Comput. **13** (1984), 156-166.
18. F. Klein and R. Fricke, *Vorlesungen über die Theorie der elliptische Modulfunktionen, vol. I*, Johnson Reprint Corp., New York, 1966.
19. N. Lebedev, *Special Functions and Their Applications*, Dover, New York, 1972.
20. W. Li, *Abelian Ramanujan graphs*, preprint.
21. A. Lubotsky, *Discrete Groups, Expanding Graphs and Invariant Measures*, Lecture Notes, U. of Oklahoma, 1989.

22. A. Lubotsky, R. Phillips, and P. Sarnak, *Ramanujan graphs*, Combinatorica **8** (1988), 261-277.
23. I. Macdonald, *Spherical Functions on a Group of p-Adic Type*, Madras, India, 1971.
24. F. J. MacWilliams and N.J.A. Sloane, *The Theory of Error-Correcting Codes*, North-Holland, Amsterdam, 1988.
25. I. I. Piatetski-Shapiro, *Complex Representations of GL(2, K) for Finite Fields K*, Contemporary Math. **16** (1983).
26. S. Poulos, *Ph.D. Thesis*, U.C.S.D., 1991.
27. W. Schmidt, *Equations over Finite Fields: An Elementary Approach*, Springer-Verlag, NY, 1976, Lecture Notes in Math **536**.
28. A. Selberg, *Collected Papers, Vol. I*, Springer-Verlag, New York, 1989.
29. D. Stanton, *An introduction to group representations and orthogonal polynomials*, in: Orthogonal Polynomials, Theory and Practice (P. Nevai and M. E .H. Ismail, eds.), pp. 419-433, Kluwer, Dordrecht, 1990.
30. H. Stark, *Modular forms and related objects*, Canadian Math. Soc. Conf. Proc. **7** (1987), 421-455.
31. T. Sunada, *Fundamental groups and Laplacians*, Springer-Verlag, New York, Lecture Notes in Math **1339** (1988), 248-277.
32. A. Terras, *Eigenvalue problems related to finite analogues of upper half planes*, in: Forty More Years of Ramifications: Spectral Asymptotics and its Applications (S. A. Fulling and F. J. Narcowich, eds.) Math. Dept., Texas A & M, College Station, TX, Discourses in Math. **1** (1991).
33. A. Terras, *Harmonic Analysis on Symmetric Spaces and Applications, I, II*, 1985, 1988.
34. E. Velasquez, *Ph.D. Thesis*, U.C.S.D., 1991.
35. J. Soto-Andrade, *Geometrical Gelfand models, tensor quotients, and Weil representations*, Amer. Math. Soc., Providence, RI, Proc. Symp. Pure Math. **47** (1987), 305-316.

DEPARTMENT OF MATHEMATICS, UNIVERSITY OF CALIFORNIA - SAN DIEGO, LA JOLLA, CA 92093-0112

Contemporary Mathematics
Volume **138**, 1992

Some Special Values for the BC Type Hypergeometric Function

R. J. BEERENDS

1. Introduction

Most of the material in this paper was inspired by the pioneering work of Koornwinder and Sprinkhuizen-Kuyper in a series of papers culminating in [13, 14]. In these papers they study in great detail orthogonal polynomials in two variables associated with the root system of type BC_2, which can be considered as two-variable analogues of the classical Jacobi polynomials. The last published paper [14] in the series deals with the real $_2F_1$ hypergeometric function of matrix argument. It is the main purpose of the present paper to extend some of their results to several variables and/or more general parameters. These generalizations have one feature in common: they concern special values of the variables and/or parameters involved.

In recent years there has been growing interest in the case of arbitrary many variables, the most general results for root systems being obtained by Heckman and Opdam [10, 11]. They have developed the basic theory of the hypergeometric function associated with arbitrary root systems, as well as the related Jacobi polynomials. Many interesting problems remain open and by specializing to specific cases one should be able to obtain more detailed results, which in turn could yield important information about the general case.

In the present paper we will deal mainly with the root system of type BC_n. For this case the generalized Jacobi polynomials were introduced by Vretare [23], although special cases had been studied earlier by James & Constantine (in connection with Grassmann manifolds and zonal spherical functions for $GL(n, \mathbb{R})$)

1991 *Mathematics Subject Classification*. Primary 33C50, 33C80, Secondary 22E30, 42C05.

Key words and phrases. Root systems of type BC, Kampé de Fériet functions, hypergeometric function of matrix argument, Jacobi polynomials of matrix argument, Harish-Chandra c-function, Appell-Lauricella functions, Jack polynomials, Gauss' summation formula.

Research supported by the Netherlands Organization for Scientific Research (NWO).

This paper is in final form and no version of it will be submitted for publication elsewhere .

and Lidl, Ricci, Dunn, and Eier (in the "Chebyshev" cases). (However, in some of these cases it is not quite obvious that definitions agree - see e.g. [3] and [4, §5] for more details and references.) The study of the BC_n Jacobi polynomials in the sense of Vretare was continued by Debiard (cf. [7]) and very recently in the sense of James & Constantine by Macdonald [19] and Lassalle [16].

Using his theory of the real $_2F_1$ hypergeometric function of matrix argument, Herz defined certain Jacobi polynomials of matrix argument which seemed to be unrelated to the root system BC_n. Indeed, quite separate from the BC_n analysis there was a further development of the theory of the hypergeometric function of matrix argument by James & Constantine and more recently by Gross & Richards and Faraut & Korányi (see e.g [4] for more details and references). This ultimately led to a generalized hypergeometric function of matrix argument as defined independently by Korányi [15], Macdonald [19] and Kaneko [12]. In [4] it was shown that the generalized $_2F_1$ hypergeometric function of matrix argument is a special case of the hypergeometric function associated with BC_n and that a similar result holds for the Jacobi polynomials of Herz, James & Constantine, Macdonald and Lassalle. In this paper we show how these relationships can be used to obtain results for the BC_n hypergeometric function.

Of course, there is yet another well developed theory of hypergeometric functions in several variables: the classical theory of the Appell-Lauricella functions and its many generalizations (cf. [9]). In [14] Koornwinder & Sprinkhuizen-Kuyper link the real $_2F_1$ hypergeometric function of matrix argument in two variables to the Appell F_4 function and it remains one of the few results in this direction. In section 3 we show that their proof can easily be extended to the generalized $_2F_1$ hypergeometric function of matrix argument in two variables. In particular we can relate this result to a classical result of Watson on the product of two Gauss hypergeometric functions. Another result has been obtained very recently by Debiard & Gaveau. In [8] they connect the n variable Lauricella F_D to a special case of the $_2F_1$ generalized hypergeometric function of matrix argument and in section 3 we use their result to obtain a reduction formula for the BC_n hypergeometric function on a wall of the Weyl chamber.

From the Jacobi polynomial case and the case of the generalized hypergeometric function of matrix argument we obtain in section 4 for some specific values of the spectral parameter the value of the BC_n hypergeometric function at a special point (see [13, §2] for the two variable polynomial case). We conjectured that the result would hold for generic values of the spectral parameter; independently Opdam [21] has shown that this is indeed the case.

In section 5 we combine results of Macdonald [19] and Stanley [22] to get an explicit expression for the BC_n Jacobi polynomials for yet another specialization of the spectral parameter. This result can also be expressed in terms of a generalized Kampé de Fériet function. Using different methods this was previously obtained by Debiard & Gaveau [8], although a constant, depending on the normalization of the Jacobi polynomials, is missing there. The two-variable case was first established in [13, §6]. We then prove that the result remains true for the non-polynomial case. Using a reduction formula for the generalized Kampé de Fériet function we arrive at a reduction formula for the BC_n hypergeometric function on the same wall of the Weyl chamber mentioned earlier. It is an intriguing matter whether a reduction formula for generic spectral parameter can be found on this (or any other) wall of the BC_n Weyl chamber. As before, the two-variable case treated by Koornwinder & Sprinkhuizen-Kuyper may serve as inspiration.

2. Notations and preliminaries

In this section we recall definitions of and results on Jack polynomials and the generalized hypergeometric function and Jacobi polynomials of matrix argument (subsection a), the hypergeometric function and the Jacobi polynomials associated with the BC_n root system (subsection b) and the relation between these functions (subsection c).

a. The generalized hypergeometric function and Jacobi polynomials of matrix argument.

For unexplained results concerning partitions and symmetric functions we refer to [17, Chapter I]. A partition λ is any sequence $\lambda = (\lambda_1, \lambda_2, \dots, \lambda_n)$ of nonnegative integers such that $\lambda_1 \geq \lambda_2 \geq \cdots \geq \lambda_n \geq 0$. The number of $\lambda_i \neq 0$ is called the length of λ and is denoted by $l(\lambda)$. The weight $|\lambda|$ of λ is defined as $|\lambda| = \sum_{i \geq 1} \lambda_i$. Given a partition $\lambda = (\lambda_1, \lambda_2, \dots, \lambda_n)$ we define the dual partition $\lambda' = (\lambda'_1, \lambda'_2, \dots, \lambda'_m)$ by $\lambda'_i = \mathrm{Card}\{\, j \mid \lambda_j \geq i \,\}$. Furthermore we put

$$(2.1) \qquad n(\lambda) = \sum_{i \geq 1}(i-1)\lambda_i = \sum_{i \geq 1}\binom{\lambda'_i}{2}.$$

The diagram of a partition λ is the set of points $(i,j) \in \mathbb{Z}^2$ such that $1 \leq j \leq \lambda_i$; we will simply write $(i,j) \in \lambda$ if (i,j) belongs to the diagram of λ. We write $\mu \leq \lambda$ if $|\mu| = |\lambda|$ and $\sum_{i=1}^k \mu_i \leq \sum_{i=1}^k \lambda_i$ for all $k \geq 1$. This is a partial ordering on the set of partitions of a given weight.

Denote by Λ_n the ring of symmetric polynomials in n independent variables with integer coefficients. If $\alpha = (\alpha_1, \alpha_2, \dots, \alpha_n) \in (\mathbb{Z}^+)^n$ ($\mathbb{Z}^+$ the set of nonnegative integers) then we put $|\alpha| = \alpha_1 + \alpha_2 + \cdots + \alpha_n$ and we let x^α denote the monomial $x_1^{\alpha_1} x_2^{\alpha_2} \dots x_n^{\alpha_n}$. For a partition λ of length $l(\lambda) \leq n$ we define the monomial symmetric polynomial m_λ in Λ_n by

$$(2.2) \qquad m_\lambda = m_\lambda(x_1, x_2, \dots, x_n) = \sum x^\alpha,$$

where the sum is taken over all distinct permutations α of λ. If $l(\lambda) > n$ we put $m_\lambda = 0$. The power sums p_r are defined for each integer $r \geq 1$ by $p_r = \sum_{i=1}^n x_i^r = m_{(r)}$. For each partition $\lambda = (\lambda_1, \lambda_2, \dots, \lambda_n)$ define $p_\lambda = p_{\lambda_1} p_{\lambda_2} \dots p_{\lambda_n}$. It is well-known that the set $\{m_\lambda\}$ forms a $\mathbb{Z}$-basis for Λ_n.

We now follow Stanley's introduction of the Jack polynomials [22, §1]. For a partition λ write $z_\lambda = \prod_{i \geq 1} i^{m_i} m_i!$, where m_i is the number of parts of λ equal to i. Let k be a parameter and $\mathbb{Q}(k)$ the field of all rational functions of k with coefficients in $\mathbb{Q}$. For convenience we will work temporarily with symmetric functions in infinitely many variables. In particular we let m_λ and p_λ denote the symmetric functions in infinitely many variables in the ring of symmetric functions Λ as defined in [17, §2]. The m_λ form a $\mathbb{Z}$-basis of Λ and the p_λ form a $\mathbb{Q}$-basis of $\Lambda \otimes \mathbb{Q}$, the ring of symmetric functions with coefficients in $\mathbb{Q}$. Let $\Lambda \otimes \mathbb{Q}(k)$ denote the ring of symmetric functions with coefficients in $\mathbb{Q}(k)$ and define a scalar product on $\Lambda \otimes \mathbb{Q}(k)$ by the condition

$$\langle p_\lambda, p_\mu \rangle_k = \delta_{\lambda\mu} z_\lambda k^{-l(\lambda)},$$

where $\delta_{\lambda\mu}$ is the Kronecker delta. We quote the following theorem from [22, §1].

THEOREM 2.1. *There are unique symmetric functions* $J_\lambda = J_\lambda(x; k^{-1})$ *in* $\Lambda \otimes \mathbb{Q}(k)$, *indexed by partitions* λ, *such that*
1. $\langle J_\lambda, J_\mu \rangle_k = 0$ *if* $\lambda \neq \mu$;
2. *if* $J_\lambda(x; k^{-1}) = \sum_\mu v_{\lambda\mu}(k^{-1}) m_\mu(x)$, *then* $v_{\lambda\mu}(k^{-1}) = 0$ *unless* $\mu \leq \lambda$;
3. *if* $|\lambda| = l$ *then the coefficient* $v_{\lambda,(1^l)}$ *of* $x_1 x_2 \ldots x_l$ *in* J_λ *equals* $l!$.

If we set $x_{n+1} = x_{n+2} = \cdots = 0$ in J_λ then we obtain for any partition of length $l(\lambda) \leq n$ a symmetric polynomial $J_\lambda(x_1, \ldots, x_n; k^{-1})$ in $\Lambda_n \otimes \mathbb{Q}(k)$, homogeneous of degree $|\lambda|$. The $J_\lambda(x_1, \ldots, x_n; k^{-1})$ vanish for $l(\lambda) > n$ and are linearly independent otherwise ([22, Proposition 2.5]). These symmetric polynomials are called Jack polynomials.

REMARK 1. Theorem 2.1 is a special case of a more general existence theorem due to Macdonald ([18, Theorem 2.3] and also Chapter VI, §4 of the new edition of [17]). Since we only need the case of the Jack polynomials, we refer to [22], where many more details and results on Jack polynomials can be found.

REMARK 2. In [22], [18, §1.7] and [19] the parameter $\alpha = k^{-1}$ is used. We will use α differently later on. To avoid confusion we prefer to write $J_\lambda(x; k^{-1})$ instead of $J_\lambda(x; k)$. This explains the rather awkward notation.

REMARK 3. In [4, Proposition 3.3] it was shown that Jack polynomials and Jacobi polynomials for the root system A_{n-1} are closely related: Jack polynomials are homogeneous versions of the A_{n-1} Jacobi polynomials (up to normalization).

We write 1_n to mean $x_1 = x_2 = \cdots = x_n = 1$. Stanley [22, Theorems 5.4 and 5.8] has shown that

$$J_\lambda(1_n; k^{-1}) = \prod_{(i,j) \in \lambda} (n - i + 1 + k^{-1}(j-1)),$$

(2.3)

$$\langle J_\lambda, J_\lambda \rangle_k = h_*(\lambda) h^*(\lambda),$$

where, for any partition λ of length $l(\lambda) \leq n$, $h_*(\lambda)$ and $h^*(\lambda)$ are defined by

(2.4)
$$h_*(\lambda) = \prod_{(i,j)\in\lambda} (\lambda'_j - i + 1 + k^{-1}(\lambda_i - j)),$$

$$h^*(\lambda) = \prod_{(i,j)\in\lambda} (\lambda'_j - i + k^{-1}(\lambda_i - j + 1)).$$

The factors in $h_*(\lambda)$ are the so-called "lower hook-lengths" at $(i, j) \in \lambda$ while the factors in $h^*(\lambda)$ are the so-called "upper hook-lengths" at $(i, j) \in \lambda$ (cf. [22, §5]).

Next we recall the Pochhammer symbol

$$(a)_s = a(a+1)\ldots(a+s-1), \qquad a \in \mathbb{C}, \quad s \in \mathbb{Z}^+,$$

and define for a partition $\lambda = (\lambda_1, \lambda_2, \ldots, \lambda_n)$ of length $l(\lambda) \leq n$

(2.5)
$$(a)_\lambda = \prod_{i=1}^{n}(a - k(i-1))_{\lambda_i}, \qquad a \in \mathbb{C}.$$

Note that we suppress the dependence on k of $(a)_\lambda$ in the notation.

We are now ready to define the generalized $_2F_1$ hypergeometric function of matrix argument. For $a, b, c \in \mathbb{C}$ such that $(c)_\lambda \neq 0$ for all λ we define

(2.6)
$$_2F_1(a, b; c; x; k^{-1}) = \sum_{\lambda} \frac{(a)_\lambda (b)_\lambda}{(c)_\lambda k^{|\lambda|}} \frac{J_\lambda(x; k^{-1})}{\langle J_\lambda, J_\lambda \rangle_k},$$

where the sum is over all partitions λ of length $\leq n$. This definition, and its extension to $_pF_q$'s, was given independently by Korányi [15, §§3–4], Macdonald [19, (6.4)] and Kaneko [12]. For more details see e.g. [4, §2a].

The $_2F_1$ as defined in (2.6) satisfies a system of n partial differential equations. To describe the result we define for $i = 1, 2, \ldots, n$ the operators $\triangle_i(a, b, c; k)$ by

(2.7)
$$\begin{aligned} \triangle_i(a, b, c; k) = &x_i(1 - x_i)\partial_{x_i}^2 \\ &+ (c - k(n-1) - (a + b + 1 - k(n-1))x_i)\partial_{x_i} \\ &+ k\sum_{j=1; j\neq i}^{n} \frac{x_i(1-x_i)}{(x_i - x_j)}\partial_{x_i} - k\sum_{j=1; j\neq i}^{n} \frac{x_j(1-x_j)}{(x_i - x_j)}\partial_{x_j}, \end{aligned}$$

where we have written $\partial_{x_i} = \partial/\partial x_i$, $\partial_{x_i}^2 = \partial^2/\partial x_i^2$, etc. One now has the following result ([15, §4], [25, Thm. A], [12, §2]).

THEOREM 2.2. *The hypergeometric function* $_2F_1(a, b; c; x; k^{-1})$ *is the unique symmetric function in the* n *variables* $x_1, \ldots, x_n$ *that satisfies*

(2.8)
$$\triangle_i(a, b, c; k) F = ab F, \qquad i = 1, 2, \ldots, n$$

and which is analytic at $x_1 = \cdots = x_n = 0$ and normalized by $F(0) = 1$.

Another result we will need later on is the generalized Gauss summation formula

$$(2.9) \qquad {}_2F_1(a, b; c; 1_n; k^{-1}) = \prod_{i=1}^{n} \frac{\Gamma(c - k(i-1))\Gamma(c - a - b - k(i-1))}{\Gamma(c - a - k(i-1))\Gamma(c - b - k(i-1))},$$

which will certainly hold when $\mathrm{Re}(b - k(n-1)) > 0$, $\mathrm{Re}(c - b - k(n-1)) > 0$, $\mathrm{Re}(c - b - a - k(n-1)) > 0$ and

$$\mathrm{Re}\ k > -\min(\frac{1}{n}, \frac{\mathrm{Re}(b - k(n-1))}{n-1}, \frac{\mathrm{Re}(c - b - a - k(n-1))}{n-1}).$$

The Gauss summation formula (2.9) is due to Macdonald [19, §7] (see [4] for some more details).

Finally we introduce the generalized Jacobi polynomials of matrix argument. These were defined independently by Macdonald [19, §9] and Lassalle [16]. We briefly describe how they are defined. Let $E(\alpha, \beta, k)$ be the differential operator

$$-E(\alpha, \beta, k) = \sum_{i=1}^{n} x_i(1 - x_i)\partial_{x_i}^2 + 2k \sum_{\substack{i,j=1 \\ i \neq j}}^{n} \frac{x_i(1 - x_i)}{(x_i - x_j)}\partial_{x_i}$$

$$+ \sum_{i=1}^{n} (\alpha + 1 - (\alpha + \beta + 2)x_i)\partial_{x_i}.$$

For $\alpha, \beta > 0$ and any partition λ define the generalized Jacobi polynomials $G_\lambda^{(\alpha,\beta)}(x; k^{-1})$ of matrix argument as the polynomials of the form

$$(2.10) \qquad G_\lambda^{(\alpha,\beta)}(x; k^{-1}) = \sum_{\mu \subset \lambda} u_{\lambda\mu}(\alpha, \beta, k) \frac{J_\mu(x; k^{-1})}{J_\mu(1_n; k^{-1})}$$

that satisfy

$$(2.11) \qquad \begin{aligned} &E(\alpha, \beta, k)\, G_\lambda^{(\alpha,\beta)} \\ &= \left((\alpha + \beta + 2 + 2k(n-1))|\lambda| + 2(n(\lambda') - kn(\lambda))\right) G_\lambda^{(\alpha,\beta)}, \end{aligned}$$

where $n(\lambda)$ is given by (2.1) and $\mu \subset \lambda$ means that $\mu_i \leq \lambda_i$ for all i. Macdonald [19] normalizes the $G_\lambda^{(\alpha,\beta)}(x; k^{-1})$ by the requirement that the coefficient $u_{\lambda\lambda}$ in (2.10) equals $(-1)^{|\lambda|}$. Lassalle [16] gives exactly the same defining properties (2.10) and (2.11) (with (α, β) replaced by (a, b), $1 + k(n-1) = q$ and $n(\lambda') - kn(\lambda) = \rho_\lambda/2$), but normalizes his Jacobi polynomials $P_\lambda^{\alpha,\beta}(x)$ by $P_\lambda^{\alpha,\beta}(0) = 1$. Hence Lassalle's $P_\lambda^{\alpha,\beta}$ and Macdonald's $G_\lambda^{(\alpha,\beta)}$ differ by $G_\lambda^{(\alpha,\beta)}(0)$. We return to this constant in section 2c.

b. BC-type Jacobi polynomials and hypergeometric functions.

Let $(e_1, e_2, \ldots, e_n)$ be the standard basis in $\mathbb{R}^n$ and let $\langle \cdot, \cdot \rangle$ denote the usual inner product for which this basis is orthonormal. We identify the dual space of $\mathbb{R}^n$ with $\mathbb{R}^n$ by means of the inner product. We consider $\mathbb{R}^n$ as the standard real form of $\mathbb{C}^n$ and extend $\langle \cdot, \cdot \rangle$ to a complex bilinear form on $\mathbb{C}^n$. Again we identify the dual space by means of $\langle \cdot, \cdot \rangle$. We use $t_1, t_2, \ldots, t_n$ as coordinates with respect to $(e_1, \ldots, e_n)$.

In $\mathbb{R}^n$ we consider the set of vectors

$$R_B = \{ \pm e_i, \pm 2e_i, \pm(e_k \pm e_l) \mid i = 1, \ldots, n; 1 \le k < l \le n \}$$

which forms a root system of type BC_n. We choose $S_B = \{ e_1 - e_2, e_2 - e_3, \ldots, e_{n-1} - e_n, e_n \}$ as basis for R_B and let R_B^+ denote the corresponding set of positive roots. The Weyl group of R_B will be denoted by W_B. Let P_B be the weight lattice of R_B. So

$$P_B = \{ \lambda \in \mathbb{R}^n \mid 2\langle \lambda, \alpha \rangle / \langle \alpha, \alpha \rangle \in \mathbb{Z} \quad \forall \alpha \in R_B \}.$$

The set of dominant weights will be denoted by P_B^+:

$$P_B^+ = \{ \lambda \in P_B \mid \langle \lambda, \alpha \rangle \ge 0 \quad \forall \alpha \in R_B^+ \}.$$

If we write $R_B^0 = \{ \alpha \in R_B \mid 2\alpha \notin R_B \}$ then R_B^0 is a root system of type C_n. Denote by $\omega_1, \omega_2, \ldots, \omega_n$ the fundamental weights of C_n, so that for $i = 1, 2, \ldots, n$ one has

(2.12) $$\omega_i = e_1 + e_2 + \cdots + e_i.$$

Then

$$P_B = \mathbb{Z}\omega_1 + \mathbb{Z}\omega_2 + \cdots + \mathbb{Z}\omega_n = \mathbb{Z}e_1 + \cdots + \mathbb{Z}e_n = \mathbb{Z}^n$$

and

$$P_B^+ = \mathbb{Z}^+\omega_1 + \mathbb{Z}^+\omega_2 + \cdots + \mathbb{Z}^+\omega_n.$$

Let Q_B be the root lattice $\mathbb{Z}R_B$ and put $Q_B^+ = \mathbb{Z}^+R_B^+$. Then $Q_B = P_B = \mathbb{Z}^n$. Let T_B be the compact torus defined by

$$T_B = i\mathbb{R}^n / 2\pi i Q_B = i\mathbb{R}^n / 2\pi i \mathbb{Z}^n.$$

We will also consider the complexification H_B, i.e.

$$H_B = \mathbb{C}^n / 2\pi i Q_B = \mathbb{C}^n / 2\pi i \mathbb{Z}^n,$$

and use the map

$$\sum_{j=1}^{n} t_j e_j \quad (\mathrm{mod}\, 2\pi i \mathbb{Z}^n) \to (e^{t_1}, \ldots, e^{t_n}) \qquad (t_j \in i\mathbb{R} \; \forall j)$$

to identify T_B with $\{ (x_1, \ldots, x_n) \in \mathbb{C}^n \mid |x_j| = 1 \; \forall j \}$.

We now introduce the Jacobi polynomials associated with the root system R_B. Our reference for this material is [11, §§2–3], [10, §8], although BC_n Jacobi polynomials were first introduced by Vretare [23]. For $\lambda \in P_B^+$ we write $\Pi_B(\lambda)$ for the convex hull of the orbit $W_B \cdot \lambda$ intersected with $\lambda + Q_B$. Note that $\Pi_B(\lambda)$ is equal to the set of all $\mu \in P_B^+$ with $\lambda - \mu \in Q_B^+$ and the W_B-conjugates of such μ. Let m be the number of W_B-orbits in R_B. Define $\mathcal{K} \cong \mathbb{C}^m$ as the vector space of W_B-invariant functions on R_B with values in $\mathbb{C}$. We call elements of $\mathcal{K}$

multiplicity functions on R_B and we denote by k_α the value of $\kappa \in \mathcal{K}$ on $\alpha \in R_B$. There are three W_B-orbits and we write

$$\begin{align}
k_1 &= k_{e_i}, \\
(2.13) \qquad k_2 &= k_{2e_i}, \\
k_3 &= k_{e_i \pm e_j}.
\end{align}$$

Let $L_B(\kappa)$ be the so-called generalized radial part of the Laplace-Beltrami operator given by

$$L_B(\kappa) = L_B(0) + \sum_{\alpha \in R_B^+} k_\alpha \coth \tfrac{1}{2}\alpha \cdot \partial_\alpha,$$

where $L_B(0)$ denotes the ordinary Laplacian on $\mathbb{R}^n$ and ∂_α denotes differentiation in the direction of α. Define the vector $\rho_B(\kappa)$ by $\rho_B(\kappa) = \tfrac{1}{2} \sum_{\alpha \in R_B^+} k_\alpha \alpha$. For $\kappa \in \mathcal{K}$ such that $k_\alpha \geq 0$ for all $\alpha \in R_B$ we now define the Jacobi polynomials $P_B(\lambda, \kappa)$ on H_B ($\lambda \in P_B^+$) associated with R_B and multiplicity function $\kappa \in \mathcal{K}$ by means of the following two properties:

1. $P_B(\lambda, \kappa) = \sum_{\mu \in \Pi_B(\lambda)} \Gamma_\mu(\lambda, \kappa) e^\mu$ with $\Gamma_\lambda(\lambda, \kappa) = 1$ and $\Gamma_{w\mu}(\lambda, \kappa) = \Gamma_\mu(\lambda, \kappa)$ for all $w \in W_B$.
2. $P_B(\lambda, \kappa)$ satisfies $L_B(\kappa) P_B(\lambda, \kappa) = \langle \lambda, \lambda + 2\rho_B(\kappa) \rangle P_B(\lambda, \kappa)$.

Here e^λ denotes the function $e^\lambda(\dot t) = e^{\langle \lambda, t \rangle}$ where $t \in \mathbb{C}^n$ and $\dot t$ denotes the equivalence class in H_B (when restricted to T_B, e^λ is the character corresponding to λ). In particular $e^{e_i}(\dot t) = e^{t_i}$ if $t = (t_1, \ldots, t_n) \in \mathbb{C}^n$. (In [10, §8], [11, §3] and [20, §2] the Jacobi polynomials are parametrized by $\mu = -\lambda \in P_B^-$ instead of $\lambda \in P_B^+$.)

Next we define for $\lambda \in \mathbb{C}^n$ and $\kappa \in \mathcal{K}$ the generalized Harish-Chandra c-function by

$$(2.14) \qquad c_B(\lambda, \kappa) = \prod_{\alpha \in R_B^+} \frac{\Gamma(\langle \lambda, \alpha^{\check{}} \rangle + \tfrac{1}{2}k_{\frac{\alpha}{2}}) \Gamma(\langle \rho_B(\kappa), \alpha^{\check{}} \rangle + \tfrac{1}{2}k_{\frac{\alpha}{2}} + k_\alpha)}{\Gamma(\langle \lambda, \alpha^{\check{}} \rangle + \tfrac{1}{2}k_{\frac{\alpha}{2}} + k_\alpha) \Gamma(\langle \rho_B(\kappa), \alpha^{\check{}} \rangle + \tfrac{1}{2}k_{\frac{\alpha}{2}})}$$

where Γ denotes the usual gamma-function and $\alpha^{\check{}} = 2\alpha/\langle \alpha, \alpha \rangle$ (we should warn the reader that this definition differs from [11, §6] and [20, §2] by a change of λ into $-\lambda$). For specific values of the k_α this is the well-known product formula of Gindikin & Karpelevich for the classical Harish-Chandra c-function. Later it will be convenient to have $\rho_B(\kappa)$ explicit; with the convention (2.13) one has

$$(2.15) \qquad 2\rho_B(\kappa) = (k_1 + 2k_2)(\sum_{i=1}^n e_i) + 2k_3 \sum_{i=1}^n (n-i)e_i.$$

The value of the Jacobi polynomial at the identity element $e \in H_A$ has been determined by Opdam [20, §5] and (in this case) is given by

$$(2.16) \qquad P_B(\lambda, \kappa)(e) = 1/c_B(\lambda + \rho_B(\kappa), \kappa).$$

Finally we let $F_{BC_n}(\lambda, \kappa; h)$ denote the hypergeometric function associated with the root system BC_n as defined in [11, §6] or [10, §7]. Here $\lambda \in \mathbb{C}^n$, $\kappa \in \mathcal{K}$

and $h \in H_B$. (We use the notation F_{BC_n} instead of the simpler F_B to avoid confusion with the Lauricella F_B function.) Recall from [10] or [11] that for $\lambda \in P_B^+$ one has

$$(2.17) \quad F_{BC_n}(-(\lambda + \rho_B(\kappa)), \kappa; \dot{\imath}) = c_B(\lambda + \rho_B(\kappa), \kappa) \, P_B(\lambda, \kappa)(\dot{\imath}), \qquad t \in \mathbb{C}^n.$$

c. Relation between the matrix and the root system cases.

We now summarize the results of [4]. For $i = 1, \ldots, n$ put $y_i = \cosh t_i$ ($t_i \in \mathbb{C}$) and let z_j be the jth elementary symmetric polynomial in $y_1, \ldots, y_n$ for $j = 1, \ldots, n$. Furthermore we let

$$(2.18) \quad x_i = \tfrac{1}{2}(1 - y_i) = \tfrac{1}{2} - \tfrac{1}{4}(e^{t_i} + e^{-t_i}) = -\sinh^2 \tfrac{1}{2} t_i, \qquad i = 1, \ldots, n.$$

Recall from section 2b that $e^{t_i} = e^{e_i}(\dot{\imath})$ for $t = (t_1, \ldots, t_n) \in \mathbb{C}^n$.

To a partition $\lambda = (\lambda_1, \lambda_2, \ldots, \lambda_n)$ of length $l(\lambda) \leq n$ we associate the vector $\lambda = \lambda_1 e_1 + \lambda_2 e_2 + \cdots + \lambda_n e_n$ in $\mathbb{R}^n$. By (2.12) we have $\lambda = (\lambda_1 - \lambda_2)\omega_1 + (\lambda_2 - \lambda_3)\omega_2 + \cdots + (\lambda_{n-1} - \lambda_n)\omega_{n-1} + \lambda_n \omega_n$ so that $\lambda \in P_B^+$. On the other hand, if $\lambda \in P_B^+$ then $\lambda = a_1\omega_1 + \cdots + a_n\omega_n$ with $a_i \in \mathbb{Z}^+$ so that by (2.12) we have $\lambda = \lambda_1 e_1 + \cdots + \lambda_n e_n$ with $\lambda_i = a_1 + \cdots + a_i$, which determines a unique partition $(\lambda_1, \ldots, \lambda_n)$. This gives a one-to-one correspondance between partitions and dominant weights of BC_n. Note that the weight $l\omega_r$ corresponds to the partition (l^r) $(r = 1, \ldots, n)$.

We first relate the hypergeometric functions. From (2.18) it is clear that from a symmetric function in the variables x_i one obtains a W_B-invariant function on H_B. In this sense we can consider the hypergeometric function of matrix argument $_2F_1(a, b; c; x; k^{-1})$ as a W_B-invariant function on H_B. We now quote the following theorem from [4, §4].

THEOREM 2.3. *Let* $_2F_1(a, b; c; x; k^{-1})$ *be defined by (2.6) and let* $F_{BC_n}(\lambda, \kappa; h)$ *be the hypergeometric function associated with the root system* BC_n. *Then*

$$(2.19) \qquad _2F_1(a, b; c; x; k^{-1}) = F_{BC_n}(-(\lambda + \rho_B(\kappa)), \kappa; \dot{\imath}),$$

where $x = (x_1, \ldots, x_n)$, $\kappa = (k_1, k_2, k_3)$, $t = (t_1, \ldots, t_n) \in \mathbb{C}^n$ *and*

$$x_i = \tfrac{1}{2} - \tfrac{1}{4}(e^{t_i} + e^{-t_i}), \qquad i = 1, \ldots, n,$$
$$\lambda = -a\omega_n,$$
$$(2.20) \qquad a + b = k_1 + 2k_2 + k_3(n - 1),$$
$$c = k_1 + k_2 + k_3(n - 1) + \tfrac{1}{2},$$
$$k = k_3.$$

Now recall from sections 2a and 2b the Jacobi polynomials $G_\lambda^{(\alpha, \beta)}$ of matrix argument and $P_B(\lambda, \kappa)$ associated with BC_n. The following result was established in [4, §5].

THEOREM 2.4. *For any partition λ one has*

$$(2.21) \qquad G_\lambda^{(\alpha,\beta)}(x;k^{-1}) = \frac{2^{-2|\lambda|}h_*(\lambda)}{J_\lambda(1_n;k^{-1})} \, P_B(\lambda,\kappa)(t),$$

where $\kappa = (k_1,k_2,k_3) = (\alpha-\beta, \beta+\frac{1}{2}, k)$ and $x_i = \frac{1}{2} - \frac{1}{4}(e^{e_i} + e^{-e_i})(t)$ for $i = 1, \ldots, n$.

In section 2a we noted that Macdonald's $G_\lambda^{(\alpha,\beta)}$ and Lassalle's $P_\lambda^{\alpha,\beta}$ differed by the constant $G_\lambda^{(\alpha,\beta)}(0)$. Since $J_\lambda(1_n;k^{-1})$, $h_*(\lambda)$ and $P_B(\lambda,\kappa)(e)$ are all known explicitly (cf. (2.3), (2.4) and (2.16)), the value $G_\lambda^{(\alpha,\beta)}(0)$ follows from (2.21).

As was shown in [4, Corollary 5.2] one can combine (2.21) and the polynomial case of (2.19) (so $-a = l \in \mathbb{Z}^+$) to obtain as a corollary an explicit expression for $G_\lambda^{(\alpha,\beta)}$ in the case $\lambda = l\omega_n$. This result was also obtained independently by Lassalle [16] and Kaneko [12, §6]. The normalizing constants are quite nice in this case since

$$(2.22) \qquad P_B(l\omega_n,\kappa)(e) = c_B(l\omega_n + \rho_B(\kappa),\kappa)^{-1} = 2^{2nl}\,(c)_{(l^n)}/(b)_{(l^n)},$$

as was observed in [4, Proposition 4.3]. Here (b,c) are related to (k_1,k_2,k_3) as in (2.20) with $a = -l$. In section 4 we describe a similar result for the special value $\lambda = l\omega_1$.

3. Connection with Appell-Lauricella functions

Theorem 2.3 shows how the BC_n hypergeometric function is related to the generalized hypergeometric function of matrix argument for the special value $\lambda = a\omega_n$ of the spectral parameter; in particular one obtains for $k = 1/2$ and $k = 1$ a relation with the classical hypergeometric function of matrix argument of James & Constantine. It is an intriguing question whether other classical hypergeometric functions in several variables (or results on such functions) are related to the root system hypergeometric function. In this section we will consider this question in the setting of (2.19): the BC_n root system and the special value $\lambda = a\omega_n = a(e_1 + e_2 + \ldots + e_n)$ ($a \in \mathbb{C}$) for the spectral variable.

A nice result in this direction has been obtained by Debiard & Gaveau in [8]. To describe their result we recall that the F_D hypergeometric function in n complex variables is defined for $|x_1| < 1, \ldots, |x_n| < 1$ by the convergent power series

$$F_D(a, b_1, \ldots, b_n; c; x_1, \ldots, x_n)$$
$$= \sum_{m_1,\ldots,m_n \geq 0} \frac{(a)_{m_1+\ldots+m_n}(b_1)_{m_1} \cdots (b_n)_{m_n}}{(c)_{m_1+\ldots+m_n} \, m_1! \cdots m_n!} x_1^{m_1} \cdots x_n^{m_n},$$

where $a, b_1, \ldots, b_n, c$ are complex and $c \neq 0, -1, -2, \ldots$. For $n = 1$ the F_D function reduces to the Gauss hypergeometric function. For $n = 2$ the F_D function is the Appell function F_1, while for $n > 2$ it is called Lauricella function. The F_D function satisfies a system of n second order linear partial differential equations (see e.g. [1, p. 117] or [8, (2.5)]). In [8, II.3 and III.5] it is shown that, after some simple manipulations, the system for the F_D function in the case $b_1 = \cdots = b_n = b$ is equivalent to the system

$$\triangle_i(a, b, c; b) - ab = 0,$$

$$\frac{\partial^2}{\partial x_i \partial x_j} + \frac{b}{x_i - x_j}\left(\frac{\partial}{\partial x_j} - \frac{\partial}{\partial x_i}\right) = 0,$$

where $\triangle_i$ is given by (2.7). Since $F_D(a, b, \ldots, b; c; x_1, \ldots, x_n)$ is a symmetric function in $x_1, \ldots, x_n$ which is analytic at $x_1 = \cdots = x_n = 0$ and is normalized by $F(0) = 1$, we conclude from theorem 2.2 that

$$(3.1) \qquad {}_2F_1(a, b; c; x; b^{-1}) = F_D(a, b, \ldots, b; c; x),$$

where $x = (x_1, \ldots, x_n)$ (in [8] this result is derived differently). Combining (3.1) with (2.19) gives a connection between a special case of the Lauricella F_D and the case $k = b$ and $\lambda = a\omega_n$ of the BC_n hypergeometric function. From (3.1) one can obtain a result which seems to be unnoticed in the literature up till now. If we take $x_i = x$ for $i = 1, \ldots, n$ then it is known that

$$F_D(a, b_1, \ldots, b_n; c; x, \ldots, x) = {}_2F_1(a, b_1 + \cdots + b_n; c; x)$$

(see e.g [9, (4.7.2)] and [1, p. 23, (25)] or [2, 9.5(1)] for the case $n = 2$). Hence we obtain from (3.1) that

$$_2F_1(a, b; c; x, \ldots, x; b^{-1}) = {}_2F_1(a, nb; c; x).$$

Since $x_i = x$ for $i = 1, \ldots, n$ means that $(t_1, \ldots, t_n)' = (t, \ldots, t)'$ we obtain from Theorem 2.3 that the hypergeometric function associated with BC_n reduces for $\lambda = -a\omega_n$ and $k = b$ to a Gauss hypergeometric function on the wall of the Weyl chamber given by $t_1 = \cdots = t_n$. Hence we have the following result, which is the first known case of a reduction formula for the hypergeometric function associated with root systems.

PROPOSITION 3.1. *On the wall of the BC_n Weyl chamber given by $t_1 = \cdots = t_n = t$ one has that*

$$F_{BC_n}(a\omega_n - \rho_B(\kappa), \kappa; (t, \ldots, t)') = {}_2F_1(a, nk_3; k_1 + k_2 + \tfrac{1}{2} + k_3(n-1); -\sinh^2 \tfrac{1}{2}t),$$

where $\kappa = (k_1, k_2, k_3)$ and $a = k_1 + 2k_2 + k_3(n-2)$.

In section 5 we return to this problem. Note that one obtains from proposition 3.1 in particular that for $a = k_1 + 2k_2 + k_3(n-2)$ one has

$$
\begin{aligned}
& F_{BC_n}(a\omega_n - \rho_B(\kappa), \kappa; (\pi i, \ldots, \pi i)') \\
(3.2) \qquad & = \frac{\Gamma(k_1 + k_2 + \tfrac{1}{2} + k_3(n-1))\Gamma(\tfrac{1}{2} - k_2 - k_3(n-1))}{\Gamma(k_1 + k_2 + \tfrac{1}{2} - k_3)\Gamma(\tfrac{1}{2} - k_2 + k_3)},
\end{aligned}
$$

by the Gauss summation formula (2.9) for $n = 1$. We return to this formula in section 4.

We now specialize to the case $n = 2$ and will relate F_{BC_n} for $\lambda = a\omega_n$ and arbitrary k to the Appell F_4 function. This extends a result by Koornwinder & Sprinkhuizen-Kuyper and in addition we will show how it is connected with a classical result due to Watson [24] which states that

$$
\begin{aligned}
(3.3) \quad & _2F_1(a,b;c;x_1)\,{}_2F_1(a,b;c;x_2) \\
&= \frac{\Gamma(c)\Gamma(c-a-b)}{\Gamma(c-a)\Gamma(c-b)}\, F_4(a,b;c,a+b-c+1;x_1x_2,(1-x_1)(1-x_2)) \\
&\quad + \frac{\Gamma(c)\Gamma(a+b-c)}{\Gamma(a)\Gamma(b)}((1-x_1)(1-x_2))^{c-a-b} \\
&\qquad \cdot F_4(c-a,c-b;c,c-a-b+1;x_1x_2,(1-x_1)(1-x_2)),
\end{aligned}
$$

where the two-variable Appell F_4 function is defined for $|\sqrt{z_1}| + |\sqrt{z_2}| < 1$ by the convergent power series

$$
F_4(a,b;c,c';z_1,z_2) = \sum_{m,n\geq 0} \frac{(a)_{m+n}(b)_{m+n}}{(c)_m(c')_n\,m!n!}\, z_1^m z_2^n.
$$

Watson used contour integration to prove this result. Later Bailey gave two proofs based on simple series manipulations (see [2, §9.6]). Burchnall [5] gave yet another proof using the differential equations for the Appell F_4 function. This proof is the most interesting for our purposes since it can also be used in a more general setting. So let us give a sketch of his proof. First note that in (3.3) we have two Appell functions $F_4(\alpha,\beta;\gamma,\gamma';\cdot;\cdot)$ with

$$
(3.4) \qquad\qquad \gamma + \gamma' = \alpha + \beta + 1.
$$

Under the assumption that the parameters α, β, γ and γ' satisfy (3.4), the system of partial differential equations satisfied by $F_4(\alpha,\beta;\gamma,\gamma';u_1;u_2)$ may be written as

$$
(3.5) \qquad
\begin{aligned}
A_1\,F &= \alpha\beta\,F, \\
A_2\,F &= \alpha\beta\,F,
\end{aligned}
$$

where

$$
\begin{aligned}
A_1 &= u_1(1-u_1)\frac{\partial^2}{\partial u_1^2} - 2u_1u_2\frac{\partial^2}{\partial u_1\partial u_2} - u_2^2\frac{\partial^2}{\partial u_2^2} \\
&\quad + (\gamma - (\gamma+\gamma')u_1)\frac{\partial}{\partial u_1} - (\gamma+\gamma')u_2\frac{\partial}{\partial u_2}, \\
A_2 &= u_2(1-u_2)\frac{\partial^2}{\partial u_2^2} - 2u_1u_2\frac{\partial^2}{\partial u_1\partial u_2} - u_1^2\frac{\partial^2}{\partial u_1^2} \\
&\quad + (\gamma' - (\gamma+\gamma')u_2)\frac{\partial}{\partial u_2} - (\gamma+\gamma')u_1\frac{\partial}{\partial u_1}.
\end{aligned}
$$

As Burchnall [5, p. 146] observed, "the results obtained by Watson and Bailey suggest that it may be profitable to transform the system by the substitution":

$$(3.6) \qquad \begin{aligned} u_1 &= x_1 x_2, \\ u_2 &= (1 - x_1)(1 - x_2). \end{aligned}$$

Always under the assumption of (3.4), Burchnall [5, (7)] obtains from (3.5) the equivalent system

$$\left[x_1(1 - x_1)\frac{\partial^2}{\partial x_1^2} + (\gamma - (\gamma + \gamma')x_1)\frac{\partial}{\partial x_1} - \alpha\beta \right] F = 0,$$

$$\left[x_2(1 - x_2)\frac{\partial^2}{\partial x_2^2} + (\gamma - (\gamma + \gamma')x_2)\frac{\partial}{\partial x_2} - \alpha\beta \right] F = 0,$$

which is a system of two Gauss hypergeometric differential equations in the separate variables x_1 and x_2. It is then not very hard to get (3.3) from this (see [5] for details). (For completeness we mention that (3.3) can also be obtained from a more general result of Burchnall & Chaundy [6] relating an arbitrary F_4 to a series of products of Gauss hypergeometric functions.)

Some 40 years later Koornwinder & Sprinkhuizen-Kuyper [14] used exactly the same proof (but transforming in reverse direction) to obtain that

$$(3.7) \qquad \begin{aligned} &_2F_1(a, b; c; x_1, x_2; 2) \\ &= \frac{\Gamma(c)\Gamma(c - a - b)}{\Gamma(c - a)\Gamma(c - b)} \, F_4(a, b; c - \tfrac{1}{2}, a + b - c + 1; x_1 x_2, (1 - x_1)(1 - x_2)) \\ &\quad + \frac{\Gamma(c)\Gamma(a + b - c)}{\Gamma(a)\Gamma(b)}((1 - x_1)(1 - x_2))^{c - a - b} \\ &\qquad \cdot F_4(c - a, c - b; c - \tfrac{1}{2}, c - a - b + 1; x_1 x_2, (1 - x_1)(1 - x_2)). \end{aligned}$$

Recall that $_2F_1(a, b; c; x_1, x_2; 2)$ is the real hypergeometric function of matrix argument (so $k = 1/2$). Note that in (3.7) we have two Appell functions $F_4(\alpha, \beta; \gamma, \gamma'; \cdot; \cdot)$ with

$$(3.8) \qquad \gamma + \gamma' = \alpha + \beta + \tfrac{1}{2}.$$

Koornwinder & Sprinkhuizen-Kuyper were unaware of Watson's result (3.3), and Burchnall's proof of it, and hence of the remarkable similarities between (3.3) and (3.7). A simple explanation for these similarities can be given using some recent results on the generalized hypergeometric function of matrix argument. First we observe the following. Write $_2F_1(a, b; c; x_1, \ldots, x_n; \infty)$ for the generalized hypergeometric function of matrix argument for $k = 0$.

PROPOSITION 3.2.

$$_2F_1(a, b; c; x_1, \ldots, x_n; \infty) = \prod_{i=1}^{n} {_2F_1}(a, b; c; x_i)$$

PROOF. The proof is extremely simple if we use theorem 2.2 since for $k = 0$ all operators $\triangle_i(a, b, c; 0)$ reduce to the Gauss hypergeometric differential equation in one variable x_i. The function $\prod_{i=1}^n {}_2F_1(a, b; c; x_i)$ is a symmetric function in the x_i that satisfies $\triangle_i(a, b, c; 0) \, F = ab \, F$ for $i = 1, \ldots, n$, is analytic at $x_1 = \cdots = x_n = 0$, and is normalized by $F(0) = 1$. Hence proposition 3.2 follows immediately from theorem 2.2.

Let us also give an elementary proof which does not depend on theorem 2.2 (which is usually stated only for $k > 0$). From the series expansion of the Gauss hypergeometric function we obtain that

$$\prod_{i=1}^n {}_2F_1(a, b; c; x_i) = \sum_{m_1, \ldots, m_n \geq 0} \frac{(a)_{m_1} \cdots (a)_{m_n} (b)_{m_1} \cdots (b)_{m_n}}{(c)_{m_1} \cdots (c)_{m_n} m_1! \cdots m_n!} x_1^{m_1} \cdots x_n^{m_n}$$

$$= \sum_{\lambda_1 \geq \cdots \geq \lambda_n \geq 0} \frac{(a)_{\lambda_1} \cdots (a)_{\lambda_n} (b)_{\lambda_1} \cdots (b)_{\lambda_n}}{(c)_{\lambda_1} \cdots (c)_{\lambda_n} \lambda_1! \cdots \lambda_n!} \sum x_1^{m_1} \cdots x_n^{m_n},$$

where the last sum is over all distinct permutations $(m_1, \ldots, m_n)$ of $(\lambda_1, \ldots, \lambda_n)$. Hence

$$\prod_{i=1}^n {}_2F_1(a, b; c; x_i) = \sum_\lambda \frac{(a)_\lambda (b)_\lambda}{(c)_\lambda} \frac{m_\lambda}{\lambda!},$$

where $(a)_\lambda$ is given by (2.5) with $k = 0$, λ runs over all partitions and $\lambda! = \lambda_1! \cdots \lambda_n!$. We now claim that

$$(3.9) \qquad \left[\frac{J_\lambda(x; k^{-1})}{\langle J_\lambda, J_\lambda \rangle_k \, k^{|\lambda|}} \right]_{k=0} = \frac{m_\lambda}{\lambda!},$$

which would prove the proposition. To prove the claim we combine the remarks in the beginning of §2 of [18] and [18, (3.13) and (6.3)] to obtain that

$$[k^{|\lambda|} J_\lambda(x; k^{-1})]_{k=0} = [k^{|\lambda|} h_*(\lambda)]_{k=0} \, m_\lambda(x)$$

(similar results can be found in Chapter IV of the new edition of [17]). From (2.3) then follows that

$$\left[\frac{J_\lambda(x; k^{-1})}{\langle J_\lambda, J_\lambda \rangle_k \, k^{|\lambda|}} \right]_{k=0} = [k^{|\lambda|} h^*(\lambda)]_{k=0}^{-1} \, m_\lambda(x)$$

and since $[k^{|\lambda|} h^*(\lambda)]_{k=0} = \lambda!$ we obtain (3.9). $\square$

Proposition 3.2 shows in particular that we can replace the left-hand side of (3.3) by ${}_2F_1(a, b; c; x_1, x_2; \infty)$ which explains the similarities between (3.3) and (3.7). Moreover, reading through Burchnall's $k = 0$ proof (or the proof in [14]) it is clear how to generalize the result to arbitrary k.

THEOREM 3.3. *In a sufficiently small neighbourhood of the origin* $(x_1, x_2) = (0,0)$ *one has that*

(3.10)
$$
\begin{aligned}
{}_2 & F_1(a, b; c; x_1, x_2; k^{-1}) \\
& = \frac{\Gamma(c)\Gamma(c-a-b)}{\Gamma(c-a)\Gamma(c-b)} F_4(a, b; c-k, a+b-c+1; x_1 x_2, (1-x_1)(1-x_2)) \\
& \quad + \frac{\Gamma(c)\Gamma(a+b-c)}{\Gamma(a)\Gamma(b)}((1-x_1)(1-x_2))^{c-a-b} \\
& \qquad \cdot F_4(c-a, c-b; c-k, c-a-b+1; x_1 x_2, (1-x_1)(1-x_2)),
\end{aligned}
$$

provided $c \neq 0, -1, -2, \ldots$ *and* $c - k$ *and* $a + b - c$ *are non-integer.*

REMARK. In (3.10) we have two Appell functions $F_4(\alpha, \beta; \gamma, \gamma'; \cdot; \cdot)$ with

(3.11)
$$
\gamma + \gamma' = \alpha + \beta + 1 - k
$$

which reduces to (3.8) for $k = \frac{1}{2}$ and (3.4) for $k = 0$.

PROOF. We follow Burchnall's $k = 0$ proof and transform the system (3.5) by the substitution (3.6), but now assuming (3.11) instead of (3.4). As in [5, p. 146] one obtains from (3.5) the equivalent system

(3.12)
$$
\begin{aligned}
(1 - x_1)D_1 - (1 - x_2)D_2 &= 0, \\
x_1\left(D_1 + k\frac{\partial}{\partial x_1}\right) - x_2\left(D_2 + k\frac{\partial}{\partial x_2}\right) &= 0,
\end{aligned}
$$

where for $i = 1, 2$ we have put

$$
D_i = x_i(1 - x_i)\frac{\partial^2}{\partial x_i^2} + (\gamma - (\gamma + \gamma' + k)x_i)\frac{\partial}{\partial x_i} - \alpha\beta.
$$

If we add up these two equations we obtain that $D_1 + kx_1\frac{\partial}{\partial x_1} = D_2 + kx_2\frac{\partial}{\partial x_2}$ and if we substitute this in (3.12) (and use that $(1-x_1)/(x_1 - x_2) = (1 - x_2)/(x_1 - x_2) - 1$), then we see that the system (3.5) is equivalent with

$$
\begin{aligned}
\triangle_2(\alpha, \beta, \gamma + k; k) - \alpha\beta &= 0, \\
\triangle_1(\alpha, \beta, \gamma + k; k) - \alpha\beta &= 0,
\end{aligned}
$$

under the assumption that (3.11) holds (for $\triangle_i$ see (2.7)). Hence the system for Appell's F_4 function is equivalent with the system (2.8) for the generalized hypergeometric function of matrix argument. The remainder of the proof is exactly the same as in [5, §4] or [14, proof of theorem 3.2], the essential point being that one has a precise description of all solutions of both system (3.5) [1, p. 51-52] and system (2.8) (theorem 2.2). $\square$

Since the system (3.12) can also be written as $(x_2 - x_1)(A_1 - \alpha\beta) = 0$ and $(x_1 - x_2)(A_2 - \alpha\beta) = 0$, we can run through the proof again to obtain the following relation between the operators $\triangle_i$ and A_i:

$$
\begin{aligned}
\triangle_1(\alpha, \beta, \gamma + k; k) &= x_2 A_1 + (1 - x_2)A_2, \\
\triangle_2(\alpha, \beta, \gamma + k; k) &= x_1 A_1 + (1 - x_1)A_2.
\end{aligned}
$$

In [14] the transformation of variables is carried out in reverse order (worth noting is a misprint in [14, theorem 3.1]: in the second relation $x - y$ should be replaced by $x + y$). Of course the theory of the generalized hypergeometric function of matrix argument, and especially theorem 2.2, was not available when [14] was published. In fact theorem 3.3 has been obtained by Sprinkhuizen-Kuyper, using completely different methods, in a manuscript which has never been published, and one could well say that part of the theory of the two variable generalized hypergeometric function and Jacobi polynomials of matrix argument was developed in that manuscript as well as in [13].

4. Value at a special point

In this section we give the value of $F_{BC_n}(-(\lambda + \rho_B(\kappa)), \kappa; t)$ at the point $t_0 = (\pi i, \dots, \pi i)$, so $x_1 = \cdots = x_n = 1$, for two special values of the spectral parameter: $\lambda \in P_B^+$ and $\lambda = a\omega_n$ ($a \in \mathbb{C}$). We start with the case $\lambda \in P_B^+$ so that we are in the Jacobi polynomial case (cf. (2.17)). As was observed by Vretare [23, Remark 5.4] one has that

$$P_B(\lambda, (\alpha, \beta, \gamma))(-y_1, \dots, -y_n) = (-1)^{|\lambda|} P_B(\lambda, (\beta, \alpha, \gamma))(y_1, \dots, y_n).$$

The parameters (k_1, k_2, k_3) have the following expression in terms of the (α, β, γ):

$$(4.1) \qquad \begin{aligned} k_1 &= \alpha - \beta, \\ k_2 &= \beta + \tfrac{1}{2}, \\ k_3 &= \gamma + \tfrac{1}{2}. \end{aligned}$$

Since the relation between the variables y_i and t_i is given by (2.18) we obtain that in the (t, κ) variables one has that

$$(4.2) \qquad P_B(\lambda, \kappa)(t + t_0) = (-1)^{|\lambda|} P_B(\lambda, \kappa')(t),$$

where

$$\kappa' = (-k_1, k_1 + k_2, k_3)$$

if $\kappa = (k_1, k_2, k_3)$ is as always. From the well-known Selberg integral it follows that the BC_n Jacobi polynomials are defined for all $\alpha, \beta > -1$, $\gamma > -\tfrac{1}{2} - \tfrac{1}{n}$, $\alpha + (\gamma + \tfrac{1}{2})(n-1) > -1$ and $\beta + (\gamma + \tfrac{1}{2})(n-1) > -1$, which by (4.1) means that

$$(4.3) \qquad \begin{aligned} k_1 + k_2, k_2 &> -\tfrac{1}{2}, \\ k_1 + k_2 + k_3(n-1) &> -\tfrac{1}{2}, \\ k_2 + k_3(n-1) &> -\tfrac{1}{2}, \\ k_3 &> -\tfrac{1}{n}. \end{aligned}$$

(This is an improvement of the condition $k_1, k_2, k_3 \geq 0$ in section 2b.) Note that the conditions (4.3) are invariant under $\kappa \leftrightarrow \kappa'$ so that $P_B(\lambda, \kappa')$ is still well-defined. From (2.16) we obtain that $P_B(\lambda, \kappa')(e) = 1/c_B(\lambda + \rho_B(\kappa'), \kappa')$ and hence it follows from (4.2) that

$$P_B(\lambda, \kappa)(\dot{t}_0) = (-1)^{|\lambda|} c_B(\lambda + \rho_B(\kappa'), \kappa')^{-1},$$

which in terms of the BC_n hypergeometric function means that

$$(4.4) \qquad F_{BC_n}(-(\lambda + \rho_B(\kappa)), \kappa; \dot{t}_0) = (-1)^{|\lambda|} \frac{c_B(\lambda + \rho_B(\kappa), \kappa)}{c_B(\lambda + \rho_B(\kappa'), \kappa')},$$

as follows from (2.17). For $\lambda = l\omega_n$ we get from (2.22) the particularly nice results

$$P_B(l\omega_n, \kappa)(\dot{t}_0) = 2^{2nl} \frac{(c-b)_{(l^n)}}{(b)_{(l^n)}},$$

$$F_{BC_n}(-(l\omega_n + \rho_B(\kappa)), \kappa; \dot{t}_0) = \frac{(c-b)_{(l^n)}}{(c)_{(l^n)}},$$

where (b, c) are related to (l, κ) as in (2.20) with $a = -l$.

Next we get the value of $F_{BC_n}(-(\lambda + \rho_B(\kappa)), \kappa; \dot{t}_0)$ for $\lambda = -a\omega_n$ $(a \in \mathbb{C})$ from (2.19) and (2.9) since $\dot{t}_0$ corresponds to $x = (1, \ldots, 1)$. This results in:

$$(4.5) \quad F_{BC_n}(a\omega_n - \rho_B(\kappa), \kappa; \dot{t}_0) = \prod_{i=1}^{n} \frac{\Gamma(c - k(i-1))\Gamma(c - a - b - k(i-1))}{\Gamma(c - a - k(i-1))\Gamma(c - b - k(i-1))},$$

under the conditions stated in section 2a. Using (2.20) these conditions are equivalent to

$$(4.6) \qquad \begin{aligned} &\mathrm{Re}(k_1 + 2k_2 - a) > 0, \\ &\mathrm{Re}(\tfrac{1}{2} - k_2 - k_3(n-1) + a) > 0, \\ &\mathrm{Re}(\tfrac{1}{2} - k_2 - k_3(n-1)) > 0, \\ &\mathrm{Re}\, k_3 > -\min(\frac{1}{n}, \frac{\mathrm{Re}(k_1 + 2k_2 - a)}{n-1}, \frac{\mathrm{Re}(\tfrac{1}{2} - k_2 - k_3(n-1))}{n-1}). \end{aligned}$$

Formulas (4.4) and (4.5) should have a common generalization and in order to find it we investigate for arbitrary λ the quotient $c_B(\lambda + \rho_B(\kappa), \kappa)/c_B(\lambda + \rho_B(\kappa'), \kappa')$ from (4.4).

PROPOSITION 4.1. *For $\kappa = (k_1, k_2, k_3)$ let $\kappa' = (-k_1, k_1 + k_2, k_3)$ and write $\lambda = \sum \lambda_i e_i$. Then*

$$\frac{c_B(\lambda + \rho_B(\kappa), \kappa)}{c_B(\lambda + \rho_B(\kappa'), \kappa')} = \prod_{i=1}^{n} \frac{\Gamma(k_1 + k_2 + \tfrac{1}{2} + k_3(n-i))\Gamma(\lambda_i + k_2 + \tfrac{1}{2} + k_3(n-i))}{\Gamma(\lambda_i + k_1 + k_2 + \tfrac{1}{2} + k_3(n-i))\Gamma(k_2 + \tfrac{1}{2} + k_3(n-i))}.$$

PROOF. Let us write $\kappa' = (k_1', k_2', k_3')$ so that $k_1' = -k_1$, $k_2' = k_1 + k_2$ and $k_3' = k_3$. Then $k_1' + 2k_2' = k_1 + 2k_2$ and hence $\rho_B(\kappa) = \rho_B(\kappa')$, as follows from

(2.15). Also $\langle\rho_B(\kappa), e_i\rangle = \frac{1}{2}k_1+k_2+k_3(n-i) = \frac{1}{2}k_1'+k_2'+k_3'(n-i) = \langle\rho_B(\kappa'), e_i\rangle$ and similarly $\langle\rho_B(\kappa), e_i\pm e_j\rangle = \langle\rho_B(\kappa'), e_i\pm e_j\rangle$. It then follows from (2.14) that

$$\frac{c_B(\lambda+\rho_B(\kappa),\kappa)}{c_B(\lambda+\rho_B(\kappa'),\kappa')} = \prod_{i=1}^n \frac{\Gamma(2(\langle\rho_B(\kappa),e_i\rangle+\frac{1}{2}k_1))\Gamma((\lambda+\rho_B(\kappa),e_i\rangle+\frac{1}{2}k_1)}{\Gamma(\langle\rho_B(\kappa),e_i\rangle+\frac{1}{2}k_1)\Gamma(2(\langle\lambda+\rho_B(\kappa),e_i\rangle+\frac{1}{2}k_1))}$$
$$\cdot\frac{\Gamma((\rho_B(\kappa'),e_i\rangle-\frac{1}{2}k_1)\Gamma(2(\langle\lambda+\rho_B(\kappa'),e_i\rangle-\frac{1}{2}k_1))}{\Gamma(2(\langle\rho_B(\kappa'),e_i\rangle-\frac{1}{2}k_1))\Gamma((\lambda+\rho_B(\kappa'),e_i\rangle-\frac{1}{2}k_1)},$$

since all other terms in the quotient cancel. Using the duplication formula $\Gamma(2z) = (2\pi)^{-\frac{1}{2}}2^{2z-\frac{1}{2}}\Gamma(z)\Gamma(z+\frac{1}{2})$ the result follows. $\square$

We can now state the results (4.4) and (4.5) in a unified way and obtain a conjecture for generic λ.

PROPOSITION 4.2. *For* $t_0 = (\pi i,\ldots,\pi i)$ *one has*

$$F_{BC_n}(-(\lambda+\rho_B(\kappa)),\kappa;i_0) =$$
(4.7)
$$\prod_{i=1}^n \frac{\Gamma(k_1+k_2+\frac{1}{2}+k_3(n-i))\Gamma(\frac{1}{2}-k_2-k_3(n-i))}{\Gamma(k_1+k_2+\frac{1}{2}+k_3(n-i)+\lambda_i)\Gamma(\frac{1}{2}-k_2-k_3(n-i)-\lambda_i)},$$

in either of the following two cases:
i. $\lambda\in P_B^+$ *and* (k_1,k_2,k_3) *satisfy (4.3)*
ii. $\lambda = -a\omega_n$ $(a\in\mathbb{C})$ *and* (k_1,k_2,k_3) *satisfy (4.6)*

PROOF. If $\lambda = a\omega_n$ then $\lambda_i = -a$ for $i = 1,\ldots,n$ and using (2.20) the result for $\lambda = a\omega_n$ follows from (4.5).

Since $\Gamma(1-z) = \pi(\Gamma(z)\sin\pi z)^{-1}$ we obtain that

$$\prod_{i=1}^n \frac{\Gamma(\lambda_i+k_2+\frac{1}{2}+k_3(n-i))}{\Gamma(k_2+\frac{1}{2}+k_3(n-i))} = \prod_{i=1}^n \frac{\sin\pi(k_2+\frac{1}{2}+k_3(n-i))}{\sin\pi(k_2+\frac{1}{2}+k_3(n-i)+\lambda_i)}$$
$$\cdot\frac{\Gamma(\frac{1}{2}-k_2-k_3(n-i))}{\Gamma(\frac{1}{2}-k_2-k_3(n-i)-\lambda_i)},$$

and if λ_i is an integer for all i then this reduces to

$$(-1)^{|\lambda|}\prod_{i=1}^n \frac{\Gamma(\frac{1}{2}-k_2-k_3(n-i))}{\Gamma(\frac{1}{2}-k_2-k_3(n-i)-\lambda_i)}.$$

Hence for $\lambda\in P_B^+$ the result follows from proposition 4.1 and (4.4). $\square$

It is easy to check that (4.7) reduces to (3.2) for $a = k_1+2k_2+k_3(n-2)$. From proposition 5.5 one obtains that (4.7) also holds for $\lambda = -a\omega_1$ $(a\in\mathbb{C})$.

It is now clear that one conjectures that the result of proposition 4.2 remains valid for generic λ (and with certain restrictions on κ). Using his thorough analysis of the hypergeometric function associated with root systems, Opdam [21] has independently shown this conjecture to be true. In [21, Corollary 6.10] the parameter λ is replaced by $\lambda-\rho_B(\kappa)$, which gives a more symmetrical formula since $(\lambda-\rho_B(\kappa))_i = \lambda_i-(\frac{1}{2}k_1+k_2+k_3(n-i))$.

5. The case of the first fundamental weight

For the two-variable case an explicit expression of the Jacobi polynomials $P_B(\lambda, \kappa)$ for $\lambda = l\omega_1 = le_1$ was given in [13, corollary 6.8]. One of the main theorems of Debiard & Gaveau in [8] is an explicit expression in the n variable case for these Jacobi polynomials $P_B(l\omega_1, \kappa)$. However, in the statement of [8, Théorème 5] a constant, which depends on the normalization of the Jacobi polynomials, is missing. Their result, including the value of the constant, can also be obtained by combining a result of Stanley with one of Macdonald. To formulate it we introduce the so-called generalized Kampé de Fériet function $F_{2:0}^{2:1}$ which is defined for $|x_i| < 1$ $(i = 1, \ldots, n)$ by the convergent power series

(5.1)
$$F_{2:0}^{2:1} \left[\begin{matrix} \alpha_1, \alpha_2 : \beta_1 ; \cdots ; \beta_n ; \\ \gamma_1, \gamma_2 : - ; \cdots ; - ; \end{matrix} \; x_1, \ldots, x_n \right]$$
$$= \sum_{m_1, \ldots, m_n \geq 0} \frac{(\alpha_1)_{m_1 + \cdots + m_n} (\alpha_2)_{m_1 + \cdots + m_n} (\beta_1)_{m_1} \cdots (\beta_n)_{m_n}}{(\gamma_1)_{m_1 + \cdots + m_n} (\gamma_2)_{m_1 + \cdots + m_n} m_1! \cdots m_n!} x_1^{m_1} \cdots x_n^{m_n},$$

where all parameters are complex and $\gamma_1, \gamma_2 \neq 0, -1, -2, \ldots$. For $n = 2$ the functions $F_{q:s}^{p:r}$ were introduced by Kampé de Fériet (cf. [1, p. 150] or [9, §1.5]), while their n variable generalizations were given by Karlsson (cf. [9, §3.7]).

Let us denote the partition $(l, 0, \ldots, 0)$ by (l) and remark that $(a)_{(l)} = (a)_l$, where $(a)_{(l)}$ is defined as in (2.5). Recall from section 2a the generalized Jacobi polynomials of matrix argument $G_{(l)}^{(\alpha, \beta)}(x; k^{-1})$.

PROPOSITION 5.1. *Let* $k = \gamma + \frac{1}{2}$, $c = \alpha + 1 + k(n - 1)$ *and* $b_1 = \alpha + \beta + 1 + 2k(n - 1) + l$. *Then*

(5.2)
$$\frac{(b_1)_l}{(c)_l} G_{(l)}^{(\alpha, \beta)}(x; k^{-1}) = \sum_{\mu \in (\mathbb{Z}^+)^n, |\mu| = s \leq l} \frac{(-l)_s (b_1)_s}{(c)_s (nk)_s} \prod_{i=1}^{n} \frac{(k)_{\mu_i}}{\mu_i!} x_i^{\mu_i}$$
$$= F_{2:0}^{2:1} \left[\begin{matrix} -l, b_1 : k ; \cdots ; k ; \\ c , nk : - ; \cdots ; - ; \end{matrix} \; x_1, \ldots, x_n \right].$$

PROOF. In [19, §9] Macdonald obtains the explicit expression

(5.3) $$G_{(l)}^{(\alpha, \beta)}(x; k^{-1}) = (c)_l \sum_{s=0}^{l} (-1)^s \binom{l}{s} ((c)_s (b_1 + s)_{l-s})^{-1} \frac{J_{(s)}(x; k^{-1})}{J_{(s)}(1_n; k^{-1})},$$

with c and b_1 as in the proposition, while from Stanley's paper [22] we quote from proposition 2.2 the result

(5.4) $$J_{(s)}(x; k^{-1}) = \sum_{|\lambda| = s} s! k^{-s} \left(\prod_{i=1}^{n} \frac{(k)_{\lambda_i}}{\lambda_i!} \right) m_\lambda,$$

where the sum is taken over all partitions λ with weight s ((5.3) and (5.4) are generalizations to n variables of [13, (6.2)] and [13, (4.7)] respectively). Note that $J_\lambda(0;k^{-1}) = 0$ except when $\lambda = (0)$, so that one obtains from (5.3) that $G_{(l)}^{(\alpha,\beta)}(0;k^{-1}) = (c)_l/(b_1)_l$. If we now combine (5.3) and (5.4) and use that $(b_1)_l/(b_1+s)_{l-s} = (b_1)_s$ then we arrive at

$$\frac{(b_1)_l}{(c)_l}G_{(l)}^{(\alpha,\beta)}(x;k^{-1}) = \sum_{|\lambda|=s\le l} (-1)^s \binom{l}{s} s! \frac{(b_1)_s}{k^s J_{(s)}(1_n;k^{-1})(c)_s} \left(\prod_{i=1}^{n} \frac{(k)_{\lambda_i}}{\lambda_i!}\right) m_\lambda.$$

Now $k^s J_{(s)}(1_n;k^{-1}) = (nk)_s$, as follows easily from (2.3), and $(-1)^s\binom{l}{s}s! = (-l)_s$. Together with the definitions of the m_λ in (2.2) and the generalized Kampé de Fériet function in (5.1) one then obtains (5.2). □

Recall that in (2.22) we gave a simple expression for $c_B(l\omega_n + \rho_B(\kappa),\kappa)$. For $\lambda = l\omega_1$ ($l \in \mathbb{Z}^+$) something similar happens. Note that from (4.1) follows that $c = k_1 + k_2 + k_3(n-1) + \frac{1}{2}$ and $b_1 = k_1 + 2k_2 + 2k_3(n-1) + l$.

COROLLARY 5.2. *For $l \in \mathbb{Z}^+$ and with $\kappa = (k_1,k_2,k_3)$ one has that*

$$(5.5)\qquad c_B(l\omega_1 + \rho_B(\kappa),\kappa) = 2^{-2l}\frac{(k_3)_l}{(nk_3)_l}\cdot\frac{(k_1+2k_2+2k_3(n-1)+l)_l}{(k_1+k_2+k_3(n-1)+\frac{1}{2})_l}.$$

PROOF. As observed in the proof of proposition 5.1, one has that

$$G_{(l)}^{(\alpha,\beta)}(0;k^{-1}) = (c)_l/(b_1)_l$$

with c and b_1 as before. Hence one obtains from (2.21) and (2.16) that

$$c_B(l\omega_1 + \rho_B(\kappa),\kappa) = \frac{2^{-2l}h_*(l\omega_1)(b_1)_l}{J_{l\omega_1}(1_n;k^{-1})(c)_l},$$

where $\kappa = (k_1,k_2,k_3) = (\alpha-\beta,\beta+\frac{1}{2},k)$. An easy calculation (using (2.3) and (2.4)) shows that $h_*(l\omega_1) = k^{-l}(k)_l$ and $J_{l\omega_1}(1_n;k^{-1}) = k^{-l}(nk)_l$. Putting these results together proves (5.5). □

Of course (5.5) can also be obtained from (2.14) by direct calculations, in which case one might as well take $\lambda = -a\omega_1$ with $a \in \mathbb{C}$. Seeing the simple expressions for $\lambda = -a\omega_n$ and $\lambda = -a\omega_1$ one realizes that similar expressions should exist for $\lambda = -a\omega_r$ for any $r = 1,\ldots,n$. Indeed one has the following result.

PROPOSITION 5.3. *For $\kappa = (k_1,k_2,k_3)$ and $a \in \mathbb{C}$ we write $c = k_1 + k_2 + \frac{1}{2} + k_3(n-1)$ and $b_r = k_1 + 2k_2 + k_3(2n-r-1) - a$. Then*

$$c_B(-a\omega_r + \rho_B(\kappa),\kappa) = 2^{2ar}\prod_{i=1}^{r}\frac{\Gamma(c-k_3(i-1))\Gamma(b_r-a-k_3(i-1))}{\Gamma(b_r-k_3(i-1))\Gamma(c-a-k_3(i-1))}$$

$$\cdot\frac{\Gamma(nk_3-k_3(i-1))\Gamma(ik_3-a)}{\Gamma(nk_3-a-k_3(i-1))\Gamma(ik_3)}.$$

In particular one obtains for $-a = l \in \mathbb{Z}^+$ that

$$(5.6)\qquad c_B(l\omega_r + \rho_B(\kappa),\kappa) = 2^{-2lr}\frac{(b_r)_{(l^r)}}{(c)_{(l^r)}}\prod_{i=1}^{r}\frac{(ik_3)_l}{((n-i+1)k_3)_l}.$$

The proof uses (2.14) and is a straightforward calculation comparable with the proof for the case $r = n$ in [4, Proposition 4.3]. For $r = 1$ the notation b_r is consistent with the earlier use of b_1, while for $r = n$ we have that $b_n = b$ with b as in (2.20). For $-a = l \in \mathbb{Z}^+$ identity (5.6) reduces to (5.5) if $r = 1$ and to (2.22) if $r = n$.

Let us return to the result of proposition 5.1 and relate (5.2) to the BC_n Jacobi polynomials $P_B(\lambda, \kappa)$. Since $F_{2:0}^{2:1}(0) = 1$ one obtains immediately from (2.17), (2.21) and (5.2) that

$$(5.7) \quad F_{2:0}^{2:1}\begin{bmatrix} -l, b_1 : k \; ; \cdots ; k \; ; \\ c \;, nk : - ; \cdots ; - ; \end{bmatrix} x_1, \ldots, x_n = c_B(l\omega_1 + \rho_B(\kappa), \kappa) P_B(l\omega_1, \kappa)(t)$$
$$= F_{BC_n}(-(l\omega_1 + \rho_B(\kappa)), \kappa; t),$$

where $l \in \mathbb{Z}^+$. We can now prove that (5.2) remains true for the non-polynomial case, i.e. that one has the following analogue of theorem 2.3.

THEOREM 5.4. *Let $F_{2:0}^{2:1}$ be defined by (5.1) and let F_{BC_n} be the hypergeometric function associated with the root system BC_n. Then*

$$F_{2:0}^{2:1}\begin{bmatrix} a \,, b_1 \; : k_3 \; ; \cdots ; k_3 \; ; \\ c \,, nk_3 : - \; ; \cdots ; - \; ; \end{bmatrix} x_1, \ldots, x_n = F_{BC_n}(-(\lambda + \rho_B(\kappa)), \kappa; t),$$

where $x = (x_1, \ldots, x_n)$, $\kappa = (k_1, k_2, k_3)$, $t = (t_1, \ldots, t_n) \in \mathbb{C}^n$ and

$$x_i = \tfrac{1}{2} - \tfrac{1}{4}(e^{t_i} + e^{-t_i}), \qquad i = 1, \ldots, n,$$
$$\lambda = -a\omega_1,$$
$$a + b_1 = k_1 + 2k_2 + 2k_3(n - 1),$$
$$c = k_1 + k_2 + k_3(n - 1) + \tfrac{1}{2}.$$

PROOF. As in the case of the hypergeometric function of matrix argument in theorem 2.3 it is clear from (2.18) that the function $F_{2:0}^{2:1}$, which is a symmetric function in the variables $x_1, \ldots, x_n$, can be considered as a W_B-invariant function on H_B. This function $F_{2:0}^{2:1}$ is analytic in $|x_i| < 1$ and in proposition 5.1 we have shown that the theorem holds for $a = -l$ with $l \in \mathbb{Z}^+$. Moreover, in $F_{2:0}^{2:1}$ the parameter a occurs polynomially in the coefficients of each monomial $x_1^{m_1} \cdots x_n^{m_n}$. As was shown in [4], one can then conclude from [11, Theorem 6.9] that

$$F_{2:0}^{2:1}\begin{bmatrix} a \,, b_1 \; : k_3 \; ; \cdots ; k_3 \; ; \\ c \,, nk_3 : - \; ; \cdots ; - \; ; \end{bmatrix} x_1, \ldots, x_n = d(\lambda, \kappa) F_{BC_n}(-(\lambda + \rho_B(\kappa)), \kappa; t),$$

where $d(\lambda, \kappa)$ is a constant depending meromorphically on λ, κ. Now Opdam has shown in [21, Theorem 6.1] that $F_{BC_n}(\lambda, \kappa; e) = 1$, and since $F_{2:0}^{2:1}(0) = 1$ the theorem follows. $\square$

As in the case of the connection with the Lauricella F_D function one can obtain from Theorem 5.4 a case of reducibility which again seems to be unnoticed up till now, even in the polynomial case (5.2). In fact, if $x_1 = \ldots, x_n = x$ one can use [9, (4.9.3)] to obtain that

$$F_{2:0}^{2:1}\left[\begin{array}{c}\alpha_1, \alpha_2 : \beta_1; \cdots; \beta_n; \\ \gamma_1, \gamma_2 : -; \cdots; -;\end{array} x, \ldots, x\right] = {}_3F_2(\alpha_1, \alpha_2, \beta_1 + \cdots + \beta_n; \gamma_1, \gamma_2; x)$$

and hence

$$F_{2:0}^{2:1}\left[\begin{array}{c}a, \ b_1 : k_3; \cdots; k_3; \\ c, nk_3 : -; \cdots; -;\end{array} x, \ldots, x\right] = {}_3F_2(a, b_1, nk_3; c, nk_3; x).$$

Since nk_3 occurs as upper and as lower parameter in the ${}_3F_2$ we obtain the following result.

PROPOSITION 5.5. *On the wall of the BC_n Weyl chamber given by $t_1 = \cdots = t_n = t$ one has that*

$$F_{BC_n}(a\omega_1 - \rho_B(\kappa)), \kappa; (t, \ldots, t)') =$$
$$\quad {}_2F_1(a, k_1 + 2k_2 + 2k_3(n-1) - a; k_1 + k_2 + \tfrac{1}{2} + k_3(n-1); -\sinh^2 \tfrac{1}{2}t),$$

where $\kappa = (k_1, k_2, k_3)$ and $a \in \mathbb{C}$.

Note that by the Gauss summation formula for the classical hypergeometric function one obtains from proposition 5.5 that (4.7) also holds for $\lambda = -a\omega_1$ ($a \in \mathbb{C}$).

As we have stated in the introduction, it is an intriguing matter whether a reduction formula as in propositions 3.1 and 5.5 can be found on this (or any other) wall of the BC_n Weyl chamber for generic spectral parameter λ.

REFERENCES

[1] P. Appell & J. Kampé de Fériet, *Fonctions hypergéométriques et hypersphériques - polynomes d'Hermite*, Gauthiers-Villars, Paris, 1926.

[2] W. N. Bailey, *Generalized hypergeometric series*, Cambridge Univ. Press, London, 1935.

[3] R. J. Beerends, *Chebyshev polynomials in several variables and the radial part of the Laplace-Beltrami operator*, Trans. Amer. Math. Soc. **328** (1991), 779–814.

[4] R. J. Beerends & E. M. Opdam, *Certain hypergeometric series related to the root system BC*, Trans. Amer. Math. Soc. (to appear).

[5] J. L. Burchnall, *The differential equations of Appell's function F_4*, Quart. J. Math. **10** (1939), 145–150.

[6] J. L. Burchnall & T. W. Chaundy, *The hypergeometric identities of Cayley, Orr, and Bailey*, Proc. London Math. Soc. (2) **50** (1948), 56–74.

[7] A. Debiard, *Système différentiel hypergéométrique et parties radiales des opérateurs invariants des espaces symétriques de type BC_p*, in: Séminaire d'algèbre Paul Dubreil et Marie-Paule Malliavin, Lect. Notes in Math. **1296** (1987), Springer, Berlin.

[8] A. Debiard & B. Gaveau, *Représentation intégrale de certaines séries de fonctions sphériques d'un système de racines BC*, J. Funct. Anal. **96** (1991), 256–296.

[9] H. Exton, *Multiple hypergeometric functions and applications*, Ellis Horwood, Chichester, 1976.

[10] G. J. Heckman, *Root systems and hypergeometric functions II*, Compositio Math. **64** (1987), 353–374.

[11] G. J. Heckman & E. M. Opdam, *Root systems and hypergeometric functions I*, Compositio Math. **64** (1987), 329–352.

[12] J. Kaneko, *Selberg integrals and hypergeometric functions associated with Jack polynomials*, preprint, Department of Mathematics, Kyushu University.

[13] T. H. Koornwinder & I. G. Sprinkhuizen-Kuyper, *Generalized power series expansions for a class of orthogonal polynomials in two variables*, SIAM J. Math. Anal. **9** (1978), 457–483.

[14] _____, *Hypergeometric functions of 2 × 2 matrix argument are expressible in terms of Appell's function F_4*, Proc. Amer. Math. Soc. **70** (1978), 39–42.

[15] A. Korányi, *Hua-type integrals, hypergeometric functions and symmetric polynomials*, in: International symposium in memory of Hua Loo Keng, vol. II Analysis, S. Gong et. al. (eds.), Science Press, Beijing and Springer Verlag, Berlin, (1991), 169–180.

[16] M. Lassalle, *Polynômes de Jacobi généralisés*, C. R. Acad. Sci. Paris Sér. I **312** (1991), 425–428.

[17] I. G. Macdonald, *Symmetric functions and Hall polynomials*, Oxford University Press, Oxford, 1979.

[18] _____, *A new class of symmetric functions*, Publ. IRMA Strasbourg, Séminaire Lotharingien (1988), 131–171.

[19] _____, *Hypergeometric functions*, unpublished manuscript.

[20] E. M. Opdam, *Some applications of hypergeometric shift operators*, Invent. Math. **98** (1989), 1–18.

[21] _____, *An analogue of the Gauss summation formula for hypergeometric functions related to root systems*, Report W 91-19, Math. Inst. Univ. Leiden, 1991.

[22] R. P. Stanley, *Some combinatorial properties of Jack symmetric functions*, Adv. Math. **77** (1989), 76–115.

[23] L. Vretare, *Formulas for elementary spherical functions and generalized Jacobi polynomials*, SIAM J. Math. Anal. **15** (1984), 805–833.

[24] G. N. Watson, *The product of two hypergeometric functions*, Proc. London Math. Soc. (2) **20** (1922), 189–195.

[25] Z. Yan, *Generalized hypergeometric functions*, C. R. Acad. Sci. Paris Sér. I **310** (1990), 349–354.

DEPARTMENT OF MATHEMATICS, UNIVERSITY OF LEIDEN, P.O. BOX 9512, 2300 RA LEIDEN, THE NETHERLANDS

Contemporary Mathematics
Volume **138**, 1992

On Certain Spaces of Harmonic Polynomials

N. BERGERON AND A. M GARSIA*

ABSTRACT. Our object of study here is the linear span of translates of the standard Garnir elements of a fixed shape μ. This remarkable space of polynomials, which we denote by $\mathbf{G}_\mu$, is an S_n-module whose graded character may be expressed in terms of the Kostka-Foulkes polynomials $K_{\lambda\mu}(q)$. This may be the simplest and most elementary graded S_n-module yielding the integrality and positivity of the coefficients of the $K_{\lambda\mu}(q)$. Another graded S_n-module with the same property is the polynomial quotient ring $\mathbf{R}_\mu$ studied by DeConcini-Procesi in [8] and by Garsia-Procesi in [12]. We prove here that $\mathbf{G}_\mu$ and $\mathbf{R}_\mu$ have the same graded character. The particular case when $\mu = 1^n$ is specially interesting since $\mathbf{G}_\mu$ then reduces to the linear span of translates of the Vandermonde determinant. A number of properties of the spaces $\mathbf{G}_\mu$ are stated without proof in a paper of Lascoux [18]. We show here how some of these properties may be derived from joint work of Garsia-Procesi [12].

Introduction

If A is an ordered subset of $X_n = \{x_1, x_2, \ldots, x_n\}$ we let $\Delta(A)$ denote the Vandermonde determinant in the elements of A. To be precise (since order counts), if $A = \{x_{a_1}, x_{a_2}, \ldots, x_{a_m}\}$ we set

$$\Delta(A) \;=\; \det \|x_{a_i}^{j-1}\|_{i,j=1}^m \;, \qquad\qquad I.1$$

with the convention that $\Delta(A) = 1$ if A is an empty subset. Let $\mu = \{0 \leq \mu_1 \leq \mu_2 \leq \cdots \leq \mu_n\}$ be a partition of n and T be an injective tableau T of shape μ with entries $1, 2, \ldots, n$. Recalling that we use the French notation of depicting tableaux, let $C_n, C_{n-1}, \ldots, C_1$ denote the columns of T ordered from left to right. Note, that with our conventions concerning partitions, if $\mu' = \{0 \leq \mu'_1 \leq \mu'_2 \leq \cdots \leq \mu'_n\}$ is the conjugate of μ, then C_i has length μ'_i. Of course, if the diagram of μ has only k columns then $C_{n-k}, \ldots, C_2, C_1$, are

1991 *Mathematics Subject Classification*. Primary 05E05, 15A52.

* Work carried out under NSF grant support.

This paper is in final form and no version of it will be submitted for publication elsewhere .

necessarily empty. Now let A_i be the ordered subset of X_n obtained by selecting the variables x_j with indices in C_i in the order occurring in T (from bottom to top). This given, we set

$$\Delta_T(x) \;=\; \Delta(C_n)\Delta(C_{n-1})\ldots\Delta(C_1) \;, \qquad\qquad I.2$$

and refer to it as the *Garnir element* corresponding to T, we shall also say that $\Delta_T(x)$ is *standard, of shape* μ,...etc, if the same holds true for T itself.

For instance if

$$T \;=\; \begin{matrix} 7 & & \\ 2 & 3 & 6 \\ 1 & 4 & 5 \end{matrix}$$

then

$$\Delta_T(x) \;=\; \det\begin{pmatrix} 1 & x_1 & x_1^2 \\ 1 & x_2 & x_2^2 \\ 1 & x_7 & x_7^2 \end{pmatrix} \times \det\begin{pmatrix} 1 & x_4 \\ 1 & x_3 \end{pmatrix} \times \det\begin{pmatrix} 1 & x_5 \\ 1 & x_6 \end{pmatrix} \;.$$

These polynomials were introduced by Garnir in [9], in his reconstruction of Young's natural representation. It will be convenient here and after to denote by $\lambda(T)$ the shape of a tableau T and by $ST(\mu)$ the collection of all standard tableux of shape μ. It is not difficult to show that the space

$$\Gamma_\mu \;=\; \mathcal{L}\{\Delta_T \,:\, T \in ST(\mu)\} \;, \qquad\qquad I.3$$

linear span of the standard Garnir elements of shape μ, is an irreducible S_n-module with character given by χ^μ in the Young indexing. To be precise, the S_n-action we refer to here is the usual action by permutation of variables. That is for a polynomial $P(x) = P(x_1, x_2, \ldots, x_n)$ and a permutation $\sigma \in S_n = (\sigma_1, \sigma_2, \ldots, \sigma_n)$ we set

$$\sigma P(x) \;=\; P(x_{\sigma_1}, x_{\sigma_2}, \ldots, x_{\sigma_n}) \;. \qquad\qquad I.5$$

Our object of study here is the space

$$\mathbf{G}_\mu \;=\; \mathcal{L}\{\Delta_T(x+t) \,:\, T \in ST(\mu)\} \qquad\qquad I.6$$

linear span of *translates* of the standard Garnir elements of shape μ. It is easy to see (by Taylor's theorem) that $\mathbf{G}_\mu$ is also the linear span of derivatives of Garnir elements of shape μ. Thus we can also write

$$\mathbf{G}_\mu \;=\; \mathcal{L}\{\partial^p \Delta_T(x) \,:\, T \in ST(\mu)\} \qquad\qquad I.7$$

with

$$\partial^p = \partial_1^{p_1} \partial_2^{p_2} \cdots \partial_n^{p_n}$$

denoting a generic monomial in the partial derivatives $\partial_i = \partial/\partial_{x_i}$.

The space Γ_μ may be viewed as the linear closure of the S_n-orbit of a Garnir element of shape μ. In the same vein $\mathbf{G}_\mu$ may be viewed as the linear closure of the orbit of a Garnir element of shape μ under the action of the *affine* version of S_n.

The presentation in I.5 makes it clear that $\mathbf{G}_\mu$ is a graded vector space, since any polynomial in $\mathbf{G}_\mu$ has all of its homogeneous components in $\mathbf{G}_\mu$ as well. We thus have the direct sum decomposition

$$\mathbf{G}_\mu = \bigoplus_{m \geq 0} H_m(\mathbf{G}_\mu) \ ,$$

where $H_m(\mathbf{G}_\mu)$ denotes the subspace of $\mathbf{G}_\mu$ consisting of all the homogeneous elements of degree m. Since the action in I.5 preserves degree, each $H_m(\mathbf{G}_\mu)$ yields a representation of S_n. Let us denote by p_m^μ its character and set

$$p^\mu(q) = \sum_{m \geq 0} q^m p_m^\mu \ . \qquad\qquad I.8$$

We shall refer to $p^\mu(q)$ as the *graded* character of $\mathbf{G}_\mu$. In particular the decomposition of $p^\mu(q)$ into its irreducible constituents

$$p^\mu(q) = \sum_{\lambda \vdash n} \chi^\lambda \, c_{\lambda,\mu}(q) \qquad\qquad I.9$$

yields that the polynomials $c_{\lambda,\mu}(q)$ necessarily have non-negative integer coefficients.

Remarkably, the $c_{\lambda,\mu}(q)$ are very closely related to the q-Kostka coefficients. The precise relation is

$$c_{\lambda,\mu}(q) = K_{\lambda\mu}(1/q) q^{n(\mu)} \qquad\qquad I.10$$

with

$$n(\mu) = \sum_{i=0}^{n-1} i \, \mu_{n-i} \qquad\qquad I.11$$

This relation yields a powerful way of studying the polynomials $K_{\lambda\mu}(q)$. In fact, as we shall see, the representation theoretical model leads to new combinatorial interpretations of their coefficients.

The main object of this paper is a proof of I.10. We shall show by elementary representation theoretical arguments that $\mathbf{G}_\mu$ is, as a graded S_n-module, equivalent to the module $\mathbf{R}_\mu$ studied by DeConcini-Procesi in [8], and by Garsia-Procesi in [12]. A proof of I.10 for the module $\mathbf{R}_\mu$ was given by Garsia-Procesi in [12]. Thus the desired result here follows in an elementary manner by combining the present arguments with those in [12].

We should mention, (see [17] II §3 ex. 1 p. 92 and III §7 ex. 9 p.136), that the first proof of the positivity result for the $K_{\lambda\mu}(q)$ is essentially contained in early work of algebraic geometers. Particularly important in this connection are the papers of Kostant [16], Steinberg [29], Hotta-Springer [14], Bohro-Kraft [5], Kraft [17], DeConcini-Procesi [8]. Unfortunately, to obtain the identity in I.10 for $\mathbf{R}_\mu$ through these works one needs deep tools (such as *Sheaves, Schemes, Cohomology Theory, etc...*) which are only accessible to a handful of experts. By contrast the basic arguments here and in [12] use only elementary tools of linear algebra and representation theory. We shall see that our approach yields an alternate way to study also the combinatorial properties of the $K_{\lambda\mu}(q)$. This is to be contrasted with the Lascoux-Schützenberger approach ([19],[20],[21],& [22]) which, though elementary, is considerably more painful (see [6]) than the one we follow here.

This paper consists of 4 sections. We have made our treatment as self-contained as possible, with exception of some facts whose proof has already been given in [12]. In the first section we present some general facts about homogeneous spaces of polynomials and graded representations. In the second section we review the definition of the DeConcini-Procesi module $\mathbf{R}_\mu$ as the quotient

$$\mathbf{R}_\mu \; = \; \mathbf{Q}[x_1, x_2, \ldots, x_n]/\mathbf{I}_\mu$$

where $\mathbf{I}_\mu$ is a suitable S_n-invariant homogeneous ideal. We introduce there the graded module $\mathbf{H}_\mu$ as simply the orthogonal complement of $\mathbf{I}_\mu$ with respect to the natural S_n-invariant scalar product of polynomials. It is then immediate that $\mathbf{H}_\mu$ is equivalent to $\mathbf{R}_\mu$ as a graded S_n-module. In section 4. we show that $\mathbf{H}_\mu$ and $\mathbf{G}_\mu$ are one and the same. In section 3. we treat in detail the special case of $\mathbf{H}_{1^n}$ since it has several combinatorial and algebraic aspects which are interesting in their own right. Several properties of the $K_{\lambda\mu}(q)$, the nesting properties of the spaces $\mathbf{H}_\mu$ and some conjectures are also to be found in sections 2. and 4.

ACKNOWLEDGEMENT. The authors are deeply indebted to Lascoux and Schützenberger for making available to us their preprints [18],[23]. These works contain a number of extremely interesting assertions without proof. The present work is the result of our efforts to provide proofs for some of them.

1. Homogeneous subspaces of polynomials

Throughout this writing we shall denote by $\mathbf{R} \; = \; \mathbf{Q}[x_1, x_2, \ldots, x_n]$ the ring of polynomials

$$P(x) \; = \; P(x_1, x_2, \ldots, x_n) \; = \; \sum_p c_p \, x^p \; ,$$

with rational coefficients c_p. Here and in the following we let $x^p = x_1^{p_1} x_2^{p_2} \cdots x_n^{p_n}$ and $\partial^p = \partial_1^{p_1} \partial_2^{p_2} \cdots \partial_n^{p_n}$ where $\partial_1 = \partial/\partial x_1$, $\partial_2 = \partial/\partial x_2, \ldots \partial_n = \partial/\partial x_n$ denote partial derivatives with respect to the given variables.

Given a formal power series $f(x) = \sum_q f_q \, x^q$ we let

$$f(\partial) \; = \; f(\partial_1, \partial_2, \ldots, \partial_n) \; = \; \sum_q f_q \, \partial^q \qquad\qquad 1.1$$

denote the corresponding differential operator. Of course the action of $f(\partial)$ on a polynomial is simply a linear extension of the action of the monomial ∂^q on the monomial x^p. That is we set

$$\partial_1^{q_1} \partial_2^{q_2} \cdots \partial_1^{q_n} \, x_1^{p_1} x_2^{p_2} \cdots x_1^{p_n} \; = \; \prod_{i=1}^n p_i (p_i - 1) \cdots (p_i - q_i + 1) \, x_i^{p_i - q_i} \; .$$

Note that if P is a polynomial of degree M and f is given by 1.1 then

$$f(\partial)P \;=\; \sum_{|q|\leq M} f_q\,\partial^q\,P \;,$$

where we have set $|q| = q_1 + q_2 + \cdots + q_n$. Thus no questions of convergence arise when dealing with formal power series as differential operators acting on polynomials.

In particular if we set

$$e^{t_1\partial_1+t_2\partial_2+\cdots+t_n\partial_n} \;=\; \sum_q t^q\frac{\partial^q}{q!} \qquad\qquad (\text{where } q! = q_1!q_2!\cdots q_n! \,)$$

then it is an immediate consequence of Taylor's theorem that for any polynomial P we have

$$e^{t_1\partial_1+t_2\partial_2+\cdots+t_n\partial_n}P(x_1,x_2,\ldots,x_n) \;=\; P(x_1+t_1,x_2+t_2,\ldots,x_n+t_n) \;.$$

It will be convenient here and in the following to denote by π_m the linear operator on polynomials defined by setting

$$\pi_m\,x^p \;=\; \begin{cases} x^p & \text{if } |p| = m, \\ 0 & \text{if } |p| \neq m. \end{cases}$$

Clearly, π_m extracts the homogeneous component of degree m out of a polynomial. In particular, given a collection $\mathbf{V}$ of polynomials we shall let $\mathcal{H}_m(\mathbf{V})$ denote the image of $\mathbf{V}$ by π_m. In other words $\mathcal{H}_m(\mathbf{V})$ simply consists of the collection of homogeneous components of degree m from elements of $\mathbf{V}$.

This given, we have the following

DEFINITION 1.1. A subspace $\mathbf{V} \subseteq \mathbf{R}$ is said to be *homogeneous* if and only if $\mathcal{H}_m(\mathbf{V}) \subseteq \mathbf{V}$ for all $m \geq 0$.

We can easily see that every homogeneous subspace $\mathbf{V}$ admits the direct sum decomposition

$$\mathbf{V} \;=\; \mathcal{H}_0(\mathbf{V}) \;\oplus\; \mathcal{H}_1(\mathbf{V}) \;\oplus\; \mathcal{H}_2(\mathbf{V}) \;\oplus\; \cdots \qquad\qquad 1.2$$

A typical example of a homogeneous subspace is the collection of solutions of a fixed system of homogeneous differential equations. More precisely, if we set

$$\mathbf{S}(f_1,f_2,\ldots,f_k) \;=\; \{P \in \mathbf{R} \;:\; f_1(\partial)P = 0, f_2(\partial)P = 0,\ldots,f_k(\partial)P = 0\}$$

then, if each of the f_i is a homogeneous polynomial, it is easily seen that the solution space $\mathbf{S}(f_1,f_2,\ldots,f_k)$ itself is a homogeneous subspace. Indeed, if

$$P \;=\; \sum_{m=0}^{M} P_m \qquad\qquad (\text{ with } P_m = \pi_m P \,)$$

and degree$(f_i) = m_i$ then degree$(f_i(\partial)P_m) = m - m_i$, and the equation

$$0 = f_i(\partial)P = \sum_{m=0}^{M} f_i(\partial)P_m$$

implies that the individual summands $f_i(\partial)P_m$ must be zero as well.

Note further that every $P \in \mathbf{S}(f_1, f_2, \ldots, f_k)$ will also be annihilated by every differential operator of the form

$$A_1(\partial)f_1(\partial) + A_2(\partial)f_2(\partial) + \cdots + A_k(\partial)f_k(\partial) \qquad 1.3$$

with each A_i an arbitrary differential operator. The collection of polynomials given by 1.3 with the A_i restricted to be also polynomials, is usually denoted by

$$(f_1, f_2, \ldots, f_k)$$

and is referred to as the *ideal generated by* $f_1, f_2, \ldots, f_k$. It is easy to see that, when each f_i is homogeneous then $(f_1, f_2, \ldots, f_k)$ is also a homogeneous subspace of $\mathbf{R}$.

A homogeneous subspace $\mathbf{V} \subseteq \mathbf{R}$, although possibly infinite-dimensional, behaves in many ways like a finite-dimensional subspace. Indeed, each of its homogeneous components $\mathcal{H}_m(\mathbf{V})$ is finite-dimensional. In fact,

$$\dim(\mathcal{H}_m(\mathbf{V})) \leq \dim(\mathcal{H}_m(\mathbf{R})) = \binom{n+m-1}{n-1}$$

This circumstance yields, as we shall see, a number of properties that are not available for more general subspaces. First of all, we can write down the generating function of dimensions of the components $\mathcal{H}_m(\mathbf{V})$. This is the formal series

$$F_{\mathbf{V}}(q) = \sum_{m \geq 0} q^m \dim(\mathcal{H}_m(\mathbf{V})) \qquad 1.4$$

which is usually referred to as the *Hilbert series* of $\mathbf{V}$. Secondly, because of 1.2, we can always produce a basis $\mathcal{B}(\mathbf{V})$ for $\mathbf{V}$ consisting of homogeneous polynomials. We shall also have the decomposition

$$\mathcal{B}(\mathbf{V}) = \sum_{m \geq 0} \mathcal{B}_m(\mathbf{V})$$

where $\mathcal{B}_m(\mathbf{V})$ denotes the collection of elements of degree m in $\mathcal{B}(\mathbf{V})$, and "$\sum$" here simply means disjoint union. This gives us the useful formula

$$F_{\mathbf{V}}(q) = \sum_{b \in \mathcal{B}(\mathbf{V})} q^{\text{degree}(b)} . \qquad 1.5$$

In particular we see that

$$F_{\mathbf{R}}(q) = \sum_p q^{\text{degree}(x^p)} = \sum_p q^{p_1 + p_2 + \cdots + p_n} = \frac{1}{(1-q)^n} . \qquad 1.6$$

There is a natural scalar product on $\mathbf{R}$ which brings to light another important property of homogeneous subspaces. To define it, let us denote by L_o the linear operation of evaluating a polynomial at 0. That is

$$L_o\, P(x_1, x_2, \ldots, x_n) \;=\; P(x_1, x_2, \ldots, x_n)\,|_{x_1 = x_2 = \cdots x_n = 0} \qquad 1.7$$

This given we set

$$\langle P\,,\,Q \rangle \;=\; L_o P(\partial) Q(X_n)\;. \qquad\qquad 1.8$$

It is easy to see that this scalar product makes the monomials $\{x^p\}_p$ into an orthogonal basis. In fact, we have

$$\langle x^p\,,\,x^q \rangle \;=\; \begin{cases} 0 & \text{if } p \neq q \\ p! & \text{if } p = q. \end{cases}$$

This scalar product has a number of useful properties. In particular, it is S_n-invariant. That is, for any $\sigma \in S_n$ and any pair of polynomials $P, Q \in \mathbf{R}$ we have

$$\langle \sigma P\,,\,\sigma Q \rangle \;=\; \langle P\,,\,Q \rangle\;. \qquad\qquad 1.9$$

Indeed, under the action I.5

$$\sigma\, P(\partial) Q(X_n) \;=\; P(\partial_{\sigma_1} \partial_{\sigma_2}, \ldots, \partial_{\sigma_n}) Q(x_{\sigma_1}, x_{\sigma_2}, \ldots, x_{\sigma_n}) \qquad 1.10$$

and 1.9 is obtained by evaluating at the origin.

Another facet of the finite-dimensional qualities of homogeneous subspaces can be stated as follows.

PROPOSITION 1.1. *The orthogonal complement* $\mathbf{U} = \mathbf{V}^{\perp}$ *of a homogeneous subspace* $\mathbf{V}$ *(with respect to the scalar product in 1.8) is also homogeneous. Moreover, we have*

$$a)\quad \mathbf{V}^{\perp\perp} \;=\; \mathbf{V}$$
$$\qquad\qquad\qquad\qquad\qquad\qquad 1.11$$
$$b)\quad \mathbf{R} \;=\; \mathbf{U}\,\oplus\,\mathbf{V}.$$

PROOF. Let $\mathcal{B}_m$ be an orthonormal basis for $\mathcal{H}_m(\mathbf{V})$. Let us then apply the Gram-Schmidt orthonormalization procedure to the ordered collection consisting of the elements of $\mathcal{B}_m$ arranged in any order followed by the monomials $\{x^p\}_{|p|=m}$ in lex order. This produces an orthonormal basis $\mathcal{C}_m$ for $\mathcal{H}_m(\mathbf{R})$ which consists of $\mathcal{B}_m$ followed by an orthonormal set $\mathcal{A}_m$. This given, any polynomial $P \in \mathbf{R}$ may be written in the form

$$P \;=\; \sum_{m \geq 0} \Big(\sum_{a \in \mathcal{A}_m} \langle P\,,\,a \rangle\, a \;+\; \sum_{b \in \mathcal{B}_m} \langle P\,,\,b \rangle\, b \Big)\;. \qquad 1.12$$

Thus the polynomials P that are orthogonal to $\mathbf{V}$ are those of the form

$$P \;=\; \sum_{m \geq 0} \sum_{a \in \mathcal{A}_m} \langle P\,,\,a \rangle\, a\;.$$

In other words $\{\mathcal{A}_m\}_{m \geq 0}$ is an orthonormal basis for $\mathbf{U} \;=\; \mathbf{V}^{\perp}$. This implies that $\mathbf{U}$ is a homogeneous subspace of $\mathbf{R}$ and that 1.11 a) & b) are immediate consequences of 1.12. $\square$

We should also note

PROPOSITION 1.2. *The solution space* $S(f_1, f_2, \ldots, f_k)$ *is the orthogonal complement of the ideal* $(f_1, f_2, \ldots, f_k)$. *In particular, if* $f_1, f_2, \ldots, f_n$ *are homogeneous then we also have*

$$(f_1, f_2, \ldots, f_n) \ = \ S(f_1, f_2, \ldots, f_n)^{\perp} \ . \qquad\qquad 1.13$$

PROOF. It is clear that $S(f_1, f_2, \ldots, f_k)$ is contained in $(f_1, f_2, \ldots, f_k)^{\perp}$. Thus we need only show the converse. To this end note that if $f \in (f_1, f_2, \ldots, f_k)$ then also the product $x^q f$ is in $(f_1, f_2, \ldots, f_k)$ for any monomial x^q. This implies that if $P \in (f_1, f_2, \ldots, f_k)^{\perp}$ then P must be orthogononal not only to f but to $x^q f$ as well . In other words we must have

$$L_o \partial^q P(\partial) Q(X_n) \ = \ 0$$

for arbitrary q. That is, the polynomial $P(\partial)Q(X_n)$ and all its derivatives must be equal to zero at the origin. From Taylor's theorem we deduce then that $P(\partial)Q(X_n)$ must be identically zero itself. This proves the reverse inclusion

$$(f_1, f_2, \ldots, f_k)^{\perp} \ \subseteq \ S(f_1, f_2, \ldots, f_k) \ .$$

The last assertion follows from proposition 1.1.

We can proceed a bit more generally and define an action of $GL(n)$ on $\mathbf{R}$ by setting, for any $n \times n$ matrix $A = \|a_{ij}\|_{i,j=1}^{n}$ and any monomial x^p

$$T_A \ x_1^{p_1} x_2^{p_2} \cdots x_n^{p_n} \ = \ \prod_{j=1}^{n} (T_A x_j)^{p_j} \ , \qquad\qquad 1.14$$

where, for each $j = 1 \ldots n$ we let

$$T_A \ x_j \ = \ \sum_{i=1}^{n} x_i a_{ij} \ .$$

This action extends, by linearity, to all polynomials $P(x_1, x_2, \ldots, x_n) \in \mathbf{R}$. Clearly, T_A preserves the degree of a polynomial and thus it leaves invariant each of the spaces $\mathcal{H}_m(\mathbf{R})$. From 1.14 we derive that for a monomial x^q of degree m we have

$$T_A \ x^q \ = \ \sum_{|p|=m} x^p \ (T_A x^q) \ |_{x^p} \ .$$

Thus the action of T_A on the subspace $\mathcal{H}_m(\mathbf{R})$, when expressed in terms of the basis

$$\mathcal{M}_m = \{x^p : |p| = m\} \ ,$$

is given by the matrix

$$A^{(m)} \ = \ \|(T_A x^q) \ |_{x^p}\| \qquad\qquad (\ |p|, |q| = m \ ) \ . \qquad\qquad 1.15$$

This is what is usually referred to as the mth *symmetric power* of A.

An important special case of 1.14 is obtained when A is taken to be the matrix which corresponds to the action of a permutation. To be precise, note that if for a permutation $\sigma \in S_n$ we set

$$P(\sigma) \;=\; \|\chi(i = \sigma_j)\|_{i,j=1}^{n} \qquad\qquad 1.16$$

then

$$T_{P(\sigma)} x_1^{p_1} x_2^{p_2} \cdots x_n^{p_n} \;=\; x_{\sigma_1}^{p_1} x_{\sigma_2}^{p_2} \cdots x_{\sigma_n}^{p_n} \;.$$

Thus in this case 1.14 reduces to I.5.

Now let G be a finite group of $n \times n$ matrices and and let $\mathbf{V}$ be a homogeneous subspace of $\mathbf{R}$ which is invariant under the action of G. More precisely, we assume that for each $A \in G$ and $P \in \mathbf{V}$ we have $T_A P \in \mathbf{V}$. Since the action in 1.14 preserves degree, each subspace $\mathcal{H}_m(\mathbf{V})$ is also invariant. Thus the action of G on $\mathbf{V}$ decomposes into a direct sum of finite-dimensional representations. Let then $\chi_m^{\mathbf{V}}(q)$ denote the character of the representation corresponding to $\mathcal{H}_m(\mathbf{V})$ and set

$$\chi^{\mathbf{V}}(q) \;=\; \sum_{m \geq 0} q^m \chi_m^{\mathbf{V}}(q) \;.$$

This is in complete analogy with what we did in the definition 1.5. Indeed, $F_{\mathbf{V}}(q)$ is none other than the evaluation of $\chi^{\mathbf{V}}(q)$ at the identity element of G. We shall refer to $\mathbf{V}$ as a *graded G-module*, and to $\chi^{\mathbf{V}}(q)$ as the corresponding *graded* character. When convenient we shall also use the symbols *char* and *char$_q$* to denote characters and graded characters respectively. In particular we may write *char$_q$* $\mathbf{V}$ instead of $\chi^{\mathbf{V}}(q)$.

Note that if $\mathcal{B}$ is any homogeneous basis of $\mathbf{V} \subseteq \mathbf{R}$ we may obtain the corresponding graded character at an element A of G, by the formula

$$\chi^{\mathbf{V}}(A; q) \;=\; \sum_{b \in \mathcal{B}} q^{\deg(b)}\, T_A b \,|_b \;. \qquad\qquad 1.17$$

We shall also refer to the representation of G corresponding to the basis $\mathcal{B}$ as a graded representation of G.

The celebrated *Master Theorem* of MacMahon [25] may be viewed as a formula for the generating function of the traces of the successive symmetric powers of a given matrix. In other words, the Master Theorem yields a closed form for the graded character of the action of the operator T_A defined in 1.14. Using the notation 1.15 we can state

THEOREM 1.1. *(MacMahon)*

$$\sum_{m \geq 0} q^m \operatorname{trace} A^{(m)} \;=\; \sum_p \det q^{|p|}\, (T_A x^p)\,|_{x^p} \;=\; \frac{1}{\det(I - qA)} \;. \qquad 1.18$$

PROOF. Although 1.18 is a purely combinatorial fact (see [7]), in our context we should be contented with the following somewhat excessive argument. We note first that the definitions 1.14 and 1.15 immediately imply that for any two $n \times n$ matrices A, B we have

$$A^{(m)} B^{(m)} \;=\; (AB)^{(m)} \;. \qquad\qquad 1.19$$

This given, since the trace function is invariant under similarity, we only have to check the validity of 1.18, for triangular matrices. However, if A is triangular and $\lambda_1, \lambda_2, \ldots, \lambda_n$ are its diagonal elements, then 1.18 reduces to the trivial identity

$$\sum_p q^{|p|} \lambda_1^{p_1} \lambda_2^{p_2} \cdots \lambda_n^{p_n} \;=\; \prod_{i=1}^{n} \frac{1}{1 - q\lambda_i} \;. \qquad\qquad 1.20$$

This proves the theorem. $\square$

By letting $A = P(\sigma)$ and setting $\chi^{\mathbf{R}}(P(\sigma); q) = \chi^{\mathbf{R}}(\sigma; q)$ we derive the important corollary:

THEOREM 1.2. *For any permutation* $\sigma \in S_n$

$$\chi^{\mathbf{R}}(\sigma, q) \;=\; \sum_p q^{|p|} \, \sigma x^p \,|_{x^p} \;=\; \prod_{i=1}^{n} \frac{1}{(1 - q^i)^{m_i(\sigma)}} \;=\; p_{\lambda(\sigma)}(1, q, q^2, \ldots)$$
$$1.21$$

where $m_i(\sigma)$ *denotes the number of cycles of length* i *in* σ, *and* $\lambda(\sigma)$ *denotes the* shape *of* σ; *that is,*

$$\lambda(\sigma) \;=\; \prod_{i=1}^{n} i^{\,m_i(\sigma)} \;.$$

PROOF. For the second equality it is sufficient to note that the characteristic polynomial of the matrix corresponding to a cycle of length i is precisely given by $1 - q^i$. We recall here that p_λ denotes the power symmetric function indexed by partition $\lambda = (\lambda_1, \lambda_2, \ldots, \lambda_n)$. That is

$$p_\lambda(x_1, x_2, x_3, \ldots) \;=\; \prod_{j=1}^{n} p_{\lambda_j}(x_1, x_2, x_3, \ldots) \;. \qquad\qquad 1.22$$

with

$$p_k(x_1, x_2, x_3, \ldots) \;=\; \sum_{i \geq 1} x_i^k \;.$$

Thus the last equality in 1.21 is obtained by making the replacements $\lambda \to \lambda(\sigma)$ and $x_i \to q^{i-1}$ in 1.22. $\square$

Making use of the Frobenius expansion of the power symmetric function we can transform 1.21 into a combinatorial identity which will turn out to be very useful for us here:

THEOREM 1.3.

$$\chi^{\mathbf{R}}(\sigma, q) \;=\; \sum_{\lambda \vdash n} \chi^\lambda(\sigma) \, \frac{\sum_{T \in ST(\lambda)} q^{c(T)}}{(1 - q)(1 - q^2) \cdots (1 - q^n)} \qquad\qquad 1.23$$

where $c(T)$ denotes the "cocharge" of T.

PROOF. The Frobenius expansion [24] gives that

$$p_{\lambda(\sigma)}(1, q, q^2, \dots) = \sum_{\lambda \vdash n} \chi^\lambda(\sigma)\, S_\lambda(1, q, q^2, \cdots) \qquad 1.24$$

and 1.23 follows from the well known ([24] ex. 14 p. 49) identity

$$S_\lambda(1, q, q^2, \cdots) = \frac{\sum_{T \in ST(\lambda)} q^{c(T)}}{(1 - q)(1 - q^2) \cdots (1 - q^n)}. \qquad 1.25$$

□

REMARK 1.1. Given a standard tableau T the *cocharge tableau* $C(T)$ corresponding to T is the column-strict tableau of same shape obtained by the following procedure. We first replace the entry 1 of T by 0. Then, inductively, having replaced the entry i of T by c, the entry $i + 1$ is replaced by c again if $i + 1$ is east of i in T and by $c + 1$ otherwise. The cocharge $c(T)$ may then be defined as the sum of the entries in $C(T)$. This given, it is easy to see that the smallest possible value of $c(T)$ as T varies among all standard tableaux of shape λ is given by the integer

$$n(\lambda) = \sum_{i=0}^{n-1} i\, \lambda_{n-i}\,, \qquad 1.26$$

which is the cocharge of the so called *superstandard* tableau of shape λ. This is the tableau obtained by filling the successive rows of the shape λ with the integers $1, 2, \dots, n$ from left to right starting with the longest row. We see then from 1.23 that the smallest m for which $H_m(\mathbf{R})$, as S_n-module, contains the irreducible representation with character χ^λ is precisely given by $n(\lambda)$.

2. Harmonic polynomials.

We have another way of obtaining graded representations. We start by giving a G-invariant homogeneous ideal $\mathbf{I}$ and then let $\mathbf{R}_I = \mathbf{R}/\mathbf{I}$ denote the corresponding quotient polynomial ring. The homogeneity of $\mathbf{I}$ makes $\mathbf{R}_I$ into a graded ring and the G-invariance of $\mathbf{I}$ allows us to define an action of G on $\mathbf{R}_I$ as well as on each of its homogeneous subspaces $H_m(\mathbf{R}_I)$. In complete analogy to what we did before we can then define the graded character of $\mathbf{R}_I$ by setting

$$char_q \mathbf{R}_I = \sum_m q^m char\, H_m(\mathbf{R}_I)\,.$$

This approach is in a sense *dual* to what we did in the previous section. To see this, for any two polynomials P, Q set

$$\langle P, Q \rangle_G = \frac{1}{|G|} \sum_{A \in G} \langle T_A P, T_A Q \rangle\,,$$

and let $\mathbf{H}_I$ be the orthogonal complement of $\mathbf{I}$ with respect to this scalar product. This is the standard construction of a G-invariant complement of a G-invariant subspace. We can easily show that $\mathbf{H}_I$ turns out to be a graded G-module with the same graded character as $\mathbf{R}_I$. More precisely, we have

PROPOSITION 2.1. *Let $H_m(\mathbf{H}_I)$ be the homogeneous component of degree m of $\mathbf{H}_I$ and let $\mathcal{B}_m = \{P_1, P_2, \ldots, P_M\}$ be any one of its bases. Then $\mathcal{B}_m$ is also a basis for $H_m(\mathbf{R}_I)$ and*

$$char_q \, \mathbf{R}_I \;=\; char_q \, \mathbf{H}_I \;=\; char_q \, \mathbf{R} \;-\; char_q \, \mathbf{I} \,. \qquad 2.1$$

PROOF. From the direct sum decomposition

$$H_m(\mathbf{R}) \;=\; H_m(\mathbf{H}_I) \;\oplus\; H_m(\mathbf{I})$$

we derive that every homogeneous polynomial P of degree m can be uniquely written in the form

$$P \;=\; \sum_{i=1}^{M} c_i \, P_i \;+\; E$$

with E an element of $\mathbf{I}$. This shows that $\mathcal{B}_m$ spans $H_m(\mathbf{R}_I)$. On the other hand the relation

$$\sum_{i=1}^{M} c_i \, P_i \;=\; E \in \mathbf{I}$$

can only hold for $E = 0$. Thus $P_1, P_2, \ldots, P_M$ are also independent as elements of $H_m(\mathbf{R}_I)$. This implies that $\mathcal{B}_m$ is a basis for $H_m(\mathbf{R}_I)$. Finally, the identity in 2.1 must hold true because the action of G on $\mathcal{B}_m$ is expressed by one and the same matrix whether we consider $\mathcal{B}_m$ in $H_m(\mathbf{H}_I)$ or in $H_m(\mathbf{R}_I)$. $\square$

The DeConcini-Procesi module $\mathbf{R}_\mu$ and the module $\mathbf{H}_\mu$ we referred to in the introduction are special cases of the pairs $\mathbf{R}_I$ and $\mathbf{H}_I$ defined above when the group G is S_n and $\mathbf{I}_\mu$ is a certain S_n-invariant ideal. To define $\mathbf{I}_\mu$ we need some notation.

Given a partition $\mu = (\mu_1, \mu_2, \ldots, \mu_n)$ $(0 \leq \mu_1 \leq \mu_2 \leq \ldots \leq \mu_n)$ we shall denote by $k(\mu)$ the number of positive components of μ. If $\mu' = (\mu'_1, \mu'_2, \ldots, \mu'_n)$ $(0 \leq \mu'_1 \leq \mu'_2 \leq \ldots \leq \mu'_n)$ is the partition conjugate of μ, we also set $d_k(\mu) = \mu'_1 + \mu'_2 + \cdots + \mu'_k$ $(1 \leq k \leq n)$. Given a subset S of the alphabet $X_n = \{x_1, x_2, \ldots, x_n\}$ we let $|S|$ denote its cardinality and $e_r(S)$ denote the rth elementary symmetric function of the variables in S. This given, we let $\mathcal{C}_\mu$ denote the collection of *partial* elementary symmetric functions

$$\mathcal{C}_\mu \;=\; \big\{ \, e_r(S) \,:\, k \geq r > k - d_k(\mu) \,,\, |S| = k \,,\, S \subseteq X_n \big\} \,. \qquad 2.2$$

The DeConcini-Procesi ideal $\mathbf{I}_\mu$ is simply the ideal generated by the elements of $\mathcal{C}_\mu$. That is we set

$$\mathbf{I}_\mu \;=\; (\mathcal{C}_\mu) \,, \qquad 2.3$$

and

$$\mathbf{R}_\mu \;=\; \mathbf{Q}[x_1, x_2, \ldots, x_n]/\mathbf{I}_\mu \,. \qquad 2.4$$

We also let

$$\mathbf{H}_\mu = (\mathbf{I}_\mu)^\perp \qquad\qquad 2.5$$

A few preliminary remarks are in order. First of all it is clear here that $\mathbf{I}_\mu$ is S_n-invariant for the simple reason that the collection $\mathcal{C}_\mu$ itself is S_n-invariant. Thus the considerations above, in particular Proposition 2.1, do apply with $G = S_n$. Of course in this case the scalar product $\langle P, Q \rangle_G$ reduces to that in 1.8. So that the superscript " $\perp$ " in 2.5 refers precisely to this scalar product. Proposition 1.2 then yields that $\mathbf{H}_\mu$ must be the solution space

$$\mathbf{H}_\mu = \{ Q \in \mathbf{Q}[x_1, x_2, \ldots, x_n] : e_r(S)(\partial)Q(x) = 0 \quad \forall e_r(S) \in \mathcal{C}_\mu \}. \qquad 2.6$$

The reader may easily verify that when $\mu = (0,0,1,2,2)$ then $\mu' = (0,0,0,2,3)$,

$$\big(1 - d_1(\mu), 2 - d_2(\mu), \ldots, n - d_n(\mu)\big) = (1,2,3,2,0)$$

and the collection $\mathcal{C}_{122}$ consists of the polynomials

$$e_1(x_1, x_2, x_3, x_4, x_5), e_2(x_1, x_2, x_3, x_4, x_5), \ldots, e_5(x_1, x_2, x_3, x_4, x_5)$$

$$e_3(x_1, x_2, x_3, x_4) , \; e_3(x_1, x_2, x_3, x_5) , \; e_3(x_1, x_2, x_4, x_5) ,$$

$$e_3(x_1, x_3, x_4, x_5) , \; e_3(x_2, x_3, x_4, x_5) ;$$

$$x_1 x_2 x_3 x_4 , \; x_1 x_2 x_3 x_5 , \; x_1 x_2 x_4 x_5 , \; x_1 x_3 x_4 x_5 , \; x_2 x_3 x_4 x_5 .$$

The presence of all the elementary symmetric functions in the alphabet $X_5 = \{x_1, x_2, \ldots, x_5\}$ is no accident here. This is true in full generality. That is

$$\mathcal{C}_\mu \supseteq \mathcal{C}_{1^n} = \{e_r(x_1, x_2, \ldots, x_n) : r = 1, 2, \ldots, n\} \qquad 2.7$$

and so

$$\mathbf{H}_\mu \subseteq \mathbf{H}_{1^n} = \{Q \in \mathbf{Q}[x_1, \ldots, x_n] : e_r(\partial_1, \ldots, \partial_n)Q(x) = 0, \; r = 1, 2, \ldots, n \} .$$
$$2.8$$

This means that if a polynomial Q belongs to a space $\mathbf{H}_\mu$ then it must satisfy all symmetric differential operators with zero constant term. In particular Q must also satify the Laplace equation. For this reason the elements of $\mathbf{H}_\mu$ will be referred to here as the *harmonic polynomials* corresponding to μ.

It is worthwhile noting that 2.8 is but one of the remarkable *nesting properties* of the spaces $\mathbf{H}_\mu$. To be precise let us recall that the partition λ is said to *dominate* the partition μ, and we write $\lambda \geq \mu$ if and only

$$\lambda_1 + \lambda_2 + \cdots + \lambda_k \geq \mu_1 + \mu_2 + \cdots + \mu_k \qquad (\forall \; 1 \leq k \leq n) .$$

It is well known that this partial order makes the collection of partitions of n into a lattice and that the lexicographic order of partitions (as vectors) refines the dominance order. This given, we can state

PROPOSITION 2.2. *The subspaces* $\mathbf{H}_\mu$ *are nested into each other according to their partition indexing. That is*

$$\mu \ \leq \ \lambda \quad \Rightarrow \quad \mathbf{H}_\mu \ \subseteq \ \mathbf{H}_\lambda \ . \hspace{3cm} 2.9$$

Moreover, we also have that for any pair λ, μ

$$\mathbf{H}_\lambda \cap \mathbf{H}_\mu \ = \ \mathbf{H}_{\lambda \wedge \mu} \hspace{3cm} 2.10$$

PROOF. It is well known (see [24] (1.11) p. 6) that we have $\mu \leq \lambda$ if and only if $\mu' \geq \lambda'$. This means that

$$\mu \ \leq \ \lambda \quad \Rightarrow \quad d_k(\mu) \ \geq \ d_k(\lambda) \quad \Rightarrow \quad \mathcal{C}_\mu \supseteq \mathcal{C}_\lambda \quad \Rightarrow \quad \mathbf{I}_\mu \supseteq \mathbf{I}_\lambda \ ,$$

and 2.9 immediately follows from the definition 2.5 of the space $\mathbf{H}_\mu$. To prove 2.10, we need to recall the definition of the partition $\lambda \wedge \mu$. To this end, note that because our partitions have weakly increasing components, the sequence $d(\mu) = (d_1(\mu), d_2(\mu), \ldots, d_n(\mu))$ satisfies the inequalities

$$d_k(\mu) \ - \ d_{k-1}(\mu) \ \leq \ d_{k+1}(\mu) \ - \ d_k(\mu) \hspace{2cm} (\text{for } 1 \leq k \leq n) \ ,$$

where we set $d_o(\mu) = 0$ here. Conversely, if a sequence $d = (0 = d_o, d_1, d_2, \ldots, d_n)$ satisfies

$$2d_k \leq d_{k-1} + d_{k+1} \hspace{2cm} (\forall \ 1 \leq k \leq n) \hspace{2cm} 2.11$$

then the differences $\nu'_k = d_k - d_{k-1}$ are weakly increasing and there is a partition ν giving $d_k = d_k(\nu)$. Now the partition $\nu = \lambda \wedge \mu$ is defined as the largest which is below both λ and μ. For this partition we must therefore have

$$d_k(\nu) \geq d_k(\lambda) \hspace{2cm} \& \hspace{2cm} d_k(\nu) \geq d_k(\mu) \ .$$

However, it is easy to see that the sequence

$$d_k = \max \left(d_k(\lambda), d_k(\nu) \right)$$

does satisfy the inequalities in 2.11, and we are forced to conclude that

$$d_k(\nu) \ = \ \max \left(d_k(\lambda), d_k(\nu) \right)$$

However, this implies that

$$\mathcal{C}_\nu \ = \ \mathcal{C}_\lambda \cup \mathcal{C}_\mu \ . \hspace{3cm} 2.12$$

But now, (again from 2.5) we get that

$$\mathbf{H}_\lambda \cap \mathbf{H}_\mu \ = \ \{ Q \in \mathbf{R} \ : \ e_r(S)(\partial)Q = 0 \quad \forall \quad e_r(S) \in \mathcal{C}_\lambda \cup \mathcal{C}_\mu \ \}$$

Thus 2.10 follows from 2.12 and 2.5. $\square$

REMARK 2.2. The definition 2.2 of the collection $\mathcal{C}_\mu$ might appear quite elaborate and unmotivated. The mystery is quickly dissipated if we recall where the spaces $\mathbf{R}_\mu$ originally came from. In the papers of Kraft [17] and DeConcini-Procesi [8], $\mathbf{R}_\mu$ is defined as the *coordinate ring of the diagonal matrices which are in the closure of the conjugacy class of a nilpotent of Jordan block structure given by the partition μ'.*

Note that if a matrix A is conjugate to such a nilpotent, then it is immediate that the (monic) greatest common divisor of the $k \times k$ minors of $A - tI$ is

$$t^{\mu'_1 + \mu'_2 + \cdots + \mu'_k} \quad . \tag{2.13}$$

Clearly this condition must also hold true for any A in the closure of this conjugacy class. In particular, if A is diagonal with entries $x_1, x_2, \ldots, x_n$, then for it to belong to this closure it is necessary that the polynomial

$$\prod_{i \in S}(x_i - t)$$

be divisible by the monomial in 2.13 for any subset S of $x_1, x_2, \ldots, x_n$ of cardinality k. But that simply says that

$$e_r(S) \;=\; 0 \qquad for \; r \;\geq\; k - (\mu'_1 + \mu'_2 + \cdots + \mu'_k) \;\;.$$

Thus the defining ideal of this variety must contain the collection $\mathcal{C}_\mu$. Now Kraft in [17] showed that the coordinate ring of this variety must have dimension greater or equal to the multinomial coefficient

$$\binom{n}{\mu} \;=\; \frac{n!}{\prod_{i=1}^{n} \mu_i!} \tag{2.14}$$

In [8] DeConcini-Procesi proved equality and identified the character of the corresponding S_n-module. However, in [8] DeConcini-Procesi use a different collection of generators for the ideal. The collection $\mathcal{C}_\mu$ was used by Tanisaki in [30] to simplify the DeConcini-Procesi proof. Basically, Tanisaki used $\mathcal{C}_\mu$ to show that

$$\dim \mathbf{R}_\mu \;\leq\; \binom{n}{\mu}$$

In this work, starting with 2.3 and 2.4 as definition of $\mathbf{R}_\mu$ we give a new proof of the Kraft inequality for the dimension of $\mathbf{R}_\mu$, by showing directly from 2.5 that

$$\dim \mathbf{H}_\mu \;\geq\; \binom{n}{\mu} \;\;. \tag{2.15}$$

In particular, we are thus providing an elementary way of completing the argument given by Tanisaki in [30], who referred to Kraft's paper for this side of the inequality.

3. Linear translates of the Vandermonde determinant.

In this section we shall study some algebraic and combinatorial properties of the space $\mathbf{G}_{1^n}$. It is easy to see that when $\mu = 1^n$ $\mathbf{G}_\mu$ reduces to

$$\mathbf{G}_{1^n} \;=\; \mathcal{L}[\Delta(x+t)] \;=\; \mathcal{L}[\partial^p \Delta(x)] \;, \qquad\qquad 3.1$$

where Δ is the Vandermonde determinant

$$\Delta(x) \;=\; \det \|x_i^{j-1}\| \;. \qquad\qquad 3.2$$

Note that the last two spaces in 3.1 are the same since, as we pointed out in section 1, every translation can be written as a differential operator. To see this we need only point out that each partial differentiation ∂_i is itself expressible in terms of translations. In fact, ∂_i is simply a *logarithm* of the corresponding difference operator $\Delta_i \;=\; e^{\partial_i} - 1$. To be precise we have

$$\partial_i \;=\; \sum_{p \geq 1} (-1)^{p-1} \frac{\Delta_i^{\,p}}{p} \;. \qquad\qquad 3.3$$

Our first goal is to show that the two spaces $\mathbf{G}_{1^n}$ and $\mathbf{H}_{1^n}$ are one and the same. The latter result is a particular case of a well known theorem of Steinberg [28] characterizing finite groups generated by reflections. This identification will be obtained by combining the following three propositions.

PROPOSITION 3.1. *For every symmetric polynomial P, with zero constant term, we have $P(\partial_1, \partial_2, \ldots, \partial_n)\, \Delta(x) \;=\; 0$. In particular we have the following inclusion*

$$\mathcal{L}\{\partial^p \,\Delta(x)\} \;\subseteq\; \mathbf{H}_{1^n} \;. \qquad\qquad 3.4$$

PROOF. Since P is symmetric and Δ is alternating, formula 1.10 yields that for any permutation σ we have

$$\sigma\, P(\partial)\, \Delta(x) \;=\; \epsilon(\sigma)\, P(\partial)\, \Delta(x) \;.$$

Now it is well known and easy to show that a polynomial that is alternating is either zero or a multiple of $\Delta(x)$. On the other hand, if P has zero constant term then the degree of $P(\partial)\, \Delta(x)$ will necessarily be less than the degree of $\Delta(x)$. Consequently $P(\partial)\, \Delta(x)$ cannot be a non-trivial multiple of $\Delta(x)$ and must therefore be identically zero. We conclude the proof by noting that $\Delta(x)$ and therefore all of its derivatives must be annihilated by each of the elementary symmetric function operators $e_r(\partial)$ and by each operator in the ideal generated by them. Thus, by 2.8, the inclusion in 3.4 must hold true as asserted. $\square$

PROPOSITION 3.2. *The monomials*

$$x_1^{\epsilon_1} x_2^{\epsilon_2} \cdots x_n^{\epsilon_n} \qquad\qquad with \quad 0 \leq \epsilon_i \leq i-1 \qquad\qquad 3.5$$

span $\mathbf{R}_{1^n}$. Moreover, every polynomial $P \in \mathbf{R}$ may be expressed in the form

$$P \;=\; \sum_{0 \leq \epsilon_i \leq i-1} A_\epsilon(x)\, x^\epsilon \;, \qquad\qquad 3.6$$

where the coefficients $A_\epsilon(x)$ are symmetric polynomials.

PROOF. Note, that from 2.7 it follows that

$$\mathbf{R}_{1^n} = \mathbf{Q}[x_1, x_2, \ldots, x_n]/(e_1, e_2, \ldots, e_n) \; . \qquad 3.7$$

Thus the elementary symmetric functions $e_r(x)$ are congruent to zero in $\mathbf{R}_{1^n}$. Denoting congruences *mod* $\mathbf{I}_\mu$ by the symbol $\cong_\mu$, we can write

$$(1 - tx_1)(1 - tx_2)\cdots(1 - tx_n) \;\cong_{1^n}\; 1 \; . \qquad 3.8$$

Thus for any $1 \le k \le n - 1$ we also have that

$$(1 - tx_1)(1 - tx_2)\cdots(1 - tx_k) \;\cong_{1^n}\; \frac{1}{(1 - tx_{k+1})(1 - tx_{k+2})\cdots(1 - tx_n)} \; . \qquad 3.9$$

Since the righthand side of this congruence is the generating function of the so-called *homogeneous* symmetric functions $h_s(x_{k+1}, x_{k+2}, \ldots, x_n)$, we deduce, by equating coefficients of t^r in 3.9, that

$$(-1)^r e_r(x_1, x_2, \ldots, x_k) \;\cong_{1^n}\; h_r(x_{k+1}, x_{k+2}, \ldots, x_n) \; . \qquad 3.10$$

This gives us the congruence

$$(t - x_1)(t - x_2)\cdots(t - x_k) \;\cong_{1^n}\; \sum_{r=0}^{k} t^r \, h_{k-r}(x_{k+1}, x_{k+2}, \ldots, x_n) \; .$$

By setting $t = x_k$ here we finally get the recursion

$$x_k^k \;\cong_{1^n}\; -\sum_{r=0}^{k-1} x_k^r \, h_{k-r}(x_{k+1}, x_{k+2}, \ldots, x_n) \; . \qquad 3.11$$

This immediately implies that, at the expense of an increase in the exponents of the variables $x_{k+1}, x_{k+2}, \ldots, x_n$ we can always reduce the exponent of the variable x_k to a value less than k. More precisely, using 3.11 for $k = 1$, which simply reduces to

$$x_1 = -x_2 - x_3 - \ldots - x_n$$

we can express (modulo $\mathbf{I}_{1^n}$) any monomial x^p as a linear combination of monomials in which x_1 doesn't appear. Recursively, having expressed every monomial x^p as a combination of monomials in which each x_i appears only to a power less than i, for $i = 1, 2, \ldots, k - 1$, by means of 3.11, we can complete the induction by expressing the latter monomials in terms of monomials in which x_k itself is also raised to a power less than k.

This shows that the monomials in 3.5 do span $\mathbf{R}_{1^n}$. In particular, every polynomial $P \in \mathbf{R}$ may be expanded in the form

$$P(x) = \sum_{0 \le \epsilon_i \le i-1} c_\epsilon \, x^\epsilon + \sum_{r=1}^{n} A_r(x) \, e_r(x) \; , \qquad 3.12$$

where the coefficients c_ϵ are given rational numbers and the $A_r(x)$ are suitable polynomials. Of course, nothing prevents us from applying the same expansion to each of the polynomials A_r . We can easily see then that by applying 3.12 recursively over and over, we end up expanding the original polynomial $P(x)$ as a linear combination of polynomials of the form

$$x^\epsilon \, e_1^{q_1}(x) e_2^{q_2}(x) \cdots e_n^{q_n}(x)$$

with $0 \le \epsilon_i \le i - 1$ and $q_i \ge 0$ integers. But this is in essence the final assertion of our proposition. □

PROPOSITION 3.3. *The polynomials*

$$\{ \, \partial^\epsilon \Delta(x) \, : \, 0 \le \epsilon_i \le i - 1 \, \} \qquad\qquad 3.13$$

are a basis for $\mathbf{G}_{1^n} = \mathcal{L}\{\partial^p \Delta(x)\}$.

PROOF. Note that the expansion in 3.12 applied to ∂^p yields that, modulo an element of the ideal generated by the differential operators $e_r(\partial)$, the operator ∂^p may be expressed as a linear combination of the operators ∂^ϵ with $0 \le \epsilon_i \le i - 1$. Thus proposition 3.1 gives that the polynomials in 3.13 do span $\mathbf{G}_{1^n}$. We only need to show then that they are linearly independent. But this is an immediate consequence of the fact that all the monomials appearing in a polynomial $\partial^\epsilon \Delta(x)$ are lexicographically strictly greater than the monomial $\partial^\epsilon x_1^0 x_2^1 \cdots x_n^{n-1}$. Indeed, we see that from the definition 3.2 it follows that

$$\partial^\epsilon \Delta(x) \;=\; \sum_\sigma \, \epsilon(\sigma) \, x_1^{\sigma_1 - 1 - \epsilon_1} x_2^{\sigma_2 - 1 - \epsilon_2} \cdots x^{\sigma_n - 1 - \epsilon_n} \; p(\sigma, \epsilon) \;,$$

where the $p(\sigma, \epsilon)$ are suitable integers. Now for each σ we do have

$$\sigma_1 - 1 - \epsilon_1 \,+\, \sigma_2 - 1 - \epsilon_2 \,+\, \cdots \,+\, \sigma_i - 1 - \epsilon_i$$
$$\ge 1 - 1 - \epsilon_1 + 2 - 1 - \epsilon_2 + \cdots + i - 1 - \epsilon_i,$$

and this inequality is strict, at least for one i, unless σ is the identity permutation. This simple fact makes it impossible that we should have a relation

$$\sum_{0 \le \epsilon_i \le i-1} c_\epsilon \, \partial^\epsilon \Delta(x) \;=\; 0 \qquad\qquad 3.14$$

without all of the c_ϵ equal to zero as well. Indeed, if such a relation were possible, then letting ϵ_o be the lexicographically smallest for which $c_{\epsilon_o} \ne 0$ we would reach a contradiction since the monomial corresponding to the identity permutation in $\partial^{\epsilon_o} \Delta(x)$ could not be cancelled by any of the terms produced by the remaining summands in 3.14. □

THEOREM 3.1.

$$\mathbf{G}_{1^n} \;=\; \mathbf{H}_{1^n} \qquad\qquad 3.15$$

PROOF. Proposition 3.1 gives the inclusion

$$\mathcal{L}[\partial^p \Delta(x)] \;=\; \mathbf{G}_{1^n} \;\subseteq\; \mathbf{H}_{1^n} \;=\; (e_1, e_2, \ldots, e_n)^\perp \qquad\qquad 3.16$$

On the other hand, what we have gathered so far implies the string of relations

$$n! \leq \dim \mathbf{G}_{1^n} \leq \dim \mathbf{H}_{1^n} = \dim \mathbf{R}_{1^n} \leq n! \qquad 3.17$$

Indeed the first follows from proposition 3.3, the second from proposition 3.1 the third from proposition 2.1 and the fourth from proposition 3.2. Thus equality must hold throughout in 3.17 and 3.16 must also be an equality as desired. $\square$

This proof has a number of consequences which are worth recording.

THEOREM 3.2.

(a) *The monomials*

$$\left\{ x_1^{\epsilon_1} x_2^{\epsilon_2} \cdots x_n^{\epsilon_n} \right\}_{0 \leq \epsilon_i \leq i-1} \qquad 3.18$$

form an integral basis for $\mathbf{R}_{1^n}$.

(b) *For a polynomial P we have*

$$P(\partial)\Delta(X) = 0 \qquad \text{if and only if} \qquad P \in (e_1, e_2, \ldots, e_n) \ . \qquad 3.19$$

In particular the map

$$P \Rightarrow P(\partial)\Delta(X) \qquad 3.20$$

is a vector space isomorphism of $\mathbf{R}_{1^n}$ *into* $\mathbf{H}_{1^n}$.

(c) *The polynomials*

$$x^{\epsilon} \, e_1^{p_1}(X) e_2^{p_2}(X) \cdots e_n^{p_n}(X) \qquad (\text{with } 0 \leq \epsilon_i \leq i-1 \text{ and } p_i \geq 0 \ ). \qquad 3.21$$

are an integral basis for $\mathbf{R}$. *In particular every polynomial* $P \in \mathbf{R}$ *has a unique expansion of the form*

$$P(X) = \sum_{0 \leq \epsilon_i \leq i-1} A_{\epsilon}(X) \, x^{\epsilon}$$

where the coefficients $A_{\epsilon}(X)$ *are symmetric polynomials in the variables* $x_1, x_2,$ x_n.

(d) *The Hilbert series of* $\mathbf{R}_{1^n}$ *and* $\mathbf{H}_{1^n}$ *are given by the polynomial*

$$F_{\mathbf{H}_{1^n}}(q) = (1+q)(1+q+q^2)\cdots(1+q+\cdots+q^{n-1}) \qquad 3.22$$

PROOF. The proof of theorem 3.1 gives that $\dim \mathbf{R}_{1^n} = n!$. Since the collection in 3.18 has also cardinality $n!$ and (by proposition 3.2) spans $\mathbf{R}_{1^n}$, then it must be a basis. This gives part (a).

The "if" in part (b) is proposition 1.3. To get the "only if" we note that 2.5 with $\mu = 1^n$ and 1.13 with $f_i = e_i$ gives (using 3.15)

$$(e_1, e_2, \ldots, e_n) = \mathbf{H}_{1^n}^{\perp} = \mathcal{L}[\partial^p \Delta(x)]^{\perp} \ .$$

This implies that if $P(\partial)\Delta(x) = 0$ then P must belong to $(e_1, e_2, \ldots, e_n)$. Moreover, it also shows that the kernel of the map in 3.20 is precisely the defining ideal of $\mathbf{R}_{1^n}$. This completes the proof of part (b).

The expansion in 3.6, as we have seen, simply says that the polynomials

$$x_1^{\epsilon_1} x_2^{\epsilon_2} \cdots x_n^{\epsilon_n} \, e_1^{p_1}(x) e_2^{p_2}(x) \cdots e_n^{p_n}(x) \qquad\qquad 0 \leq \epsilon_i \leq i-1 \qquad 3.23$$

span **R**. In fact, the proof of proposition 3.2 shows that if P is a polynomial with integer coefficients then its expansion in terms of these polynomials has also integer coefficients. This means that to prove part (c) we need only show that the right number of them fall in each subspace $H_m(\mathbf{R})$. In other words, (using 1.5 and 1.6) we must have

$$\sum_{0 \le \epsilon_i \le i-1} q^{\Sigma_i \epsilon_i} \sum_p q^{\Sigma_i i p_i} = \frac{1}{(1-q)^n} .$$

However, the left hand side here sums to

$$\frac{(1+q)(1+q+q^2)\cdots(1+q+\cdots+q^{n-1})}{(1-q)(1-q^2)\cdots(1-q^n)}$$

and this is easily seen to reduce to the right hand side.

Finally, part (d) is an immediate consequence of part (a) and the equality of $\mathbf{R}_{1^n}$ and $\mathbf{H}_{1^n}$ as graded modules.

There are two further useful corollaries which we leave to the reader to derive.

THEOREM 3.3. *A collection of polynomials $\mathcal{B}$ is a basis for $\mathbf{R}_{1^n}$ if and only if the collection $\{b(\partial)\Delta(X_n) : b \in \mathcal{B}\}$ is a basis for $\mathbf{H}_{1^n}$.*

Surprisingly, we have also the following remarkable fact.

THEOREM 3.4. *A collection of polynomials $\mathcal{B} \subseteq \mathbf{H}_{1^n}$ is a basis for $\mathbf{H}_{1^n}$ if and only if it is a basis for $\mathbf{R}_{1^n}$.*

We can also obtain a combinatorial interpretation for the coefficients $c_{\lambda,\mu}(q)$ when $\mu = 1^n$. More precisely, we can state:

THEOREM 3.5. *The action of S_n on $\mathbf{H}_{1^n}$ is simply a graded version of the left regular representation. In fact, we have*

$$p^{1^n}(\sigma;q) = \sum_{\lambda \vdash n} \chi^\lambda(\sigma)\, c_\lambda(q) .$$

with

$$c_\lambda(q) = \sum_{T \in ST(\lambda)} q^{c(T)} . \qquad\qquad 3.24$$

PROOF. We start by applying 1.17 for $A = P(\sigma)$, that is

$$\chi^{\mathbf{V}}(\sigma;q) = \sum_{b \in \mathcal{B}} q^{\text{degree}(b)}\, \sigma b\,|_b , \qquad\qquad 3.25$$

for $\mathbf{V} = R_{1^n}$ and $\mathcal{B}$ the basis in 3.18. To this end we note that using 3.12 we get the expansion

$$\sigma\, x^\epsilon = \sum_{\epsilon'} x^{\epsilon'}\, a_{\epsilon,\epsilon'}(\sigma) + \sum_{r=1}^n A_r(X)\, e_r(X) , \qquad\qquad 3.26$$

where (because of part (a) of Theorem 3.2) the coefficients $a_{\epsilon,\epsilon'}(\sigma)$ will necessarily be integers. This gives that in $\mathbf{R}_{1^n}$

$$\sigma\, x^\epsilon\,|_{x^\epsilon} \;=\; a_{\epsilon,\epsilon}(\sigma)\ ,$$

and therefore 3.25 yields

$$\chi^{\mathbf{R}_{1^n}}(\sigma;q) \;=\; \sum_\epsilon q^{|\epsilon|}\, a_{\epsilon,\epsilon}\ . \qquad\qquad 3.27$$

We now apply again 3.25 with $\mathbf{V} = \mathbf{R}$ and $\mathcal{B}$ the basis in 3.21. Here, 3.26 and the S_n-invariance of each e_r immediately give that

$$\sigma(x^\epsilon\, e_1^{p_1}\cdots e_n^{p_n}) \;=\; e_1^{p_1}\cdots e_n^{p_n}\, \sigma x^\epsilon$$

$$=\; \sum_{\epsilon'} x^{\epsilon'} e_1^{p_1}\cdots e_n^{p_n}\, a_{\epsilon,\epsilon'}(\sigma) \;+\; \sum_{r=1}^n A_r(X)\, e_1^{p_1}\cdots e_r^{p_r+1}\cdots e_n^{p_n}\ . \qquad 3.28$$

Note that, whatever the expansion of each polynomial $A_r(X)$ in terms of the basis in 3.21 turns out to be, the second sum in 3.28 cannot produce a term involving the basis element

$$x^\epsilon e_1^{p_1}\cdots e_n^{p_n}\ .$$

Thus we must have again

$$\sigma\, x^\epsilon e_1^{p_1}\cdots e_n^{p_n}\,|_{x^\epsilon e_1^{p_1}\cdots e_n^{p_n}} \;=\; a_{\epsilon,\epsilon}(\sigma)\ , \qquad\qquad 3.29$$

and 3.25 now yields the identity

$$\chi^{\mathbf{R}}(\sigma;q) = \sum_\epsilon \sum_p q^{|\epsilon|+p_1+2p_2+\cdots+np_n}\, a_{\epsilon,\epsilon}(\sigma)$$

$$=\; \frac{\sum_\epsilon q^{|\epsilon|}\, a_{\epsilon,\epsilon}(\sigma)}{(1-q)(1-q^2)\cdots(1-q^n)}\ . \qquad\qquad 3.30$$

Equating the right hand sides of 1.23 and 3.30 gives (using 3.27)

$$\chi^{\mathbf{R}_{1^n}}(\sigma;q) \;=\; \sum_\epsilon q^{|\epsilon|}\, a_{\epsilon,\epsilon}(\sigma) \;=\; \sum_{\lambda \vdash n} \chi^\lambda(\sigma) \sum_{T \in ST(\lambda)} q^{c(T)}$$

and the identity 3.24 follows then from 3.15 and the fact that $\mathbf{R}_{1^n}$ and $\mathbf{H}_{1^n}$ are equivalent graded modules. $\square$

REMARK 3.1. The expression for the coefficients $c_\lambda(q)$ in 3.24 suggests that there ought to be a natural basis $\{\pi[A,T]\}$ for $\mathbf{H}_{1^n}$ indexed by pairs of standard tableaux A, T of the same shape which would directly explain formula 3.24. Such a basis was recently discovered by E. Allen. To describe it we need some notation. First of all it will be convenient to have a standard way of labeling the cells of a Ferrers' diagram. We shall agree to label them by successive integers $1, 2, \ldots$ from left to right along the rows starting from the bottom row. Hereafter, the cell bearing label i will be referred to as the ith cell of the shape λ. In this manner, given a partition λ, we are able to identify a tableau A of shape λ with the function $a(x)$ which for $x = i$ gives the entry of A in the ith cell of λ. For A and B arbitrary tableau of shape $\lambda \vdash n$ let

$$m_{A,B}(x) \;=\; \prod_{i=1}^{n} x_{a(i)}^{b(i)}$$

This given, E. Allen sets, when A, T are also standard,

$$\pi[A,T] \;=\; P(A)m_{A,C(T)}(x) \qquad\qquad 3.31$$

where $P(A)$ denotes the row group of A. He then arranges the basis $\mathcal{A} = \{\,\pi[A,T]\,\}$ in blocks $\mathcal{A}_T$ defined by setting, for $T \in ST(\lambda)$:

$$\mathcal{A}_T \;=\; \{\,\pi[A,T] \;:\; A \in ST(\lambda)\,\}\,. \qquad\qquad 3.32$$

The final result in [2] is that, under a suitable total order of these blocks, the matrix which expresses the action of S_n in terms of this basis is block triangular with the diagonal block corresponding to $\mathcal{A}_T$ yielding a version of Young's natural representation with character χ^λ (in the Young indexing). Since all of the elements of $\mathcal{A}_T$ have degree $c(T)$, we easily see that the contribution of this block to the graded character is the term $q^{c(T)}\chi^{\lambda(T)}$, and formula 3.24 must necessarily follow.

The properties of $\mathbf{R}_{1^n}$ stated in Theorem 3.2 are more or less well known to the algebraic-combinatorial audience. The fact that the polynomials in 3.21 are an integral basis for $\mathbf{R}$ is proved in [3] using Galois theory. We have covered them here for sake of completeness and to prepare for our treatment of the general spaces $\mathbf{H}_\mu$. We shall present next some further facts that perhaps are not so well known. Some of these facts are also stated without proof in [18].

In [15] S.G. Hulsurkar shows, that a certain collection of translates of $\Delta(X)$ is an integral basis for $\mathcal{L}\{\partial^p \Delta(X)\}$. Hulsurkar proves this result in the more general context of Weyl groups. Now, the Hulsurkar collection, which we denote here by $\mathcal{HU}$, when specialized to the symmetric group case, is defined as follows. Let $u_1, u_2, \ldots, u_n$ be the unit coordinate vectors (that is u_i has the ith component equal to 1 and all the other equal to zero). For each permutation σ let

$$\epsilon_\sigma \;=\; \sum_{\sigma_i > \sigma_{i+1}} (u_{\sigma_1} + u_{\sigma_2} + \cdots + u_{\sigma_i}) \;-\; \mathrm{maj}(\sigma)(u_1 + u_2 + \cdots + u_n)$$

and set

$$\Delta_\sigma \;=\; \Delta(X + \epsilon_\sigma)\,. \qquad\qquad 3.33$$

We should mention, that by $\mathrm{maj}(\sigma)$ we mean the *major index* of σ, but as we shall quickly see its actual value has no effect on this formula. Then the Hulsurkar result may be stated as follows

THEOREM 3.6. *The collection $\mathcal{H}U = \{\Delta_\sigma\}_{\sigma \in S_n}$ is an integral basis for $\mathbf{H}_{1^n}$.*

To see how this theorem fits into the present context, we need a few observations and prove a few facts that will also be useful in the sequel. To this end, note first that there is also another natural way of mapping a polynomial $P \in \mathbf{R}$ into an element of $\mathbf{H}_{1^n}$. Roughly speaking, instead of making P into a differential operator and then applying it to $\Delta(X)$ as in the map

$$P \rightarrow P(\partial)\Delta(X) \qquad\qquad 3.34$$

we rather first make P into a *translation* operator. To be precise, we define a new map ψ by setting

$$\psi P = P(e^\partial)\Delta(X) = P(e^{\partial_1}, e^{\partial_2}, \ldots, e^{\partial_n})\Delta(X) . \qquad 3.35$$

This given, we can show that ψ has many properties in common with the map in 3.34. Before we can state them we need the following auxiliary fact.

PROPOSITION 3.4. *For any symmetric polynomial P we have*

$$P(e^\partial) \Delta(X) = P(1, 1, \ldots, 1) \Delta(X) . \qquad\qquad 3.36$$

PROOF. Since the identity in 3.36 is linear and multiplicative in P it is sufficient to verify it for the generators of a multiplicative basis of the symmetric polynomials. We can easily do it for the power symmetric functions $p_k(X) = \sum_{i=1}^n x_i^k$. We note then that

$$p_k(e^\partial) = \sum_{i=1}^n e^{k\partial_i} = n + \sum_{m \geq 1} \frac{k^m}{m!} p_m(\partial)$$

Thus from proposition 3.1 we derive that

$$p_k(e^\partial)\Delta(X) = n \Delta(X) = p_k(1, 1, \ldots, 1)\Delta(X) .$$

This establishes 3.36 and our proposition.

It will be convenient here and in the following to denote by $\mathbf{I}_{1^n}(1)$ the ideal generated by the symmetric polynomials that vanish at the point $(1, 1, \ldots, 1)$. We leave it to the reader to show that

$$\mathbf{I}_{1^n}(1) = \left(e_k(X) - \binom{n}{k} \ : \ for \ k = 1, 2 \ldots, n \ \right) .$$

It is clear from 3.36 that every element of $\mathbf{I}_{1^n}(1)$ is in the kernel of ψ. However, the converse is also true as we shall presently show.

THEOREM 3.7. *If a collection $\{b\}_{b\in\mathcal{B}}$ of homogeneous polynomials is a basis for $\mathbf{R}_{1^n}$ then the collection $\{b(e^\partial)\Delta(X)\}_{b\in\mathcal{B}}$ is a basis for $\mathbf{H}_{1^n}$. In particular, we have ker $\psi = \mathbf{I}_{1^n}(1)$.*

PROOF. We have shown (theorem 3.2 (c)) that each polynomial P has an expansion of the form

$$P(X) = \sum_{b\in\mathcal{B}} A_b(X)b(X) \ . \tag{3.37}$$

where the coefficients $A_b(X)$ are suitable symmetric polynomials. Combining this with proposition 3.4 we get

$$P(e^\partial)\Delta(X) = \sum_{b\in\mathcal{B}} A_b(e^\partial)b(e^\partial)\Delta(X) = \sum_{b\in\mathcal{B}} A_b(1,1,\ldots,1)b(e^\partial)\Delta(X)$$
$$\tag{3.38}$$

This gives that the $b(e^\partial)\Delta(X)$'s do span $\mathcal{H}_{1^n}$. Since, their number is equal to the dimension of $\mathbf{H}_{1^n}$, they must form a basis as well. This proves our first assertion. To complete the proof we need only show that the kernel of ψ is contained in the ideal $\mathbf{I}_{1^n}(1)$. Assume that $\psi P = 0$. Then 3.37 yields that

$$\sum_{b\in\mathcal{B}} A_b(1,1,\ldots,1)b(e^\partial)\Delta(X) = 0 \ .$$

Since we know now that the polynomials $b(e^\partial)\Delta(X)$ are independent we must conclude that each of the coefficients $A_b(1,1,\ldots,1)$ must vanish. But then 3.37 gives that $P \in \mathbf{I}_{1^n}(1)$. $\square$

We are now in a position to derive Hulsurkar's result. It was shown in [11] (see also [1] for an independent combinatorial proof) that the so-called *descent monomials* $\{m_\sigma(X)\}_{\sigma\in S_n}$ form an integral basis for $\mathbf{R}_{1^n}$. These monomials are simply defined by setting

$$m_\sigma(X) = \prod_{\sigma_i>\sigma_{i+1}} x_{\sigma_1}x_{\sigma_2}\cdots x_{\sigma_i} \ . \tag{3.39}$$

Now it is easy to see that

$$m_\sigma(e^\partial)\Delta(X) = \Delta_\sigma(X) \ . \tag{3.40}$$

In fact, the only difference between the translation effected by the operator $m_\sigma(e^\partial)$ and that given by the vector ϵ_σ in the definition of $\Delta_\sigma(X)$ is the extra translation by the vector

$$- \operatorname{maj}(\sigma)(u_1 + u_2 + \cdots + u_n) \ .$$

However, this translation is given by $\operatorname{maj}(\sigma)$ applications of the operator

$$e^{\sum_{i=1}^n \partial_i}$$

which has absolutely no effect on $\Delta(X)$. This proves 3.40 and thus the Hulsurkar result follows from Theorem 3.7.

REMARK 3.2. From proposition 3.4 we deduce that for each element $P \in \mathbf{H}_{1^n}$ and for any t we have

$$\sum_{i=1}^{n} e^{t\partial_i} P = n P$$

This identity may be rewritten in the form

$$P(x_1, x_2, \ldots, x_n) = \frac{1}{n} \sum_{i=1}^{n} P(x_1, .., x_i + t, .., x_n)$$

which may be viewed as saying that P satisfies the *mean value property* (see [10]) with respect to the measure which concentrates n unit masses at the tips of the n coordinate vectors. Thus it is also in this stronger sense that the elements of the spaces $\mathbf{H}_\mu$ may be said to be harmonic.

4. Linear translates of Garnir elements.

In this section we shall study the spaces $\mathbf{G}_\mu$ in full generality. Of course, our goal is the proof that the coefficients $c_{\lambda\mu}(q)$ in their graded character

$$char_q \, \mathbf{G}_\mu = p^\mu(q) = \sum_\lambda \chi^\lambda \, c_{\lambda\mu}(q)$$

are given by I.10. Since this identification was already carried out for $\mathbf{R}_\mu$ in [12], we need only show here that $\mathbf{G}_\mu$ and $\mathbf{R}_\mu$ are equivalent as graded S_n-modules. As we stated in the introduction we shall achieve this by showing that $\mathbf{G}_\mu$ and $\mathbf{H}_\mu$ are one and the same space. In the process of proving this we shall also derive a number of results which are interesting in their own right.

We should note first that the relation of the Garnir module Γ_μ to Young's natural is quite immediate. In fact, we only need to point out that each Garnir element $\Delta_T(x)$ may also be written in the form

$$\Delta_T(x) = N(T) \, m_T(x) = \frac{1}{\mu!} \dot{N}(T) P(T) m_T(x) \, . \qquad 4.1$$

where, as customary, $P(T)$ and $N(T)$ denote the so called *positive* and *negative*, *row* and *column groups* of T respectively and

$$m_T(x) = \prod_{i=1}^{n} x_i^{h_i(T)-1} \qquad 4.2$$

with $h_i(T)$ denoting the height of the label "i" in T. The expression $m_T(x)$ will occur quite often in this writing, we shall refer to it as the *monomial of* T. Perhaps all of this can be best understood through an example. For instance if

$$T = \begin{array}{ccc} 7 & & \\ 2 & 3 & 6 \\ 1 & 4 & 5 \end{array}$$

then as we have seen

$$\Delta_T(x) \;=\; \det \begin{pmatrix} 1 & x_1 & x_1^2 \\ 1 & x_2 & x_2^2 \\ 1 & x_7 & x_7^2 \end{pmatrix} \times \det \begin{pmatrix} 1 & x_4 \\ 1 & x_3 \end{pmatrix} \times \det \begin{pmatrix} 1 & x_5 \\ 1 & x_6 \end{pmatrix} \; .$$

Now, using Young's notation, we can rewrite these determinants in the form

$$\det \begin{pmatrix} 1 & x_1 & x_1^2 \\ 1 & x_2 & x_2^2 \\ 1 & x_7 & x_7^2 \end{pmatrix} = [1,2,7]' \, x_2 x_7^2 \; ,$$

$$\det \begin{pmatrix} 1 & x_4 \\ 1 & x_3 \end{pmatrix} = [4,3]' x_4 \; , \quad \det \begin{pmatrix} 1 & x_5 \\ 1 & x_6 \end{pmatrix} = [5,6]' x_6 \; ,$$

and this is an instance of the first equation in 4.1. Clearly the last equality in 4.1 holds true since the monomial of T is left invariant by any permutation of the row group of T.

Let now $T_1, T_2, \ldots, T_{n_\mu}$ denote the standard tableaux of shape μ (in the usual lexicographic order say) and σ_{ij} be the permutation that sends T_j into T_i, we recall that the Young *natural* representation corresponding to the partition μ may be defined as the integral matrix $A(\sigma) = \|a_{ij}(\sigma)\|$ which occurs in the relations (see Garsia-Wachs [13] eq. 2.13)

$$\sigma N(T_j) P(T_j) \;=\; \sum_{i=1}^{n_\mu} a_{ij}(\sigma) \, N(T_i) P(T_i) \, \sigma_{ij} \; . \qquad\qquad 4.3$$

Applying this group algebra element to the monomial of T_j we finally get

$$\sigma \Delta_{T_j}(x) \;=\; \sum_{i=1}^{n_\mu} \Delta_{T_i}(x) \, a_{ij}(\sigma) \; . \qquad\qquad 4.4$$

It will be convenient here to denote by $\mathcal{T}(\mu)$ the collection of injective tableaux of shape μ. We shall for a moment alter our the definition of Γ_μ given in the introduction and set

$$\Gamma_\mu \;=\; \mathcal{L}\{\Delta_T(x) \,:\, T \in \mathcal{T}(\mu)\} \; .$$

We can immediately see from 4.4 that we have not changed the space.

THEOREM 4.1. *(Garnir)* *The polynomials* $\{\Delta_{T_i}(x)\}_{i=1}^{n_\mu}$ *form a basis for* Γ_μ. *Moreover, this space is an irreducible* S_n-*module with character* χ^μ, *in the Young indexing.*

PROOF. Using 4.4 we derive that the image of a linear combination

$$P(x) \;=\; \sum_{j=1}^{n_\mu} c_j \, \Delta_{T_j}(x)$$

by the action of an element f of the group algebra $\mathbf{Q}(S_n)$ is given by acting on the coefficient sequence $(c_1, c_2, \ldots, c_n)$ by the matrix

$$A(f) = \sum_{\sigma \in S_n} f(\sigma) A(\sigma) \ .$$

Using, the irreducibility of the representation $A(\sigma)$ we deduce that, as long as the coefficients c_i are not all zero, we can choose f in such a manner that the polynomial $f\, P(x)$ reduces to one of the $\Delta_{T_i}(x)$. Since none of the Garnir elements is equal to zero, it follows that $P(x)$ itself cannot be zero. This shows the independence of the standard Garnir elements. Thus they must form a basis. Finally, we see (again from 4.4) that in terms of this basis the action of a permutation σ is precisely given by Young's natural. Thus the last assertion of the theorem holds true in the strongest possible sense. $\square$

We shall next proceed to show that $\mathbf{G}_\mu = \mathbf{H}_\mu$. We shall break up the proof into three basic steps, as we did in the previous section for $\mathbf{G}_{1^n} = \mathbf{H}_{1^n}$. Only the details will be more demanding in the general case. We prove first that each translate of a Garnir element lies in $\mathbf{I}_\mu^\perp$. Then show that the dimension of $\mathbf{H}_\mu$ is at most $\binom{n}{\mu}$ by exhibiting a spanning set of monomials for $\mathbf{R}_\mu$. Finally, we construct an independent subset of $\mathbf{G}_\mu$ which is of cardinality precisely $\binom{n}{\mu}$. We start with

PROPOSITION 4.1.

$$\mathcal{L}\{\partial^p \Delta_T(x) \ : \ T \in ST(\mu)\} \ \subseteq \ \mathbf{H}_\mu \ . \qquad 4.5$$

PROOF. Clearly, we need only show that each Garnir element $\Delta_T(x)$ is killed by the generators of the ideal $\mathbf{I}_\mu$. More, precisely, we show that for each $T \in \mathcal{T}(\mu)$ and each $e_r(S) \in \mathcal{C}_\mu$ we have

$$e_r(S)(\partial) \, \Delta_T(x) \ = \ 0 \ . \qquad 4.6$$

We recall, from the definition 2.2, that if $e_r(S) \in \mathcal{C}_\mu$ and s denotes the cardinality of S then

$$r \ > \ s \ - \ \mu_1' - \mu_2' - \cdots - \mu_s' \ . \qquad 4.7$$

Using the same conventions as in the definition of $\Delta_T(x)$, let $C_1, C_2, \ldots, C_n$ denote the columns of T and let

$$S \ = \ S_1 + S_2 + \cdots + S_n \qquad (\ S_i \ = \ S \cap C_i\ ) \qquad 4.8$$

be the corresponding disjoint decomposition of S. Accordingly we shall necessarily have

$$e_r(S)(\partial) \, \Delta_T(x) \ = \ \sum_{r_1 + r_2 + \cdots + r_n = r} \prod_{i=1}^{n} e_{r_i}(S_i)(\partial) \, \Delta(C_i) \ . \qquad 4.9$$

Note now that if in any summand of 4.9 we have an index i such that

$$r_i \ \geq \ 1 \ \ \& \ \ S_i \ = \ C_i \qquad 4.10$$

then from Proposition 3.1 (with the alphabet $X_n = \{x_1, x_2, \ldots, x_n\}$ replaced by C_i) we get

$$e_{r_i}(S_i)(\partial)\, \Delta(C_i) \;=\; 0$$

and the corresponding summand must necessarily vanish as well. We should also note that, since an elementary symmetric function of index greater than the cardinality of the alphabet is by definition equal to zero, we must also have

$$r_i \;\leq\; |S_i| \qquad (\; \forall\; i = 1, 2, \ldots, n\;) \qquad\qquad 4.11$$

Our goal is to show that the conditions in 4.7, 4.8 and 4.11 force the existence of at least one index i satisfying 4.10 for each of the summands in 4.9. To this end let $i_1 < i_2 < \cdots < i_k$ be the indices i for which S_i is a *proper* subset of the corresponding column C_i. Clearly, since each S_{i_a} $(a = 1, .., k)$ misses at least one element in its column, the set S must then altogether miss at least k elements from $[1, n]$. This gives that

$$s \;\leq\; n - k \;. \qquad\qquad 4.12$$

In other words there must be $t = n - k \geq s$ indices $j_1 < j_2 < \cdots < j_t$ such that

$$S_{j_1} = C_{j_1}\,, \; S_{j_2} = C_{j_1}\,, \; \ldots\,, \; S_{j_t} = C_{j_t}$$

Thus to show the existence of an index i satisfying 4.10 we need only show that for each summand in 4.9 we have

$$r_{j_1} \;+\; r_{j_2} \;+\; \cdots \;+\; r_{j_t} \;\geq\; 1 \;. \qquad\qquad 4.13$$

Now to show this note that from 4.7 we get

$$r_{j_1} + \cdots + r_{j_t} \;=\; r \;-\; r_{i_1} - \cdots - r_{i_k} \;>\; s \;-\; \mu_1' - \mu_2' - \cdots - \mu_s' \; - r_{i_1} - \cdots - r_{i_k} \;.$$

However, since $|S_{j_b}| = |C_{j_b}| = \mu_{j_b}'$ (*for* $b = 1..t$), this may be rewritten as

$$r_{j_1} + \cdots + r_{j_t} \;>\; \sum_{a=1}^{k} \left(|S_{i_a}| - r_{i_a}\right) \;+\; \mu_{j_1}' + \mu_{j_2}' + \cdots + \mu_{j_t}' \;-\; \mu_1' - \mu_2' - \cdots - \mu_s' \;.$$

Now, 4.11 gives

$$\sum_{a=1}^{k} \left(|S_{i_a}| - r_{i_a}\right) \;\geq 0$$

moreover we have also

$$\mu_{j_1}' + \mu_{j_2}' + \cdots + \mu_{j_t}' \;-\; \mu_1' - \mu_2' - \cdots - \mu_s' \;\geq 0 \;,$$

because $\mu_1', \mu_2', \cdots, \mu_s'$ are the s smallest parts of μ' and $t \geq s$. This establishes 4.13 and 4.6 must be true as asserted. Our proof is thus complete. $\square$

To state the next proposition we need to recall some notation an review a few auxiliary facts from a recent work of Garsia-Procesi [12].

Given a partition μ and a height i less than or equal to the height of μ we denote by $\mu^{(i)}$ the partition obtained by removing from the diagram of μ the lowest corner square that lies on or above row i. This given, we let $\mathcal{B}_\mu$ denote the collection of monomials obtained from the recursion

$$\mathcal{B}_\mu = \sum_{i=1}^{h(\mu)} x^{i-1} \, \mathcal{B}_{\mu^{(i)}} \qquad\qquad 4.14$$

together with the initial condition $\mathcal{B}_\mu = \{1\}$ for $\mu = (1)$. In 4.14 the "$\sum$" means disjoint union and the expression $x^{i-1}\mathcal{B}(\mu^{(i)})$ is to represent the set of monomials obtained by multiplying each monomial in $\mathcal{B}(\mu^{(i)})$ by x^{i-1}. It is easy to see that when $\mu = 1^n$ the collection $\mathcal{B}_\mu$ reduces to the collection $\{x^\epsilon\}_{1 \le \epsilon_i \le i-1}$ we have shown in section 3 to yield a basis for $\mathbf{R}_{1^n}$. We shall derive here a similar result for each space $\mathbf{R}_\mu$.

PROPOSITION 4.2. *The collection $\mathcal{B}_\mu$ spans the space $\mathbf{R}_\mu$. In other words*

$$\mathbf{R}_\mu = \mathcal{L}\{ \, x^\epsilon \in \mathcal{B}_\mu \, \} \qquad\qquad 4.15$$

PROOF. We shall simply give an algorithm for expanding $mod\ I_\mu$ any given monomial

$$x^p = x_1^{p_1} x_2^{p_2} \cdots x_n^{p_n} \qquad\qquad 4.16$$

in terms of the monomials in $\mathcal{B}_\mu$. We shall proceed by a double induction. We assume we have done it up to $n-1$ and for all $p_n \ge i$, then prove it for $p_n = i-1$. The case $n = 1$ being trivial we start by proving that x^p is expandable as soon as $p_n \ge k(\mu)$. Where $k(\mu) = \mu'_n$ denotes the number of rows in the diagram of μ. In fact, more than that is true. We can show that

$$x_n^{k(\mu)} \cong_\mu 0 \ . \qquad\qquad 4.17$$

To this end note that $d_{n-1}(\mu) = n - k(\mu)$. From the definition 2.2 of $\mathcal{C}_\mu$ it follows then that $e_r(X_{n-1}) \in \mathcal{C}_\mu$ for $r \ge k(\mu)$ where for convenience we set $X_m = \{x_1, x_2, \ldots, x_m\}$. This means that, $mod\ I_\mu$, the polynomial $\prod_{i=1}^{n-1}(1-tx_i)$ has no terms in $t^{k(\mu)}$. Since as we have observed all the elementary symmetric functions in the alphabet X_n belong to $\mathcal{C}_\mu$, we see that we must have as well that

$$\prod_{i=1}^{n-1}(1 - tx_i) \cong_\mu \sum_{r \ge 0} t^r \, x_n^r \ .$$

Equating coefficients of $t^{k(\mu)}$ in this identity gives 4.17 as desired.

Having established our first step, we assume by descent induction, that every polynomial $P(x_1, x_2, \ldots, x_n)$ in which x_n appears to a power at least i is expandable in terms of the monomials in $\mathcal{B}_\mu$ and proceed to show that every x^p with $p_n = i - 1$ is also expandable. To simplify our language, we shall omit repeating *in terms of $\mathcal{B}_\mu$* or *mod I_μ* whenever it is clear from the context. Since the result is assumed to be true for $n - 1$ we shall be able to expand in terms of $\mathcal{B}_{\mu^{(i)}}$ and *mod $I_{\mu^{(i)}}$* any polynomial in the alphabet X_{n-1}. In particular every monomial

$$x^p = x_1^{p_1} x_2^{p_2} \cdots x_{n-1}^{p_{n-1}} x_n^{i-1} \qquad\qquad 4.18$$

may be written as a linear combination of the monomials in $x_n^{i-1}\mathcal{B}_{\mu^{(i)}}$ with an error which is a sum of terms of the form

$$A(X_{n-1})x_n^{i-1}e_r(S) \qquad\qquad e_r(S) \in \mathcal{C}_{\mu^{(i)}} \ \& \ S \subseteq X_{n-1} \ , \qquad 4.19$$

where $A(X_{n-1})$ denotes an arbitrary polynomial in the alphabet X_{n-1}. To study such a term we assume that $k = |S|$. From the definition of $\mathcal{C}_{\mu^{(i)}}$ we then get that

$$r > k - d_k(\mu^{(i)}) \ . \qquad\qquad 4.20$$

Note now that, by our convention for writing partitions, we have

$$d_k(\mu^{(i)}) = \begin{cases} d_{k+1}(\mu) - 1 & \text{if } k \geq n - \mu_i, \\ d_{k+1}(\mu) & \text{if } k < n - \mu_i \ . \end{cases} \qquad 4.21$$

Thus in any case $d_k(\mu^{(i)}) \leq d_{k+1}(\mu)$. Using the simple identity

$$e_r(S) = e_r(S + x_n) - x_n e_{r-1}(S) \qquad\qquad 4.22$$

we can rewrite the error term in 4.19 as

$$A(X_{n-1}) \, x_n^{i-1} e_r(S + x_n) - A(X_{n-1}) \, x_n^i e_{r-1}(S) \ . \qquad 4.23$$

Now the second term here is expandable by the induction hypothesis. For the first term, we need only be concerned when

$$r = k + 1 - d_k(\mu^{(i)}) \qquad and \qquad k < n - \mu_i \ . \qquad 4.24$$

For if $r > k+1-d_k(\mu^{(i)})$ then $r > k+1-d_{k+1}(\mu)$ and $e_r(S+x_n) \in \mathcal{C}_\mu$ since $S + x_n$ is a $k + 1$-element set. Thus the first term in 4.23 is equal to zero $mod \ I_\mu$ and we needn't be concerned with it. On the other hand, if $k \geq n - \mu_i$ then the first case of 4.21 holds true and 4.20 reduces to $r > k + 1 - d_{k+1}(\mu)$ again and the same reasoning applies. Let us then assume that the conditions in 4.24 hold true. In this case we resort to the identity

$$(-1)^{i-1}x_n^{i-1}e_r(S) = e_{r+i-1}(S) - \sum_{s=0}^{i-2} (-1)^s \, x_n^s \, e_{r+i-1-s}(S + x_n) \ .$$

When we substitute this into 4.19 we see that the terms coming from the summation are again of no concern since $r + i - 1 - s \geq r + 1 > k + 1 - d_{k+1}(\mu)$. So we are left with the term

$$A(X_{n-1})e_{r+i-1}(S) \ .$$

However, this is also zero $mod \ I_\mu$ since for $k < n - \mu_i$ we have $\mu'_{k+1} < i$ (by the construction of $\mu^{(i)}$) and thus

$$d_k(\mu^{(i)}) = d_{k+1}(\mu) = d_k(\mu) + \mu'_{k+1} < d_k(\mu) + i$$

which combined with 4.24 gives

$$r + i - 1 > k + 1 - d_k(\mu) - i + i - 1 = k - d_k(\mu) \ .$$

We can thus conclude that our error term is in any case expandable. This completes our induction and the proof of the proposition. □

REMARK 4.1. The fact that $\mathcal{B}_\mu$ spans $\mathbf{R}_\mu$ was first proved by DeConcini-Procesi in [8]. Their proof was elementary but rather intricate. Tanisaki in [30] using $\mathcal{C}_\mu$ as a generating set for $\mathbf{I}_\mu$ gave a simpler proof. The expansion algorithm we give here is based on a further simplified version of the Tanisaki argument.

REMARK 4.2. Our expansion algorithm strongly suggests that the collection $\mathcal{B}_\mu$ is precisely the monomial basis we would obtain by the Gröbner algorithm when the alphabet is totally ordered by $x_1 > x_2 > \cdots > x_n$. Numerical evidence confirms this. In fact, using MAPLE, the *gbasis* command with inputs $\mathcal{C}_\mu$ and X_n, constructs a Gröbner basis $\mathcal{G}B$ for $\mathbf{I}_\mu$ which yields expansions in terms of $\mathcal{B}_\mu$ when *normalf* is used with input $\mathcal{G}B$. This in turn suggests that $\mathcal{B}_\mu$ must be a lower ideal of monomials. We can prove this fact, without using Gröbner basis methods, directly from the combinatorics of $\mathcal{B}_\mu$.

Let us recall that the natural partial order of monomials is defined by setting $x^p = x_1^{p_1} \cdots x_n^{p_n} \leq x^q = x_1^{q_1} \cdots x_n^{q_n}$ if and only if $p_i \leq q_i$ *(for $i = 1..n$)*. It develops that $\mathcal{B}_\mu$ has the following remarkable characterization.

PROPOSITION 4.3. *For each partition μ the basis $\mathcal{B}_\mu$ is the lower ideal of monomials whose maximal elements are the monomials $m_T(x)$ (see 4.2) of the standard tableaux of shape μ.*

PROOF. We shall only sketch the argument here. Details may be found in [12] (prop. 4.1 & 4.2). We first note that the bases $\mathcal{B}_\mu$ are nested into each other in the same manner as the corresponding spaces $\mathbf{H}_\mu$. That is we have

$$\mu \geq \nu \quad \rightarrow \quad \mathcal{B}_\mu \supseteq \mathcal{B}_\nu . \qquad 4.24$$

This is not difficult to verify, by a straightforward induction argument, in the case that μ is an immediate successor of ν. This done, we then observe that for any μ we have

$$\mu^{(1)} \geq \mu^{(2)} \geq \cdots \geq \mu^{(k(\mu))} ,$$

and 4.24 gives

$$\mathcal{B}_{\mu^{(1)}} \supseteq \mathcal{B}_{\mu^{(2)}} \supseteq \cdots \supseteq \mathcal{B}_{\mu^{(h(\mu))}} . \qquad 4.25$$

To show our assertion we can now proceed by induction on the number of squares of the diagram of μ. Since the elements of highest degree in of $\mathcal{B}_\mu$ are the standard tableau monomials defined in 4.2, we only need to show that $\mathcal{B}_\mu$ is a lower ideal of monomials. Thus our assertion will follow if we show that if a monomial $x_n^{i-1}m(x_1,\ldots,x_{n-1})$ occurs in $\mathcal{B}_\mu$ then also the monomial $x_n^{j-1}m(x_1,\ldots,x_{n-1})$ can be found in $\mathcal{B}_\mu$ for any $1 \leq j < i$. By the recursion 4.14, the latter monomial must come out of $x_n^{j-1}\mathcal{B}_{\mu^{(j)}}$. In other words the monomial $m(x_1,\ldots,x_{n-1})$ itself should be in $\mathcal{B}_{\mu^{(j)}}$. Since this monomial must have come out of $\mathcal{B}_{\mu^{(i)}}$, we are left to show that $\mathcal{B}_{\mu^{(i)}} \subseteq \mathcal{B}_{\mu^{(j)}}$. However, this is an immediate consequence of 4.25. □

This proposition allows us to write each monomial $x^\epsilon \in \mathcal{B}_\mu$ in the form

$$x^\epsilon = m_{T_\epsilon}(x)/x^{\rho_\epsilon} , \qquad\qquad 4.26$$

where $m_{T_\epsilon}(x)$ is a standard tableau monomial and ρ_ϵ is a nonnegative exponent vector. It will be convenient for our next argument that we assign to each $x^\epsilon \in \mathcal{B}_\mu$ unique standard tableau T_ϵ giving 4.26. For instance we can let T_ϵ be the lexicographically first such tableau. This given we have:

PROPOSITION 4.4. *The collection of polynomials* $\{\partial^{\rho_\epsilon} \Delta_{T_\epsilon}(x) : x^\epsilon \in \mathcal{B}_\mu\}$ *is an independent set in* $\mathbf{H}_\mu$.

PROOF. Note that if A is an ordered subset of x then the corresponding Vandermonde determinant $\Delta(A)$ is an alternating sum of monomials all lexicographically larger that the monomial corresponding to the diagonal term in $\Delta(A)$. Thus if T is a standard tableau, the same will be true for each of the factors $\Delta(C_i)$ occurring in the definition of $\Delta_T(x)$. From this we easily deduce that $\Delta_T(x)$ has an expansion of the form

$$\Delta_T(x) = m_T(x) + \sum_{x^p >_{lex} m_T(x)} c_p\, x^p ,$$

where "$>_{lex}$" denotes the lexicographic order of monomials. Now, the definition of lexicographic order immediately yields that the implication

$$x^p <_{lex} x^q \quad \rightarrow \quad \partial^\eta x^p <_{lex} \partial^\eta x^q$$

must hold for any differentiation ∂^η which doesn't kill x^q. However, this gives that the polynomials $\partial^{\rho_\epsilon} \Delta_{T_\epsilon}(x)$ may be be written in the form

$$\partial^{\rho_\epsilon} \Delta_{T_\epsilon}(x) = a x^\epsilon + \sum_{x^p >_{lex} x^\epsilon} a_p x^p ,$$

with suitable coefficients a, a_p, and this is all that we need to establish their independence. $\square$

We can now put together these propositions and derive the basic result of the section.

THEOREM 4.2.

$$\mathbf{G}_\mu = \mathbf{H}_\mu \qquad\qquad 4.27$$

PROOF. From the recursion 4.14 we can easily deduce that

$$\text{card } \mathcal{B}_\mu = \frac{n!}{\mu_1!\mu_2!\cdots\mu_n!} = \binom{n}{\mu} .$$

Thus propositions 4.4, 4.1 and 4.2, in that order (combined with the fact that $\mathbf{R}_\mu$ and $\mathbf{H}_\mu$ are isomorphic S_n-modules), yield the following sequence of inequalities

$$\binom{n}{\mu} = \text{card } \{\, \partial^{\rho_\epsilon} \Delta_{T_\epsilon}(x) \,\}_{x^\epsilon \in \mathcal{B}_\mu} \leq \dim \mathbf{H}_\mu = \dim \mathbf{R}_\mu \leq \binom{n}{\mu} .$$

Thus equalities must hold throughout. This forces equality in 4.5. Thus 4.27 must hold true as asserted. □

We should point out that, as a by-product of this proof, we also get the following two corollaries.

THEOREM 4.3. *The monomials in $\mathcal{B}_\mu$ form a basis for $\mathbf{R}_\mu$.*

THEOREM 4.4. *The collection $\{\partial^{\rho_\epsilon} \Delta_{T_\epsilon}(x) : x^\epsilon \in \mathcal{B}_\mu\}$ is a basis for $\mathbf{H}_\mu$.*

Remarkably the space $\mathbf{H}_\mu$ allows us to derive properties of the ideal $\mathbf{I}_\mu$ which appear difficult to approach directly. A useful example is given by the following result.

PROPOSITION 4.5. *A monomial x^p belongs to $\mathbf{I}_\mu$ if and only if it is not an S_n-image of an element of $\mathcal{B}_\mu$.*

PROOF. Since $\mathcal{B}_\mu$ is a basis for $\mathbf{R}_\mu$ none of its elements nor their images by any permutation can be in $\mathbf{I}_\mu$. Thus we need only show the sufficiency of the condition. However, this is also immediate. In fact, the equality in 1.13 specialized to the collection $\mathcal{C}_\mu$ yields

$$\mathbf{I}_\mu = \mathbf{H}_\mu^\perp = \mathcal{L}[\partial^p \Delta_T(x) : T \in \mathcal{T}(\mu)]^\perp .$$

This implies that every monomial x^p which kills all the Garnir elements $\Delta_T(x)$ of shape μ must necessarily be in $\mathbf{I}_\mu$. Now it is easy to see that if x^p is not an S_n-image of an element of $\mathcal{B}_\mu$ then ∂^p will kill *every term* of each of the polynomials $\Delta_T(x)$. This is because the latter are all images of the maximal elements of $\mathcal{B}_\mu$. □

REMARK 4.3. Let us use for a moment the English convention for partitions, and let the nonzero parts of μ be $\mu_1 \geq \mu_2 \geq \cdots \geq \mu_k > 0$. This given, we can easily deduce from proposition 4.5 that every monomial of the form

$$m_t(S) = \prod_{x_i \in S} x_i^t \qquad\qquad 4.28$$

with S a subset of x of cardinality $1 + \mu_{t+1} + \mu_{t+2} + \cdots + \mu_k$ necessarily belongs to $\mathbf{I}_\mu$. Indeed, these are the smallest monomials whose *shape* is not contained in the shape of a tableau monomial $m_T(x)$ for $T \in \mathcal{T}(\mu)$. This is easily seen since the latter shape is given by the partition $0^{\mu_1} 1^{\mu_2} 2^{\mu_3} \cdots (k-1)^{\mu_k}$ and the shapes of the monomials in 4.28 correspond to the outer corners of this partition.

REMARK 4.4. Note that the equality of $\mathbf{H}_\mu$ and $\mathbf{R}_\mu$ yields that the maximum degree of any element of $\mathbf{H}_\mu$ as well as $\mathbf{R}_\mu$ is given by $n(\mu)$. Thus from theorem 4.1 we deduce that the irreducible representation with character χ^μ in Young's indexing occurs in the top degree in $\mathbf{H}_\mu$ as well as $\mathbf{R}_\mu$. This proves a conjecture made by Kraft in [17]. In particular this means that the action of S_n on the standard tableau monomials $m_T(x)$ of shape μ, when expressed *mod I_μ*, is an irreducible with character χ^μ. In a recent paper H. Barcelo [4] shows, by a combinatorial argument that only uses the vanishing *(mod I_μ)* of the monomials in

4.28, that again the matrix which expresses this action in the standard monomial basis may be constructed by Young's straightening algorithm.

REMARK 4.5. Combining the results of this paper with those in [12] we have established that the graded character of $\mathbf{H}_\mu$ may be written in the form

$$p^\mu(q) \; = \; \sum_{\lambda \vdash n} \chi^\lambda \, K_{\lambda\mu}(1/q) q^{n(\mu)} \; . \hspace{2cm} 4.29$$

This given, we immediately derive from 2.9 that

$$\mu \leq \nu \quad \longrightarrow \quad K_{\lambda\mu}(q) \; << \; K_{\lambda\nu}(q) \; ,$$

where the symbol " $<<$ " represents coefficientwise inequality. Lascoux in [18] states that the Kostka-Foulkes polynomials $K_{\lambda\mu}(q)$ may be written in the form

$$K_{\lambda\mu}(q) \; = \; \sum_{\nu \leq \mu} A_{\lambda\nu}(q) \; , \hspace{2cm} 4.30$$

where the polynomials $A_{\lambda\mu}(q)$ also have non negative integer coefficients. The result is claimed in [18] to be provable by a combination of various facts derived by Lascoux-Schützenberger in [19],[21],[22]. Numerical evidence, suggests that 4.30 may also have an explanation in terms of the modules $\mathbf{H}_\mu$. The evidence exhibits the existence of a decomposition of the space $\mathbf{H}_{1^n}$ as a direct sum of invariant subspaces

$$\mathbf{H}_{1^n} \; = \; \bigoplus_{\mu \vdash n} \mathbf{A}_\mu \; ,$$

with the property that for every $\mu \vdash n$ we have

$$\mathbf{H}_\mu \; = \; \bigoplus_{\nu \leq \mu} \mathbf{A}_\mu \; . \hspace{2cm} 4.31$$

The S_n-invariance of the subspaces $\mathbf{A}_\mu$ would immediately yield 3.30. It should make a worthwhile project to establish 4.31 in full generality. For a result in this direction we refer to [12] (theorem 4.1).

REMARK 4.6. The basis $\mathcal{A} = \{\pi[A, T]\}$ discovered by E. Allen may be used as a tool for studying combinatorial properties of the polynomials $K_{\lambda\mu}(q)$. It can be shown that when we restrict $\mathcal{A}$ to the quotient $\mathbf{R}_\mu$ then, under a suitable total order of the blocks $\mathcal{A}_T$, if a single element of a given block $\mathcal{A}_T$ may be expressed ($mod\ I_\mu$) in terms of elements of previous blocks, the same holds true for all the other elements of $\mathcal{A}_T$. This allows us to select out of $\mathcal{A}$ a basis for $\mathbf{R}_\mu$ which consists of a certain subcollection of blocks

$$\sum_{T \in \mathcal{ST}(\mu)} \mathcal{A}_T \; ,$$

with the property that

$$\mu \leq \nu \quad \longrightarrow \quad \mathcal{ST}(\mu) \subseteq \mathcal{ST}(\nu) \; .$$

Since the action of S_n on $\mathbf{R}_\mu$ when expressed in terms of $\sum_{T \in ST_\mu} \mathcal{A}_T$ is still block triangular with irreducible diagonal blocks (as indicated in section 2.), we see that we can write the character of $\mathbf{R}_\mu$ in the form

$$p^\mu(q) \;=\; \sum_{T \in \mathcal{ST}(\mu)} \chi^{\lambda(T)}\, q^{c(T)} \;.$$

Comparing with 4.29 gives

$$K_{\lambda\mu}(1/q) q^n(\mu) \;=\; \sum_{\substack{T \in ST(\mu) \\ \lambda(T) = \lambda}} q^{c(T)} \;.$$

which expresses the Kostka-Foulkes polynomial as the q-enumerator of a certain collection of standard tableau. Thus a combinatorial characterization of the collections $\mathcal{ST}(\mu)$ yields a new interpretation of the coefficients of the $K_{\lambda\mu}(q)$. We refer to [1] for some results and conjectures on this matter.

REFERENCES

1. E. Allen, *On a conjecture of Procesi and a new basis for a graded left-regular representation of S_n*, U.C.S.D. thesis.
2. _____, *A conjecture of Procesi and the straightening algorithm of G.C. Rota*, to appear, Proc. Nat. Acad. Sci..
3. E. Artin, *Galois Theory*, Notre Dame Mathematical Lectures No. 2,, Notre Dame, IN, 1944.
4. H. Barcelo, *Young straightening in a quotient S_n-module*, to appear.
5. W. Borho, H. Kraft, *Über Bahnen und deren Deformationen bei linearen Aktionen reduktiver Gruppen*, Comment. Math. Helv. **54** (1909), 61-104.
6. L. Butler, *Doctoral thesis*, M.I.T..
7. P. Cartier and D. Foata, Lecture Notes in Math. **85** (1969), *Problèmes combinatoires de commutation et réarrangements,,* Springer, Berlin.
8. C. de Concini and C. Procesi, *Symmetric functions, conjugacy classes and the flag variety*, Invent. Math. **64** (1981), 203-230.
9. H. Garnir, *Théorie de la représentation lineaire des groupes symmétriques*, (4), Mémoires de la Soc. Royale de Liège **10** (1950).
10. A. M. Garsia, *A note on the mean value property*, Trans. A.M.S. **102** (1962), 181-186.
11. _____, *Combinatorial methods in the theory of Cohen-Macaulay rings*, Advances in Math. **38** (1980), 229-266.
12. A. M. Garsia and C. Procesi,, *On certain graded S_n-modules and the q-Kotska polynomials*, to appear, Advances in Math..
13. A. M. Garsia and M. Wachs, *Combinatorial aspects of skew representations of the symmetric group*, J. Comb. Theory, Series A **50** (1989), 47-81.
14. R. Hotta and T. A. Springer, *A specialization theorem for certain Weyl group representations and an application to Green polynomials of unitary groups*, Invent. Math. **41** (1977), 113-127.
15. S.G. Hulsulkar, *Proof of Verma's conjecture on Weyl's dimension polynomial*, Invent. Math. **27** (1974), 45-52.
16. B. Kostant, *Lie group representations on polynomial rings*, Amer. J. Math. **85** (1979), 327-404.
17. H. Kraft, *Conjugacy classes and Weyl group representations*, Proc. 1980 Torun Conf. Poland, Astérisque **87-88** (1981), 195-205.
18. A. Lascoux, *Cyclic permutations on words, tableaux and harmonic polynomials*, to appear.
19. A. Lascoux and M.P. Schützenberger, *Le monoide plaxique*, Quaderni della Ricerca Scientifica **109** (1981), 129-156.

20. _____, *Sur une conjecture de H. O. Foulkes*, C. R. Acad. Sci. Paris **286** (1978), 323-324.

21. _____, *Croissance des polynômes de Foulkes/Green*, C.R. Acad. Sc. Paris, 288 (1979), 95-98..

22. _____, *Croissance des tableaux de Young et foncteurs de Schur*, Astérisque **87-88** (1981), 191-205.

23. _____, *Cyclic permutations, tableaux and harmonic polynomials*, (preliminary version of [18] above).

24. I. G. Macdonald, *Symmetric Functions and Hall Polynomials*, Clarendon Press, Oxford, 1979.

25. P.A. MacMahon, *A certain class of generating functions in the theory of numbers*, Phil. Trans. **185** (1894), 111-160.

26. N. Spaltenstein, *On the fixed point set of a unipotent transformation on the flag manifold*, Proc. Kon. Akad. v. Wetensch. **79** (1976), 452-456.

27. T.A. Springer, Invent. Math. **44** (1978), 279-293.

28. R. Steinberg, *Differential equations invariant under finite reflection groups*, A. M. S. Transactions **112** (1964), 392-400.

29. _____, *On the desingularization of the unipotent variety*, Invent. Math. 36 (1976), 209-224.

30. T. Tanisaki, *Defining ideals of the closures of conjugacy classes and representations of the Weyl groups*, Tohoku J. Math. **34** (1982), 575-585.

31. A. Young, *On quantitative substitutional analysis (sixth paper)*,, The Collected Papers of A. Young, University of Toronto Press.

DEPARTMENT OF MATHEMATICS, U.C.S.D., LA JOLLA, CA 92093-0112

Contemporary Mathematics
Volume **138**, 1992

Identities for Generalized
Hypergeometric Coefficients*

L. C. BIEDENHARN** AND J. D. LOUCK

ABSTRACT. Generalizations of the hypergeometric functions to arbitrarily many symmetric variables are discussed, along with the associated hypergeometric coefficients and the setting within which these generalizations arose. Identities generalizing the Euler identity for $_2F_1$, the Saalschütz identity, and two generalizations of the $_4F_3$ Bailey identity, among others, are given

1. Introduction

Hypergeometric functions, denoted by $_pF_q$, are very well-known in special function theory and the associated 'hypergeometric coefficients' are simply monomial products of gamma functions appearing in the formal power series expansion of $_pF_q(z)$ over the single variable z. The 'generalized hypergeometric coefficients' in our title denotes a generalization to coefficients in a formal power series expansion over arbitrarily many symmetric variables. (The explicit definition is given in Section 2, below.) In order to emphasize that this generalization is not arbitrary or *ad hoc*, it is useful to sketch the background and setting within which this generalization arose.

There is at present no overarching general theory of special functions, but several investigations—primarily those of Wigner in the 1950's—have shown[1,2,3] that large classes of special functions arise as matrix elements of irreps (irreducible representations) of symmetry groups, such as the quantal rotation group $SU(2)$, so important in theoretical physics. There have been further generalizations of this approach to special functions via symmetry groups (a survey is

---

1991 *Mathematics Subject Classification*. Primary 33C50, 33C80, Secondary 22E30.

Key words and phrases. Tensor operators, hypergeometric functions of matrix argument, Hopf algebra, enveloping algebra, Young tableaux.

*Work performed under the auspices of the U.S. Department of Energy.

**Consultant, Department of Physics, Duke University, Durham, NC 27706.

This paper is in final form and no version of it will be submitted for publication elsewhere .

given in ref. 4), and a particularly fruitful procedure[5] focuses on *tensor operators*, rather than *irrep matrices*, as the fundamental structure. It is this approach which led to the generalization of the hypergeometric functions reported here.

Tensor operators are operators that are most easily defined as acting on a special Hilbert space, called *model space* (by Gel'fand). Consider the symmetry group of interest for the present work: $SU(3)$, the group of 3×3 unitary, unimodular matrices. The set of all unitary irreps can be labeled by the set of *Young frames* $\{\lambda\}$, with $\lambda = [\lambda_1, \lambda_2, 0]$, where the λ_i are nonnegative integers satisfying $\lambda_1 \geq \lambda_2 \geq 0$. Let us define $\mathcal{H}_\lambda$ as a vector space carrying the irrep λ. Then *model space* $\mathcal{H}$ is the direct sum: $\mathcal{H} = \sum_{\lambda \in \{\lambda\}} \oplus \mathcal{H}_\lambda$, where each irrep λ occurs *exactly once* in the direct sum. A tensor operator $\mathcal{O}$ is then a bounded map $\mathcal{O} : \mathcal{H} \to \mathcal{H}$.

It is a basic result that—by defining an equivariant action on tensor operators by the symmetry group—one may classify tensor operators as *sets of operators* identified by irrep labels of the group. For tensor operators in $SU(3)$, one accordingly labels a tensor operator by an irrep label $[M] = [M_{13}, M_{23}, 0] \in \{\lambda\}$, and a triangular pattern of integers:

$$(M) \equiv \begin{pmatrix} M_{13}\, M_{23}\, 0 \\ M_{12}\, M_{22} \\ M_{11} \end{pmatrix} ,$$

obeying the betweenness constraints $M_{ij} \geq M_{i,j-1} \geq M_{i+1,j}$. (The pattern (M) is called a *Gel'fand-Weyl pattern* and denotes a unique vector in the irrep $[M]$.) This classification is further refined by a second (inverted triangular) labeling pattern,

$$(\Gamma) \equiv \begin{pmatrix} \Gamma_{11} \\ \Gamma_{12}\, \Gamma_{22} \\ M_{13}\, M_{23}\, 0 \end{pmatrix} ,$$

whose integer entries also obey the betweenness constraints. (The pattern (Γ)—called an *operator pattern*—is structurally the same as (M) but does not have a direct group-theoretic significance.) For *unit* tensor operators (those having norm one) it has been proved[6] that: *the unit tensor operators in $SU(3)$ are a basis for all $SU(3)$ tensor operators and are canonically labeled (to within $\pm$ signs), by the two patterns (M) and (Γ).*

A second basic result is that the set of all unit tensor operators (of a given unitary symmetry group) form an algebra, in fact, a non-commutative, but co-commutative, Hopf algebra. This algebra contains as a sub-algebra the universal enveloping algebra of the group. As one might expect, matrix elements of these algebras and sub-algebras are a prolific source of special functions. For example, the group $SU(2)$ leads in this way to $_3F_2$ functions (for the unit tensor operators) and to $_4F_3$ functions (for invariant products of three coupled unit tensor operators).

The special functions that we discuss below are all associated with tensor operators in $SU(3)$, and arose in defining maps[7] of canonical unit tensor operators into invariant functions. The most interesting of these invariant functions (related to the denominator function of the canonical form) are a family of multi-variable polynomial special functions, denoted by G_q^t (see section 3), which have remarkable symmetry properties among which is a characterization by zeroes that fall into $SU(3)$ weight space patterns[7]. It was the search for a proof of these properties that led to a generalization[8] of the Gauss hypergeometric function that we now describe.

2. Symmetric Generalized Hypergeometric Functions

We begin with a generalization of the Gauss function $_2F_1\,(a, b; c; z)$. The generalization consists, in essence, of replacing the single variable z in the Gauss series by the Schur functions, $e_\mu(z)$, of symmetric function theory[9], with z denoting a set of t indeterminates $z_1, z_2 \ldots, z_t$. The Schur functions are defined in terms of standard Weyl tableaux by the formula:

$$e_\mu(z) \;=\; \sum_\alpha z_1^{\alpha_1} z_2^{\alpha_2} \ldots z_t^{\alpha_t}, \tag{2.1}$$

where $\alpha = (\alpha_1, \ldots, \alpha_t)$ is a *weight* of the partition μ into t parts. (Equivalently, α is the content of the Young frame of the irrep $[\mu]$ filled in lexically with $1, 2, \ldots, t$ according to the usual rules for a standard tableau.) The summation in eq. (2.1) is over all weights α of the $U(t)$ irrep $[\mu]$, *including repetitions.*

To define our generalized hypergeometric function, denoted $_2\mathcal{F}_1$, we generalize the Gauss series to be:

$$_2\mathcal{F}_1(a, b; c; z_1, z_2, \ldots, z_t) \;\equiv\; \sum_\mu \langle _2\mathcal{F}_1(a, b; c)|\mu\rangle e_\mu(z_1, z_2, \ldots, z_t), \tag{2.2}$$

where the generalized hypergeometric coefficients are:

$$\langle _2\mathcal{F}_1(a, b; c)|\mu\rangle \;\equiv\; M_\mu^{-1} \prod_{s=1}^{t}(a - s + 1)_{\mu_s}(b - s + 1)_{\mu_s}/(c - s + 1)_{\mu_s}. \tag{2.3}$$

In eq. (2.3), M_μ denotes the *measure* of the irrep $[\mu]$ defined by

$$M_\mu = (\dim \mu)^{-1} \prod_{i=1}^{t}(\mu_i + t - i)!/(i - 1)!, \tag{2.4}$$

where dim μ denotes the Weyl dimension formula,

$$\dim \mu = \prod_{\substack{i<j}}^{t}{}_1 (\mu_i - \mu_j + j - i)/1!2! \ldots (t - 1)!, \tag{2.5}$$

which is also the number of standard tableaux for the Young frame associated to the irrep $[\mu]$. We have also used in eq. (2.3) Pochhammer's notation for the rising factorial, that is, for $a \in \mathbf{N}$, $(x)_a = x(x+1)\cdots(x+a-1) = \Gamma(x+a)/\Gamma(x)$ with $(x)_0 = 1$.

It is easily verified that for $t = 1$ definition (2.2) reduces to the classical Gauss series.

The main results proved in Refs. 8 and 10 are the following theorem and two identities.

THEOREM 2.6. *The generalized Gauss series obeys the Euler identity:*

$$_2\mathcal{F}_1(a,b;c;z)\,_2\mathcal{F}_1(c-a-b,d;d;z) = \,_2\mathcal{F}_1(c-a,c-b;c;z).$$

An immediate consequence of this theorem are the following two identities for generalized hypergeometric coefficients:

IDENTITY 2.7. *Generalized Saalschütz identity:*

$$\sum_{\mu\nu} g(\mu\nu\lambda)\langle\,_2\mathcal{F}_1(a,b;c)|\mu\rangle\langle\,_2\mathcal{F}_1(c-a-b,d;d)|\nu\rangle = \langle\,_2\mathcal{F}_1(c-a,c-b;c)|\lambda\rangle.$$

This relation is an easy consequence of the Euler identity and the multiplicative property of the Schur functions,

$$e_\mu(z)e_\nu(z) = \sum_\lambda g(\mu\nu\lambda)e_\lambda(z), \tag{2.8}$$

where $g(\mu\nu\lambda)$ denotes the Littlewood-Richardson numbers for $GL(t,\mathbf{C})$.

For $t = 1$, we have $g(\mu\nu\lambda) = \delta_{\mu+\nu,\lambda}$, and eq. (2.7) reduces to the classic Saalschütz identity (see Bailey[11]).

IDENTITY 2.9. *Generalized Bailey Identity of the First Kind:*

$$\sum_{\mu\nu} g(\mu\nu\lambda)\langle\,_2\mathcal{F}_1(c-a,c-b;c)|\mu\rangle\langle\,_2\mathcal{F}_1(c'-a',c'-b';c')|\nu\rangle$$

$$= \sum_{\mu\nu} g(\mu\nu\lambda)\langle\,_2\mathcal{F}_1(a,b;c)|\mu\rangle\langle\,_2\mathcal{F}_1(a',b';c')|\nu\rangle,$$

where the parameters are required to satisfy $c - a - b + c' - a' - b' = 0$.

Identity (2.9) is called a generalized Bailey identity because for $t = 1$, we have

$$\sum_{\mu+\nu=\lambda} \langle\,_2\mathcal{F}_1(a,b;c)|\mu\rangle\langle\,_2\mathcal{F}_1(a',b';c')|\nu\rangle$$

$$= \frac{(a)_\lambda(b)_\lambda}{\lambda!(c)_\lambda}\,_4F_3\left(\begin{matrix}a',b',1-c-\lambda,-\lambda;\\c',1-a-\lambda,1-b-\lambda\end{matrix}\right)$$

$$= \frac{(a')_\lambda(b')_\lambda}{\lambda!(c')_\lambda}\,_4F_3\left(\begin{matrix}a,b,1-c'-\lambda,-\lambda;\\c,1-a'-\lambda,1-b'-\lambda\end{matrix}\right), \tag{2.10}$$

in which $c - a - b + c' - a' - b' = 0$. The identity between the two $_4F_3$ hypergeometric series (of unit argument) is the *reversal identity* (reverse the order of terms in the finite series expression). Using (2.10) in identity (2.9) for $t = 1$ now gives Bailey's identity[11]. (Shukla[12] independently obtained relation (2.9).)

It is straightforward to give further generalized hypergeometric functions, corresponding to the standard generalization from $_2F_1$ to $_pF_q$. These are the $_p\mathcal{F}_q$ functions, which we now define.

Let $a = (a_1, \ldots, a_p)$, $b = (b_1, \ldots, b_q)$, and $z = (z_1, \ldots, z_t)$ denote arbitrary complex *numerator* and *denominator parameters*, p and q in number, respectively, and z a set of t indeterminates. We define *generalized hypergeometric coefficients* by

$$\langle {}_pF_q(a;b)|\mu\rangle \;=\; M_\mu^{-1} \prod_{s=1}^{t} \left[\frac{\prod_{i=1}^{p}(a_i - s + 1)_{\mu_s}}{\prod_{j=1}^{q}(b_j - s + 1)_{\mu_s}} \right], \tag{2.11}$$

where $\mu = [\mu_1\mu_2\ldots\mu_t]$ is an arbitrary (lexical) partition in t parts and M_μ is the measure factor of eq. (2.4).

These coefficients are now used to define a *generalized hypergeometric function* by the formal series:

$$_pF_q(a;b;z) \;\equiv\; \sum_\mu \langle {}_pF_q(a;b)|\mu\rangle e_\mu(z), \tag{2.12}$$

where $e_\mu(z)$ denotes a Schur function in the indeterminates $z_1, z_2, \ldots, z_t$ as defined in eq. (2.1).

One of the simplest special functions that occurs in the $_pF_q$ class is $_1F_0$. (This function already occurs in the work of Littlewood[13].) It has the explicit definition given by:

$$_1F_0(a;z) \;=\; \sum_\mu \langle {}_1F_0(a)|\mu\rangle e_\mu(z) \;=\; \prod_{s=1}^{t}(1 - z_s)^a, \tag{2.13}$$

where the hypergeometric coefficient is given by

$$\langle {}_1F_0(a)|\mu\rangle \;=\; (\dim \mu) \prod_{s=1}^{t} \frac{(a - s + 1)_{\mu_s}}{(t - s + 1)_{\mu_s}}. \tag{2.14}$$

It is easily shown that one has:

IDENTITY 2.15. *Addition rule:*

$$_1F_0(a;z){}_1F_0(b;z) \;=\; {}_1F_0(a + b;z).$$

We can now state another identity for generalized hypergeometric coefficients.

IDENTITY 2.16. *Generalized Addition Rule of Binomial Type:*

$$\sum_{\mu\nu} g(\mu\nu\lambda)\langle {}_1F_0(x)|\mu\rangle\langle {}_1F_0(y)|\nu\rangle \;=\; \langle {}_1F_0(x + y)|\lambda\rangle.$$

For $t = 1$, this relation reduces to

$$\sum_{\mu+\nu=\lambda} \frac{(x)_\mu (y)_\nu}{\mu!\nu!} \;=\; \frac{(x + y)_\lambda}{\lambda!}, \tag{2.17}$$

hence, the designation of identity (2.16) as a *generalized binomial identity*.

Before we can state our next identity it is necessary to define the polynomials G_q^t mentioned in Section 1.

3. The G_q^t polynomials of $SU(3)$.

The invariant polynomials G_q^t occur in the canonical form[7] for unit tensor operators in $SU(3)$. Each such unit tensor operator is canonically labeled by the (M) and (Γ) patterns (which both contain the irrep labels $[M_{13}\ M_{23}\ 0]$ for $SU(3)$, or the labels $[M_{13}\ M_{23}\ M_{33}]$ if we generalize to $U(3)$). The operator pattern (Γ) specifies that the unit tensor operator in question induces the shifts $(\Delta_1, \Delta_2, \Delta_3)$ when acting on the irrep $[m_{13}m_{23}m_{33}]$ of model space; that is $[m_{13}m_{23}m_{33}] \rightarrow [m_{13} + \Delta_1,\ m_{23} + \Delta_2,\ m_{33} + \Delta_3]$ under the action of the operator. (The relation of the shifts to the (Γ) pattern is that $\Delta_1 = \Gamma_{11}$, $\Delta_2 = \Gamma_{12} + \Gamma_{22} - \Gamma_{11}$ and $\Delta_3 = M_{13} + M_{23} + M_{33} - \Gamma_{12} - \Gamma_{22}$.) The label t in G_q^t is also determined by the pattern (Γ) and canonically labels the multiplicity.

It is convenient in defining the G_q^t functions to use a different set of variables than those mentioned above. In place of the irrep variables $[m_{13}m_{23}m_{33}]$ in model space we first go over to the symmetric variables $p_{ij} = m_{ij} + j - i$ (the *partial hooks*) and then introduce the differences: $x_i = p_{j3} - p_{k3}$, where (ijk) is a positive permutation of $(1\ 2\ 3)$, with $x_1 + x_2 + x_3 = 0$. Thus irreps in model space are associated with integer lattice (L) points (x_1, x_2, x_3) in the Möbius plane (M), and automatically show S_3 symmetry. The parameters $(\Delta_1, \Delta_2, \Delta_3)$ now denote shifts in the Möbius plane. Finally the label M_{13} is replaced by p and M_{23} by q (with $M_{33} = 0$), as a typographic convenience.

Accordingly we have the symbol definitions:

(i) The three-tuple $(\Delta_1, \Delta_2, \Delta_3)$, such that $\Delta_i \in \mathbf{N}, 0 \leq \Delta_i \leq p$, and $\Delta_1 + \Delta_2 + \Delta_3 = p + q$, is denoted $\Delta = (\Delta_1, \Delta_2, \Delta_3)$.

(ii) A point in **L** is denoted $x = (x_1, x_2, x_3)$.

(iii) $\lambda = [\lambda_1 \lambda_2 \ldots \lambda_t]$ denotes an irrep label of $U(t)$. The symbols $\mu, \nu, \ldots$ also denote irrep labels with t parts.

(iv) $h(\lambda\mu\nu\rho)$ denotes the number of times irrep $[q - t + 1, \ldots, q - t + 1]$—this denotes the integer $q - t + 1$ repeated t times—is contained in the direct product $\lambda \times \mu \times \nu \times \rho$, and is defined to be zero if $[q - t + 1, \ldots, q - t + 1] \notin \lambda \times \mu \times \nu \times \rho$. The conjugate irrep $\bar{\lambda}$ has the partition with $\bar{\lambda}_i = q - t + 1 - \lambda_i$.

(v) The symbol A denotes the 3×3 array of variables defined by

$$A = A_t(\Delta; x) = (a_{ij})$$

$$= \begin{bmatrix} \Delta_1 - t + 1 & \Delta_2 - t + 1 + x_1 & \Delta_3 - t + 1 - x_1 \\ \Delta_2 - t + 1 & \Delta_3 - t + 1 + x_2 & \Delta_1 - t + 1 - x_2 \\ \Delta_3 - t + 1 & \Delta_1 - t + 1 + x_3 & \Delta_2 - t + 1 - x_3 \end{bmatrix}. \tag{3.1}$$

(vi) For $t = 0$ and $t = q + 1$, we define $G_q^0(\Delta; x) = G_q^{q+1}(\Delta; x) = 1$.

(vii) The notation

$$\mathcal{G}_q^t(A) = G_q^t(\Delta; x) \tag{3.2}$$

is used to signify that $\mathcal{G}_q^t$ is a polynomial in the variables a_{ij} of the array A, hence of Δ_i and x_j.

We can now give explicitly the polynomials $G_q^t(\Delta; x)$, $q \in \mathbf{N}$, $t = 1, 2, \ldots, q$, using the terms defined above and referring to Section 2 for the definitions of the hypergeometric coefficients.

$$G_q^t(\Delta; x) = \mathcal{G}_q^t(A) = \prod_{s=1}^{t} \frac{(q-s+1)!}{(s-1)!} \prod_{i=1}^{3} \prod_{s=1}^{t} (-a_{i1} - s + 1)_{q-t+1}$$

$$\times \sum_{\lambda\mu\nu\rho} h(\lambda\mu\nu\rho) \langle {}_1\mathcal{F}_0(K - \ell)|\rho\rangle \langle {}_2\mathcal{F}_1(-a_{12}, -a_{13}; a_{11} - \ell)|\lambda\rangle$$

$$\times \langle {}_2\mathcal{F}_1(-a_{22}, -a_{23}; a_{21} - \ell)|\mu\rangle$$

$$\times \langle {}_2\mathcal{F}_1(-a_{32}, -a_{33}; a_{31} - \ell)|\nu\rangle, \tag{3.3}$$

where we have defined $\ell = q - 2t + 1$, $K = \Delta_1 + \Delta_2 + \Delta_3 - 3t + 3$.

REMARK. It is of interest to note that the problem of determining $SU(3)$ tensor operators (and hence the G_q^t functions) involves generalized hypergeometric functions having arbitrarily large partitions, that is, the Young frames that enter in eq. (3.3) have t rows, corresponding to irreps $[\mu]$ of $U(t)$, with t arbitrarily large (thus *not* limited to $SU(3)$.)

Let us summarize the properties[7] of the G_q^t polynomials:

(i) *Total degree* $2t(q - t + 1)$ in x. By this we mean that $G_q^t(\Delta; x)$ is a sum of monomials of the form $x_1^\alpha x_2^\beta x_3^\gamma$, where α, β, γ are nonnegative integers such that $\alpha + \beta + \gamma \le 2t(q - t + 1)$ and the sum is over all such monomials multiplied by real coefficients that are themselves functions of Δ_i. This polynomial property is placed in evidence when eq. (3.3) is rewritten in terms of the quantities (see Ref. 7) ($k \in \mathbf{N}$):

$$F_{k,\lambda}(x, y, z) = \prod_{s=1}^{t} (x + t - k - s + 1)_k \langle {}_2\mathcal{F}_1(-y, -z; x + t - k)|\lambda\rangle. \tag{3.4}$$

(ii) *Determinantal symmetry.* This symmetry refers to the invariance of $G_q^t(\Delta; x)$ under the transformation of the six variables $(\Delta_1, \Delta_2, \Delta_3, x_1, x_2, x_3)$ induced by row interchange, column interchange, and transposition of the 3×3 array A defined by eq. (3.1). For example, under matrix transposition of A, that is $A \to A^{\mathrm{tr}}$, we have

$$(\Delta_1, \Delta_2, \Delta_3, x_1, x_2, x_3) \to (\Delta_1, \Delta_2 + x_1, \Delta_3 - x_1, -x_1, -x_3, -x_2). \tag{3.5}$$

(iii) *Weight space* $W_q^t(\Delta)$ *of zeroes.* The points in $W_q^t(\Delta)$ are in one-to-one correspondence with those of the weight space of irrep $[q - t, 0, -t + 1]$ of $U(3)$. With each point $x \in W_q^t(\Delta)$, we associate a multiplicity number $M_q^t(\Delta; x)$,

$$M_q^t(\Delta; x) \equiv \min\{(t, q - t + 1, 1 + d_t(x)\}, \tag{3.6}$$

where $d_t(x)$ is the "distance" from lattice point $x \in W_q^t(\Delta)$ to the nearest boundary point as measured along the direction of a coordinate axis (one lattice spacing = one unit of distance, with $d_t = 0$ at the boundary). The multiplicity function $M_q^t(\Delta; x)$ assigns to each point $x \in W_q^t(\Delta)$ exactly the value of the multiplicity of the weight $w = (w_1, w_2, w_3)$ of irrep $[q-t, 0, -t+1]$, where w is related to the point $x \in W_q^t(\Delta)$ by $x_1 = \Delta_3 - t + 1 - w_1$, $x_2 = -\Delta_2 - \Delta_3 + q - 1 - w_2$, $x_3 = \Delta_2 - t + 1 - w_3$. By the phrase "a polynomial has the weight space $W_q^t(\Delta)$ of zeroes", we mean that each $x \in W_q^t(\Delta)$ is a zero of the polynomial with multiplicity $M_q^t(\Delta; x)$.

Property (i) is already evident from the definition (3.3) of G_q^t, as is the invariance of $G_q^t(\Delta; x)$ under the transformation of the variables $(\Delta; x)$ corresponding to the column (and less obviously row) interchanges in the array A. Accordingly, the proof of the determinantal symmetry stated in (ii) requires only the invariance under the transformation (3.5) corresponding to transposition of the array A. This transpositional symmetry is also the key to proving that $G_q^t(\Delta; x)$ possesses the zeroes described in (iii), a result proved in Ref. 7.

It was the search for a proof of transpositional symmetry that led to the discovery of many of the special functions discussed in Section 2, although the $_2F_1$ generalized hypergeometric function and the associated Saalschütz identity came earlier in developing properties of the G_q^t polynomials for the special case $t = 1$.

The problem of proving transpositional symmetry can be made more evident if we introduce yet another polynomial function, A_λ. Define new variables a, b, c, d, e by $a = -a_{33}$, $b = -a_{32}$, $d = -a_{22}$, $e = -a_{23}$, $c = K - \ell$ and define the polynomial A_λ of these variables by

$$A_\lambda \begin{pmatrix} a, b, d, e \\ c \end{pmatrix} = \prod_{s=1}^t (a+b+c-s+1)_{\lambda_s} (d+e+c-s+1)_{\lambda_s}$$

$$\times \sum_{\mu\nu} g(\mu\nu\lambda)\langle {_2F_1}(a, b; a+b+c)|\mu\rangle$$

$$\times \langle {_2F_1}(d+c, e+c; d+e+c)|\nu\rangle. \tag{3.7}$$

We find the following expression for G_q^t in terms of the A_λ functions (using the generalized Saalschütz identity):

$$G_q^t(A) = (-1)^{t(q-t+1)} \left[\prod_{s=1}^t \frac{(q-s+1)!}{(s-1)!} \right]$$

$$\times \sum_\lambda M_\lambda^{-1} \left[\prod_{s=1}^t (-1)^{\lambda_s}(-a_{11}-s+1)_{q-t+1-\lambda_s}(-a_{12}-s+1)_{\lambda_s} \right.$$

$$\left. \times (-a_{13}-s+1)_{\lambda_s}(-a_{21}-s+1)_{\lambda_s}(-a_{31}-s+1)_{\lambda_s} \right]$$

$$\times A_{\bar\lambda} \begin{pmatrix} a, b, d, e \\ c \end{pmatrix}. \tag{3.8}$$

The summation in eq. (3.8) is over all partitions λ such that $q - t + 1 \geq \lambda_1 \geq \cdots \geq \lambda_t \geq 0$. The functions A_λ given by eq. (3.7) are defined for all partitions, hence for the conjugate partition (see definition (iv) above) that appears in eq. (3.8).

It follows from the discussion above that a *sufficient condition* for transpositional symmetry of the polynomial $\mathcal{G}_q^t$, that is, for $\mathcal{G}_q^t(A) = \mathcal{G}_q^t(A^{\mathrm{tr}})$ is the $b \leftrightarrow e$ symmetry:

$$A_\lambda \begin{pmatrix} a, b, d, e \\ c \end{pmatrix} = A_\lambda \begin{pmatrix} a, e, d, b \\ c \end{pmatrix}, \tag{3.9}$$

which constitutes Identity (4.1) below.

4. Further Identities for Generalized Hypergeometric Functions and Coefficients

Two basic identities are required to prove the determinantal symmetry of the polynomials $\mathcal{G}_q^t(A)$ defined and discussed in Section 3. These are the *generalized Saalschütz identity* (identity (2.7) above) and the *generalized Bailey identity of the second kind* which we now state.

IDENTITY 4.1. *Generalized Bailey Identity of the Second Kind:*

$$
\begin{aligned}
A_\lambda \begin{pmatrix} a, b, d, e \\ c \end{pmatrix} &= \prod_{s=1}^{t} (a + b + c - s + 1)_{\lambda_s} (d + e + c - s + 1)_{\lambda_s} \\
&\quad \times \sum_{\mu\nu} g(\mu\nu\lambda) \langle {}_2\mathcal{F}_1(a, b; a + b + c)|\mu\rangle \, \langle {}_2\mathcal{F}_1(d + c, e + c; d + e + c)|\nu\rangle \\
&= \prod_{s=1}^{t} (a + e + c - s + 1)_{\lambda_s} (b + d + c - s + 1)_{\lambda_s} \\
&\quad \times \sum_{\mu\nu} g(\mu\nu\lambda) \langle {}_2\mathcal{F}_1(a, e; a + e + c)|\mu\rangle \langle {}_2\mathcal{F}_1(b + c, d + c; b + d + c)|\nu\rangle.
\end{aligned}
$$

For $t = 1$, this relation reduces to

$$
(c + d)_n (c + e)_n \, {}_4F_3 \begin{pmatrix} a, b, 1 - c - d - e - n, -n; \\ a + b + c, 1 - c - d - n, 1 - c - e - n \end{pmatrix}
$$

$$
= (a + c + d)_n (a + c + e)_n
$$

$$
\times \, {}_4F_3 \begin{pmatrix} a, a + c, a + b + d + e + 2c + n - 1, n; \\ a + b + c, a + c + d, a + c + e \end{pmatrix}, \tag{4.2}
$$

which is again an expression of Bailey's identity[11]. Since identity (4.1) is distinct from identity (2.9), it is called a *generalized Bailey identity of the second kind*.

Let us remark that the proof of identity (4.1), given in Ref. 14 is equivalent to the proof of the b, e interchange symmetry of the A_λ coefficients defined by eq. (3.7). This proof we found to be *very* difficult. The proof was achieved by showing that eq. (3.7) could be expressed in terms of yet another generalization of the hypergeometric functions, very different from the definition in eq. (2.12). In this new form the symmetry in question becomes self-evident. This new generalization (discussed in ref. 5) involves a new type of *inhomogeneous* symmetric function, but space prevents further discussion here.

We conclude this section by noting several summation identities that have been proved for the A_λ functions:

SUMMATION IDENTITIES 4.3.

$$(a) \qquad A_\lambda \begin{pmatrix} 0, b, d, e \\ c \end{pmatrix} = \langle {}_3\mathcal{F}_0(b+c, d+c, e+c) | \lambda \rangle,$$

$$(b) \qquad A_\lambda \begin{pmatrix} a, b, 0, e \\ c \end{pmatrix} = \langle {}_3\mathcal{F}_0(a+c, b+c, e+c) | \lambda \rangle,$$

$$(c) \qquad A_\lambda \begin{pmatrix} a, b, d, e \\ -a \end{pmatrix} = \langle {}_3\mathcal{F}_0(b, d, e) | \lambda \rangle,$$

$$(d) \qquad A_\lambda \begin{pmatrix} a, b, d, e \\ -d \end{pmatrix} = \langle {}_3\mathcal{F}_0(a, b, d) | \lambda \rangle,$$

where

$$\langle {}_3\mathcal{F}_0(a, b, c) | \lambda \rangle = M_\lambda^{-1} \prod_{s=1}^{t} (a-s+1)_{\lambda_s} (b-s+1)_{\lambda_s} (c-s+1)_{\lambda_s}.$$

5. Concluding Remarks

Let us conclude by remarking that the results given above may be extended substantially, even without going beyond the group $SU(3)$. We have in mind the generalization which involves q-analogs and accordingly the q-analog extension of the generalized hypergeometric function ${}_p\mathcal{F}_q$; this is quite analogous to the familiar q-extension of ${}_pF_q$ to ${}_p\phi_q$. Recently it has been found there exists a a q-generalization of the classical Lie groups to the so-called 'quantum groups' (actually deformations of the universal enveloping algebras). There is a corresponding generalization of tensor operator theory[15], and from this result it is evident that *q-analog extensions of all the identities found above must exist as a consequence of the existence of the quantum group $SU_q(3)$.* Quantum group investigations of such extensions have only just begun, primarily for q-analogs of $SU(2)$ results (some of which were obtained by Askey and Wilson[16] prior to quantum groups); the prospect of $SU(3)$ extensions is both interesting and challenging.

Acknowledgements:

We would like to thank Professor Donald Richards for his continuing interest in our work and for the invitation to participate in this special session of the A.M.S. Tampa Meeting where these results were presented. We would also like to thank Professor Richards for pointing out the relevance of Khatri's work[17] to the generalization given in eq. (2.2).

REFERENCES

1. E. P. Wigner, *Application of Group Theory to the Special Functions of Mathematical Physics*, (Unpublished lecture notes, Princeton University, Princeton NJ, 1955).

2. J. Talman, *Special Functions: A Group Theoretical Approach*, (W. A. Benjamin, NY, 1968). This monograph is based on ref. 1.

3. N. Vilenkin, *Special Functions and the Theory of Group Representations*, (Transl. from the Russian, *Amer. Math. Soc. Transl.* **22**, Ameri. Math. Soc., Providence, RI, 1968).

4. L. C. Biedenharn, R. S. Gustafson, M. A. Lohe, J. D. Louck, and S. C. Milne, *Special Functions and Group Theory in Theoretical Physics*, in *Special Functions, Group Theoretical Aspects and Applications*, Eds. R. Askey, T. H. Koornwinder, and W. Schempp, (Riedel, NY, 1984) p. 129.

5. J. D. Louck and L. C. Biedenharn, *Special Functions Associated with* $SU(3)$ *Wigner-Clebsch-Gordan Coefficients*, Proceedings of the International School on "Symmetry and Structural Properties of Condensed Matter", (September 6-12, 1990, Poznan, Poland), Eds. W. Florek, P. Lulek and M. Nucha, World Scientific (Singapore) 1991.

6. L. C. Biedenharn and J. D. Louck, *J. Math. Phys.* **13**, (1972) 1985.

7. J. D. Louck, M. A. Lohe, and L. C. Biedenharn, *J. Math. Phys.* **16**, (1975) 2408; ibid, *J. Math. Phys.* **26**, (1985) 1458; ibid, **29** (1988) 1106.

8. J. D. Louck and L. C. Biedenharn, *J. Math. Anal.* **59** (1977) 423.

9. I. G. Macdonald, *Symmetric Functions and Hall Polynomials*, (Clarendon Press, Oxford, 1979).

10. L. C. Biedenharn and J. D. Louck *Adv. in Appl. Math.* **9** (1988) 477; ibid, **10** (1989) 396.

11. W. N. Bailey, *Generalized Hypergeometric Series*, (Cambridge Univ. Press, Cambridge, 1935), p. 56.

12. D. P. Shukla, *Indian J. Pure Appl. Math.* **12**, (1981) 994.

13. D. E. Littlewood, *The Theory of Group Characters and Matrix Representations of Groups*, (2nd ed., Oxford Univ. Press, London, 1950).

14. L. C. Biedenharn, A. M. Bincer, M. A. Lohe and J. D. Louck (to appear in *Adv. in Appl. Math.*).

15. L. C. Biedenharn and M. Tarlini, *Lett. in Math. Phys.* **20** (1990) 271.

16. R. Askey and J. A. Wilson, *Some basic hypergeometric orthogonal polynomials that generalize Jacobi polynomials*, *Memoirs Amer. Math. Soc.* **319** (1985).

17. A. T. James, *Ann. Math. Statist.*, **35** (1964) 475-501.

18. C. G. Khatri, "On the moments of traces of two matrices in three situations for complex multivariate normal populations," *Sankhya A* **32** (1970) 65-80.

THEORETICAL DIVISION, LOS ALAMOS NATIONAL LABORATORY, LOS ALAMOS, NM 87545

Contemporary Mathematics
Volume **138**, 1992

Eigen Analysis for Some Examples
of the Metropolis Algorithm

PERSI DIACONIS AND PHIL HANLON

ABSTRACT. The Metropolis algorithm allows us to sample from a given probability distribution by running a Markov chain. We derive the eigenvalues for a class of simple chains. These appear to be the first examples where such explicit computation is possible. They thus allow us to compare exact results with currently available bounds. The eigenvalues turn out to be related to families of orthogonal and symmetric polynomials; as one varies the temperature in the Metropolis algorithm one runs through the natural parameter in the family.

1. Introduction

Let X be a finite set and let $\pi(x)$ be a probability distribution on X with $\pi(x) > 0$ for all $x \in X$. The Metropolis algorithm is a strategy for sampling from π which is effective when π is only known up to a norming constant which is difficult to compute because $|X|$ is large. The algorithm proceeds by running a Markov chain with transition $M(x, y)$ and stationary distribution π. The chain is constructed by "thinning down" an easy to run "base chain." A more careful description is given below. The Metropolis algorithm is very widely used in statistical mechanics (Ising simulations), statistics, and as an ingredient in simulated annealing. Little is known about its non-asymptotic rate of convergence to stationarity.

This paper, along with Hanlon (1992), gives a class of examples where eigenvalues and sharp computations for rates of convergence can be derived. The examples involve natural random walks on a group "thinned down" by a natural distance function. The examples are interesting in two directions. They seem

---

1991 *Mathematics Subject Classification*. Primary 05E05, 60J27.

Key words and phrases. Metropolis algorithm, Markov chains, Ising models, simulated annealing, zonal polynomials, Jack polynomials, Krawtchouk polynomials, random walks, convergence to stationarity.

Research supported by grants from the NSF.

This paper is in final form and no version of it will be submitted for publication elsewhere .

to be the first examples where explicit computation can be carried out. They can thus serve as test problems for bounds on rates of convergence. Further, they produce classical families of one-parameter orthogonal polynomials such as Krawtchouk polynomials and symmetric functions such as Jack symmetric functions.

The Metropolis algorithm is explained more carefully in section 2. Our most interesting example appears in section 4. This analyzes a process on the symmetric group. Here, the base chain is based on randomly transposing pairs of cards. The chain is "thinned down" according to its distance to the identity using d – the minimum number of transpositions metric. The resulting chain has stationary distribution proportional to $\theta^{d(\pi, \pi_0)}$, a measure that arises in statistical applications. The eigenvalues are the change of basis coefficients when the Jack symmetric functions are expanded in the power sum symmetric functions. Using recent results of Stanley (1989) and Macdonald (1989), these eigenvalues are explicitly available. We determine the convergence properties and study the dependence on θ.

Section 3 carries out an easier analysis for nearest neighbor walk on the cube $\mathbb{Z}_2^d$, thinning down with Hamming distance. Now the one-parameter family of Krawtchouk polynomials appears in the eigen analysis.

In both cases, the chains are rapidly mixing: the Metropolis thinning only changes the eigenvalues in a linear way. In contrast, use of "off-the-shelf" bounds leads to exponentially perturbed eigenvalues and a poor picture of convergence rates.

The final section carries out this analysis for a twisted Markov chain on the space of matchings. Again, exact analysis and interesting special functions appear.

2. The Metropolis algorithm and eigen analysis of reversible Markov chains

This section contains a careful description of the Metropolis algorithm along with background on the use of eigenvalues to study total variation convergence for reversible Markov chains.

Let X be a finite set. Let $\pi(x)$ be a positive probability on X. Often, practical considerations allow easy access to the ratios $r_{yx} = \pi(y)/\pi(x)$. Let $S(x, y)$ be the transition matrix of a symmetric irreducible Markov chain on X. This is the base chain which is assumed to be easy to run. Let $M(x, y)$ be defined by "thinning down" $S(x, y)$ according to the following

$$(2.1) \quad M(x, y) = \begin{cases} S(x, y) r_{yx} & \text{if } r_{yx} < 1 \\ S(x, y) & \text{if } x \neq y \text{ and } r_{yx} \geq 1 \\ S(x, x) + \sum_{\substack{z \neq x \\ r_{zy}}} S(x, z)(1 - r_{zy}) & \text{if } x = y. \end{cases}$$

This definition has a simple implementation. If the chain is at x, pick y with probability $S(x, y)$. If $x \neq y$ and $r_{yx} \geq 1$, the chain moves to y. If $y \neq x$ and $r_{yx} < 1$, flip a coin with success probability r_{yx}. If the coin toss succeeds, the chain moves to y. In all other cases the chain stays at x.

It is straightforward to show that $M(x,y)$ defines an irreducible aperiodic transition matrix with stationary distribution π. For further details, see e.g. Hammersley and Handscomb (1964).

The Metropolis chain M is reversible: $\pi(x)M(x,y) = \pi(y)M(y,x)$. To conclude this section we derive bounds on rates of convergence for a general reversible chain in terms of eigenvalues. Thus for the remainder of this section, let (M,π) be a reversible Markov chain on a finite set X. Let D be a diagonal matrix having $\sqrt{\pi(x)}$ down the diagonal. Reversibility yields that $T = DMD^{-1}$ is symmetric. Thus T can be orthogonally diagonalized $T = \Gamma\beta\Gamma^t$ with Γ orthogonal and β a diagonal matrix having the eigenvalues of T, and so M, on the diagonal. Thus $M = V\beta V^{-1}$ with $V = D^{-1}\Gamma, V^{-1} = \Gamma^t D$. This implies that the right eigenvectors of M are the columns of V: $V_{xy} = \Gamma_{xy}/\sqrt{\pi(x)}$. These are orthonormal in $L^2(\pi)$. The left eigenvectors are the rows of V^{-1}: $V_{xy}^{-1} = \Gamma_{yx}\sqrt{\pi(y)}$. These are orthonormal in $L^2(1/\pi)$.

Define total variation distance as

$$\|M^k(x,\cdot) - \pi(\cdot)\| = \frac{1}{2}\sum_y |M^k(x,y) - \pi(y)|.$$

The following lemma gives bounds on total variation in terms of eigenvalues.

LEMMA 1. *Let (M,π) be a reversible Markov chain on a finite set X. Let β_y denote the eigenvalues, β^* the second largest eigenvalue in absolute value. Let $f_y(\cdot)$ be an orthonormal basis of right eigenfunctions in $L^2(\pi)$. Let $g_y(\cdot)$ be an orthonormal basis of left eigenfunctions in $L^2(\frac{1}{\pi})$. Then, for any starting state x, the total variation $4\|M(x,\cdot) - \pi(\cdot)\|^2$ is bounded above by any of the following three quantities:*

(2.2) $\displaystyle\sum_y \beta_y^{2k} f_y^2(x) - 1$

(2.3) $\displaystyle\frac{1}{\pi^2(x)}\sum_y \beta_y^{2k} g_y(x) - 1$

(2.4) $\displaystyle\frac{1}{\pi(x)}(\beta^*)^{2k}.$

PROOF. For any fixed x

$$\left(\sum_y |M^k(x,y) - \pi(y)|\right)^2 = \left(\sum_y \frac{|M^k(x,y) - \pi(y)|}{\sqrt{\pi(y)}}\sqrt{\pi(y)}\right)^2$$

$$\leq \sum_y \frac{|M^k(x,y) - \pi(y)|^2}{\pi(y)}$$

$$= \sum_y \left(\frac{M^k(x,y)}{\pi(y)}\right)^2 - 1$$

$$= \frac{M^{2k}(x,x)}{\pi(x)} - 1$$

$$= \frac{1}{\pi(x)}\sum_y \beta_y^{2k}\Gamma_{xy}^2 - 1.$$

The inequality above is Cauchy-Schwarz. The next to last equality used the formula $\pi(x)M(x,y) = \pi(y)M(y,x)$. The bounds now follow by relating Γ_{xy} to left or right eigenvectors (multiplying or dividing by $\sqrt{\pi(x)}$). The final bound follows by bounding β_y by β^*. □

The Metropolis chains in this paper are all built by thinning down a simple chain with known spectral properties. It is natural to try to compare the eigenvalues. This can be achieved by comparing the Dirichlet forms. For a reversible chain (M, π), define

$$\mathcal{E}(f, f) = \frac{1}{2} \sum_{x,y} (f(x) - f(y))^2 \pi(x) M(x, y).$$

This is just the quadratic form defined by I-M. The classical minimax characterization of eigenvalues yields

$$1 - \beta_i = \min_{W_i} \min_{f \in W_i} \frac{\mathcal{E}(f, f)}{\|f\|_\pi^2}, \qquad \dim W_i = i + 1.$$

Here β_i is the ith largest eigenvalue $1 = \beta_0 > \beta_1 \geq \beta_2 \cdots \geq \beta_{|X|-1} \geq -1$. This renders the following comparison lemma evident.

LEMMA 2. *Let* $(M, \pi), (\widetilde{M}, \widetilde{\pi})$ *be reversible Markov chains on a finite set* X. *Suppose the associated Dirichlet forms satisfy*

$$\widetilde{\mathcal{E}} \leq A\mathcal{E} \quad \text{and} \quad \widetilde{\pi} \geq a\pi \quad \text{for positive } A, a.$$

Then

$$\beta_i \leq 1 - \frac{a}{A}(1 - \widetilde{\beta}_i).$$

As will emerge, these bounds can be very far off in the Metropolis setting. For lower bounds, and examples where comparison is useful, see Diaconis and Saloff-Coste (1992).

3. A simple example: nearest neighbor walk on $\mathbb{Z}_2^d$

Let $\mathbb{Z}_2^d$ be the group of binary d-tuples under coordinatewise addition, thought of as the vertices of a cube in d-dimensions. Nearest neighbor walk has transition matrix

$$(3.1) \qquad\qquad S(x, y) = \begin{cases} \frac{1}{d} & \text{if } H(x, y) = 1 \\ 0 & \text{otherwise} \end{cases}$$

with $H(x, y)$ the Hamming distance: the number of coordinates where x and y disagree. This walk has been extensively analyzed because it gives a representation of the Ehrenfest urn. See Kac (1947), Letac and Takacs (1979), or Diaconis (1988) and references cited there. In particular, the matrix S is diagonalizable with eigenvalues $(1 - \frac{2j}{d})$ occurring with multiplicity $\binom{d}{j}$, $0 \leq j \leq d$, and eigenvectors given by the characters of $\mathbb{Z}_2^d$. Note this walk has periodicity problems since -1 is an eigenvalue. These will disappear for the Metropolis walk.

Consider next the family of measures on $\mathbb{Z}_2^d$ defined by

$$(3.2) \qquad\qquad P_\theta(x) = \frac{\theta^{H(x)}}{(1+\theta)^d}$$

with $0 < \theta \leq 1$, $H(x) = H(x,0)$. For this example, there is no problem simulating from P_θ by a variety of schemes. For example, coordinates of x can be chosen as independent 1's or 0's with probability $\theta/(1+\theta)$ and $1/(1+\theta)$ respectively. This takes order d operations.

The Metropolis algorithm can be used to generate from P_θ. Starting from $S(x,y)$ at (3.1), the recipe (2.1) results in a chain with

$$(3.3) \qquad M(x,y) = \begin{cases} \frac{1}{d} & \text{if } H(x,y) = 1, \ H(y) < H(x) \\ \frac{\theta}{d} & \text{if } H(x,y) = 1, \ H(y) > H(x) \\ (1 - \frac{H(x)}{d})(1-\theta) & \text{if } H(x,y) = 0 \\ 0 & \text{otherwise.} \end{cases}$$

For example, when $d = 2$, M appears as

$$
\begin{array}{c}
\begin{array}{cccc} 00 & 01 & 10 & 11 \end{array} \\
\begin{array}{c} 00 \\ 01 \\ 10 \\ 11 \end{array}
\begin{pmatrix}
1-\theta & \frac{\theta}{2} & \frac{\theta}{2} & 0 \\
\frac{1}{2} & \frac{1-\theta}{2} & 0 & \frac{\theta}{2} \\
\frac{1}{2} & 0 & \frac{1-\theta}{2} & \frac{\theta}{2} \\
0 & \frac{1}{2} & \frac{1}{2} & 0
\end{pmatrix}.
\end{array}
$$

The permutation group S_d operates on $\mathbb{Z}_2^d$. It is clear that $M(\pi x, \pi y) = M(x,y)$ for $\pi \in S_d$. This implies that the orbit chain, which just records $H(x)$ when the chain is in x, is again a Markov chain. This chain takes values in $(0, 1, \ldots, d)$ with transition matrix

$$(3.4) \qquad m(i,j) = \begin{cases} \frac{i}{d} & \text{if } j = i - 1 \\ (1 - \frac{i}{d})\theta & \text{if } j = i + 1 \\ (1 - \frac{i}{d})(1-\theta) & \text{if } j = i. \end{cases}$$

For example, when $d = 2$, m appears as

$$
\begin{array}{c}
\begin{array}{ccc} 0 & 1 & 2 \end{array} \\
\begin{array}{c} 0 \\ 1 \\ 2 \end{array}
\begin{pmatrix}
1-\theta & \theta & 0 \\
\frac{1}{2} & \frac{1-\theta}{2} & \frac{\theta}{2} \\
0 & 1 & 0
\end{pmatrix}.
\end{array}
$$

In general, m has $(1 - \frac{i}{d})(1 - \theta)$ down the diagonal, $0 \leq i \leq d$; $\frac{i}{d}$ below the diagonal $1 \leq i \leq d$; $(1 - \frac{i}{d})\theta$ above the diagonal $0 \leq i \leq d - 1$; and zeros elsewhere. This m has stationary distribution $\pi(i) = \binom{d}{i}\theta^i(1+\theta)^{-d}$.

THEOREM 1. *The matrix m defined at (3.4) has eigenvalues*

$$\beta_i = 1 - \frac{i}{d}(1 + \theta), \qquad 0 \le i \le d.$$

The corresponding right eigenvector is the Krawtchouk polynomial

$$P_i(j) = \left(\theta^i \binom{d}{i}\right)^{-1/2} \sum_{k=0}^{i} (-1)^k \binom{j}{k}\binom{d-j}{i-k}\theta^{i-k}.$$

These have been normalized to be orthonormal in $L^2(\pi)$.

PROOF. This is essentially contained in Krawtchouk (1929). The statement is easy to verify from known properties of Krawtchouk polynomials as given by MacWilliams and Sloane (1977, p. 150). □

For example, when $d = 2$, the eigenvalues are 1, $1 - \frac{(1+\theta)}{2}$, $-\frac{\theta}{2}$, with eigenvectors

$$\begin{pmatrix} 1 \\ 1 \\ 1 \end{pmatrix}, \qquad \frac{1}{\sqrt{2\theta}}\begin{pmatrix} 2\theta \\ \theta - 1 \\ -2 \end{pmatrix}, \qquad \frac{1}{\theta}\begin{pmatrix} \theta^2 \\ -\theta \\ 1 \end{pmatrix}.$$

The main point is that thinning down by distance leads to an interesting deformation of the eigenvalues and eigenvectors. The next result converts this into a rate of convergence result for the original chain M.

THEOREM 2. *Fix θ in $(0,1)$. For the Metropolis chain M at (3.3) started at 0, let $k = \frac{d}{2(1+\theta)}(\log d\theta + c)$. Then*

$$\|M^k(0, \cdot) - P_\theta(\cdot)\| \le f(\theta, c)$$

with $f(\theta, c)$ independent of d, tending to zero as $c \to \infty$, for each fixed θ.

PROOF. Because of invariance, the distance $\|M^k(0, \cdot) - P_\theta(\cdot)\| = \|m^k(0, \cdot) - \pi_\theta(\cdot)\|$ where the orbit chain is defined at (3.4). Using Theorem 1 and the upper bound (2.2),

$$(3.5) \qquad 4\|m^k(0, \cdot) - \pi_\theta(\cdot)\|^2 \le \sum_{j=1}^{d}\left(1 - \frac{j(1+\theta)}{d}\right)^{2k}\theta^j\binom{d}{j}.$$

Break the sum in (3.5) into $S_1 + S_2$ with S_1 summed over $1 \le j \le d/(1+\theta)$ and S_2 summed over $d/(1+\theta) < j \le d$. For S_1, use $1 - x \le e^{-x}$, $\binom{d}{j} \le d^j/j!$, and the definition of k to conclude

$$S_1 \le \sum_{j=1}^{\infty}\frac{e^{-c}}{j!} = e^{e^{-c}} - 1.$$

For S_2, replace j by $d - \ell$ and use the inequalities above to conclude

$$S_2 \le \theta^{d+2k}\sum_{\ell=1}^{d\theta(1+\theta)}\frac{1}{\ell!}\exp\{\frac{-\ell}{\theta}(\log(d\theta) + c) + \ell\log(d/\theta)\}.$$

Now $\theta < 1$ yields $\frac{-\ell}{\theta} \log d + \ell \log d < 0$. Making this replacement,

$$S_2 \leq \sum_{\ell=1}^{\infty} \frac{1}{\ell!} \exp\{-\ell[\log \theta^{\frac{\theta+1}{\theta}} + \frac{c}{\theta}]\}.$$

The bounds for S_1 and S_2 sum to give $f(\theta, c)$ with the stated properties. $\quad\square$

REMARKS. 1. The bound is sharp in the sense that if $k = \frac{d}{2(1+\theta)}(\log d\theta - c)$ the variation distance does not tend to zero. This can be shown by using the first eigenfunction as in Diaconis (1988, Chapter 3).

2. Theorem 2 shows that the Metropolis algorithm is only "off by a log." That is, its running time (order $d \log d$) is comparable to the optimal algorithm (order d) which uses the structure of the normalizing constant. Of course, the function $H(x)$ is relatively simple.

3. It is instructive to compare the second eigenvalue, $1 - \frac{(1+\theta)}{d}$, with what results from comparison of Dirichlet forms (Lemma 2 of Section 2). Passing to the chain m and comparing with the unthinned base chain ($\theta = 1$) requires $a = \left(\frac{1+\theta}{2}\right)^d$ and $A = \frac{1}{\theta}\left(\frac{1+\theta}{2\theta}\right)^d$. This yields

$$\beta_i \leq 1 - \theta^{d+1}(1 - \widehat{\beta}_i).$$

This is off by an exponential factor and virtually useless in practice.

4. Simulating from a distribution on the symmetric group

We begin with some motivation for the example to be studied. Statisticians sometimes have to work with ranked data as when a group of people are each asked to rank order 5 wines. To facilitate data analysis, a variety of metrics between permutations are employed. For example, the Cayley distance is defined as

$$d(\pi, \sigma) = \text{minimum number of transpositions required}$$
$$\text{to bring } \pi \text{ to } \sigma, \text{ for } \pi, \sigma \in S_n.$$

This is named after Cayley who discovered $d(\pi, \sigma) = n - C(\pi\sigma^{-1})$, with $C(\tau)$ the number of cycles in τ. One use of such metrics is to build probability distributions on S_n,

(4.1) $$P_\theta(\pi) = c(\theta)\theta^{d(\pi, \pi_0)}$$

where $0 < \theta \leq 1$,

$$c^{-1}(\theta) = \sum_{\pi} \theta^{d(\pi, \pi_0)} = \prod_{i=1}^{n}(1 + \theta(i - 1))$$

is a normalizing constant, and π_0 is a "location parameter." This model describes a population peaked about π_0 which falls off geometrically at rate θ. For $\theta = 1$,

P_1 becomes the uniform distribution. In statistical applications, π_0 and θ would be unknown and estimated from a sample of rankings. Further background can be found in Critchlow (1985) or Diaconis (1988, Chapter 6).

In any modern statistical work, the ability to sample efficiently from a probability distribution is crucial. In this section we analyze the Metropolis algorithm for this problem and compare it with the best available alternative. Without essential loss, take $\pi_0 = id$ throughout.

The Metropolis algorithm will be based on repeated random transpositions. Thus, consider the Markov chain on S_n with transition matrix

$$S(\sigma, \tau) = \begin{cases} 1/\binom{n}{2} & \text{if } \tau = \sigma(i,j) \text{ for some } i < j \\ 0 & \text{otherwise.} \end{cases}$$

The chain S was analyzed by Diaconis and Shahshahani (1981). As given, it has periodicity problems which will disappear when it is thinned down. Define a Metropolis chain as in (2.1) with

$$r(\sigma, \tau) = \theta^{C(\sigma) - C(\tau)}.$$

When $n = 3$, the transition matrix becomes

$$
M_3^\theta =
\begin{array}{c}
\\
id \\
(12) \\
(13) \\
(23) \\
(123) \\
(132)
\end{array}
\begin{array}{cccccc}
id & (12) & (13) & (23) & (123) & (132)
\end{array}
\left(
\begin{array}{cccccc}
1-\theta & \frac{\theta}{3} & \frac{\theta}{3} & \frac{\theta}{3} & 0 & 0 \\
\frac{1}{3} & \frac{2}{3}(1-\theta) & 0 & 0 & \frac{\theta}{3} & \frac{\theta}{3} \\
\frac{1}{3} & 0 & \frac{2}{3}(1-\theta) & 0 & \frac{\theta}{3} & \frac{\theta}{3} \\
\frac{1}{3} & 0 & 0 & \frac{2}{3}(1-\theta) & \frac{\theta}{3} & \frac{\theta}{3} \\
0 & \frac{1}{3} & \frac{1}{3} & \frac{1}{3} & 0 & 0 \\
0 & \frac{1}{3} & \frac{1}{3} & \frac{1}{3} & 0 & 0
\end{array}
\right).
$$

The stationary distribution is the left eigenvector proportional to

$$(1, \theta, \theta, \theta, \theta^2, \theta^2).$$

By construction, the matrix M_n^θ commutes with the action of S_n on itself by conjugation. This implies that if π and σ are conjugate in S_n, $M^k(id, \pi) = M^k(id, \sigma)$ for all $k = 1, 2, \ldots$. Thus the chain lumped to conjugacy classes is Markov. When $n = 3$, this lumped chain has transition matrix m_3^θ

$$
(4.2) \qquad
\begin{array}{c}
1 \\
1,2 \\
3
\end{array}
\begin{array}{ccc}
1^3 & 1,2 & 3
\end{array}
\left(
\begin{array}{ccc}
1-\theta & \theta & 0 \\
\frac{1}{3} & \frac{2}{3}(1-\theta) & \frac{2}{3}\theta \\
0 & 1 & 0
\end{array}
\right).
$$

This has stationary distribution proportional to $(1, 3\theta, 2\theta^2)$. In general, the conjugacy classes are indexed by partitions of n. The transition matrix for the lumped chain has

$$(4.3) \qquad m_n^\theta(\lambda, \mu) = \sum M_n^\theta(\pi, \sigma)$$

summed over all σ in the conjugacy class μ with π any permutation in conjugacy class λ.

Any of the standard measures of speed of convergence to stationarity are the same for M_n^θ and m_n^θ (see, e.g., Diaconis and Zabell (1982). The stationary distribution for m is calculated by summing over the conjugacy class. This gives

$$(4.4) \qquad \pi^\theta(\lambda) = \theta^{n-r} \frac{n!}{z_\lambda} \prod_{i=1}^{n} (\theta(i-1)+1)^{-1}.$$

Here, if the partition $\lambda = (\lambda_1, \ldots, \lambda_r)$ with $\lambda_1 \geq \lambda_2 \cdots \geq \lambda_r > 0$ has a_i parts equal to i,

$$(4.5) \qquad z_\lambda = \prod_{i=1}^{n} i^{a_i} a_i!$$

Hanlon (1992) has derived the eigenvalues and eigenvectors of the Markov chain m. This description involves an interesting class of symmetric functions called the Jack symmetric function. We will use the notation in Stanley (1984) and Macdonald (1979) for these functions. For each partition λ of n and each real $\alpha \neq 0$ there is a homogeneous symmetric function $J_\lambda(\mathbf{x}; \alpha)$. Here $\mathbf{x} = (x_1, \ldots, x_k)$ with $k \geq n$. For fixed α, $\{J_\lambda(\mathbf{x}; \alpha)\}_\lambda$ is a basis of the homogeneous symmetric polynomials of degree n in $x_1, x_2, \ldots, x_k$ as λ varies over partitions of n. A more familiar basis is the power sum symmetric functions defined by $P_i(\mathbf{x}) = x_1^i + \cdots + x_k^i$; $P_\lambda = \prod_{i=1}^{r} P_{\lambda_i}$. Denote the change of basis coefficients by $C(\lambda, \mu)$:

$$(4.6) \qquad J_\lambda(\mathbf{x}; \alpha) = \sum_{\mu \vdash n} C^\alpha(\lambda, \mu) P_\mu(\mathbf{x}).$$

The $C^\alpha(\lambda, \mu)$ are rational functions in α. For example, when $n = 3$,

$$J_{1^3} = P_1^3 - 3P_{1,2} + 2P_3$$
$$J_{2,1} = P_1^3 + (\alpha - 1)P_{1,2} - \alpha P_3$$
$$J_3 = P_1^3 - 3\alpha P_{1,2} + 2\alpha^2 P_3 .$$

When $\alpha = 1$, the Jack polynomials become the Schur functions. When $\alpha = 2$, the Jack polynomials become the zonal polynomials (spherical functions of GL_n/O_n)). Until now, no interpretation of other values was known.

To state the main result, one further piece of notation is needed. Define

$$(4.8) \qquad j_\lambda(\alpha) = \prod_{s \in \lambda} h_*^\lambda(s) h_\lambda^*(s)$$

where the product is over square s in the diagram of λ. If $s = (i, j)$ has a squares strictly below it and b squares strictly to its right,

$$\begin{aligned} h^*(s) &= (a+1)\alpha + b, \\ h_*(s) &= a\alpha + (b+1). \end{aligned}$$

When $\alpha = 1, h^* = h_*$ is the usual hook length. For example, when $n = 3$, the upper and lower hook lengths are as shown:

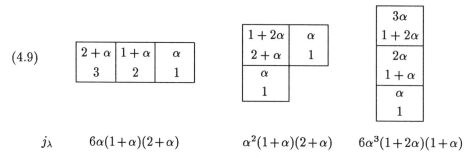

$$(4.9)$$

$$j_\lambda \qquad 6\alpha(1+\alpha)(2+\alpha) \qquad \alpha^2(1+\alpha)(2+\alpha) \qquad 6\alpha^3(1+2\alpha)(1+\alpha)$$

With this notation, our main result can be stated.

THEOREM 1. *For $0 < \theta \le 1$, the Markov chain m_n^θ defined at (4.3) with stationary distribution π_n^θ at (4.4) has an eigenvalue β_λ for each partition $\lambda = (\lambda_1, \lambda_2, \ldots, \lambda_r)$ of n. These are*

$$\beta_\lambda = (1 - \theta) + \frac{\theta n(\lambda^t) + n(\lambda)}{\binom{n}{2}}$$

with

$$n(\lambda) = \sum_{i=1}^r (i-1)\lambda_i = \sum_j \binom{\lambda_j^t}{2}.$$

The corresponding left eigenvector, normed to be orthonormal in $L^2(1/\pi_n^\theta)$ is

$$\frac{C^\theta(\lambda, \cdot)}{\{j_\lambda \Pi / (\theta^n n!)\}^{1/2}}$$

with C^θ as in (4.6), $j_\lambda(\theta)$ as in (4.8), and $\Pi = \prod_{i=1}^n (\theta(i-1) + 1)$.

EXAMPLE. The matrix m_3^θ at (4.2) has eigenvalues and eigenvectors

λ	1^3	$2, 1$	3
β_λ	$-\theta$	$\frac{2}{3}(1-\theta)$	1
$C(\lambda, \cdot)$	$(1, -3, 2)$	$(1, \theta - 1, -\theta)$	$(1, 3\theta, 2\theta^2)$.

The normalizing constants for these eigenvectors appear in (4.9).

PROOF OF THEOREM 1. Stanley (1989, Theorem 5.4) shows, in the present notation

$$(4.10) \qquad J_\lambda(1^k; \alpha) = \prod_{(i,j) \in \lambda} (k - (i-1) + \alpha(j-1)).$$

On the left, the function J_λ is evaluated when all of its arguments are equal to 1. Since this is true for all k, one can equate coefficients in

$$J_\lambda(1^k; \alpha) = \sum_{\mu \vdash n} C^\alpha(\lambda, \mu) P_\mu(1^k) = \sum_{\mu \vdash n} C^\alpha(\lambda, \mu) k^{\ell(\mu)}$$

with $\ell(\mu) = r$ if $\mu = (\mu_1, \mu_2, \ldots, \mu_r)$. Of course, this only determines $C^\alpha(\lambda, \mu)$ for certain special μ, e.g., $\mu = 1^2, \mu = 1^{n-2}, 2, \mu = n, \mu = 1, n-1$. In particular

(4.11)
$$C^\alpha(\lambda, 1^n) = 1$$

and

(4.12)
$$C^\alpha(\lambda, 1^{n-2}, 2) = \alpha \sum_j \binom{\lambda'_j}{2} - \sum_i \binom{\lambda_i}{2}$$

for all λ. The argument in Hanlon (1992) shows that the $C^\theta(\lambda, \mu)$ are left eigenvectors of the matrix m_n^θ. To determine the corresponding eigenvalue, observe that the first column of m_n^θ is $(1 - \theta, \frac{1}{\binom{n}{2}}, 0, 0, \ldots, 0)^t$. The eigenvalue equation becomes

$$C^\theta(\lambda, 1^n)(1 - \theta) + \frac{C^\theta(\lambda, 1^{n-2}, 2)}{\binom{n}{2}} = \beta_\lambda C^\theta(\lambda, 1^n).$$

Now (4.11), (4.12) gives the eigenvalue.

The orthonormality follows from Stanley (1989) who gives

$$\sum_{\nu \vdash n} C^\alpha(\lambda, \nu) C^\alpha(\mu, \nu) \alpha^{\ell(\nu)} z_\nu = j_\lambda \delta_{\lambda\mu}.$$

$\square$

The ingredients above can be combined together to yield the following reasonably sharp bounds for rates of convergence for the original chain M.

THEOREM 2. For $0 < \theta < 1$, let P_θ^k be the probability on S_n associated to the Markov chain defined from M_n^θ starting as id, with stationary distribution P_θ defined at (4.1). Let

$$k = an \log n + cn, \quad \text{with} \quad a = \frac{1}{2\theta} + \frac{1}{4\theta}(\frac{1}{\theta} - \theta) \quad \text{and} \quad c > 0.$$

Then, there is a function $f(\theta, c)$ independent of n, with $f(\theta, c) \to 0$ for $c \to \infty$ such that

$$\|P_\theta^k - P_\theta\| \le f(\theta, c).$$

PROOF. From (2.5),

$$4\|P_\theta^k - P_\theta\|^2 \le \theta^n n! \prod_{i=1}^n (1 + \theta(i - 1)) \sum_{\substack{\lambda \vdash n \\ \lambda \ne (n)}} \frac{1}{j_\lambda} \beta_\lambda^{2k}.$$

Here β_λ and j_λ are as in Theorem 1 and $C(\lambda, n) = 1$ from (4.11) was used. Now, one proceeds along the lines of Diaconis and Shahshahani (1981) or Diaconis (1988, pp. 36-43) who essentially did the case $\theta = 1$. The eigenvalue β_λ is monotone if the majorization ordering is used on $\lambda : \lambda \ge \lambda'$ implies $\beta_\lambda \ge \beta_{\lambda'}$. This holds for all $\theta \in (0, 1)$. It allows the terms in the upper bound to be grouped just as in the $\theta = 1$ case. As there, the sum is dominated by the term for $\lambda = (n - 1, 1)$. For this term,

$$\beta_{n-1,1} = 1 - \frac{2\theta}{n} - \frac{2}{n(n-1)}$$

and

$$\theta^n n! \prod_{i=1}^{n} (1 + \theta(i-1))/j_{n-1,1}$$

$$= \frac{\theta^n n!(1+\theta)(1+2\theta)\cdots(1+(n-1)\theta)}{\theta^2(n-1)!(1+\theta)(2+\theta)\cdots(n-3+\theta)(n-1+\theta)(n-2+2\theta)}$$

$$= f(\theta)n^{2+(\frac{1}{\theta}-\theta)}(1 + O(\frac{1}{n}))$$

for an explicit continuous function $f(\theta)$. It follows from this, with k as given, that

$$\frac{\theta^n n! \prod_{i=1}^{n}(1 + \theta(i-1))}{j_\lambda}\beta_\lambda^{2k} \leq f(\theta)n^{2+(\frac{1}{\theta}-\theta)}e^{-\frac{4\theta}{n}(an\log n - 2C^n)}(1 + O(\frac{1}{n}))$$

$$= f(\theta)e^{-8\theta C}(1 + O(\frac{1}{n})).$$

We omit the rest of the argument, except for consideration of the term corresponding to 1^n. For this term $\beta_{1^n} = -\theta$ and

$$\theta^n n! \prod_{i=1}^{n}(1 + \theta(i-1))/j_{1^n} = \frac{\theta^n n!(1+\theta)(1+2\theta)\cdots(1+(n-1)\theta)}{n!\theta(1+\theta)(2+\theta)\cdots(n-1+\theta)}.$$

This last quantity is bound above by a positive continuous function $g(\theta)$, for all n. It follows that the term for 1^n tends to zero. $\square$

REMARKS. 1) The function $a(\theta) = \frac{1}{2\theta} + \frac{1}{4\theta}(\frac{1}{\theta} - \theta)$ increases as θ decreases from 1 to 0 so it takes longer to converge for small θ. This seems curious since the walk starts at the identity which is also the most likely state.

2) We have not attempted to prove a matching lower bound but are morally certain that the variation distance does not tend to zero for $h = an\log n - cn$ when n is large and c is positive. The arguments in Diaconis and Shahshahani (1981) show this for $k = \frac{1}{2}n\log n - cn$ for all θ. This goes part of the way to explaining the curiosity in remark 1: for the walk to reach stationarity, the distribution of the number of fixed points has to be correct. In particular there has to be a reasonable chance of hitting every card. This already requires $\frac{1}{2}n\log n + cn$ moves be made.

5. One-factors and signed Markov chains

In this section, we will consider two other cases in which the Metropolis Algorithm produces twisted Markov chains with interesting steady states. The second of these will be obtained from the first by appropriately introducing signs in the transition matrix. In view of this, it will not technically be a Markov chain. Nevertheless, the Metropolis Algorithm still applies and gives us an interesting twisting of our "signed" Markov chain.

This section twists a chain studied by Diaconis (1986) who gave the following interpretation: consider f pairs of mathematicians who come to a party. They arrive in pairs as $\{(1, 2), (3, 4), \ldots, (2f - 1, 2f)\}$. Being mathematicians, they stand there talking to the person they arrived with. A host decides to mix things up by picking a pair, say (i_1, i_2), at random, then a second pair, say (j_1, j_2), at random, and switching, say, to $(i_1, j_1), (i_2, j_2)$ or $(i_1, j_2), (i_2, j_1)$. This defines a process on Ω_f the partitions of $2f$ into f 2-element blocks where order within or between doesn't matter.

The symmetric group S_{2f} acts transitively on Ω_f. The isotropy subgroup can be identified with $B_f = \mathbb{Z}_2^f \alpha S_f$, the hyperoctahedral group. The pair S_{2f}, B_f is a Gelfand pair with spherical functions given by the coefficients $C^2(\lambda, \mu)$ of section 4. This allowed a complete analysis when $\theta = 1$. The present section refines this analysis to allow for twisting.

5.1 Switching random pairs in one-factors.

A *one-factor* on $2f$ points is a graph with $2f$ points in which every point has degree 1. Let Ω_f denote the set of 1-factors on $(2f)$ points. It is easy to see that $|\Omega_f| = (2f)!/2^f f!$. We usually draw a one-factor $\delta \in \Omega_f$ by putting the points $1, 2, \ldots, f$ in a top row and the points $(f + 1), \ldots, 2f$ in a bottom row. For example, the one-factor Δ_0 which has an edge from i to $i + f$ for $i = 1, 2, \ldots, f$ is drawn as:

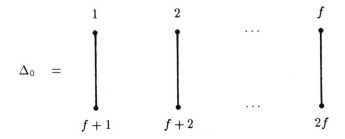

Given $\delta_1, \delta_2 \in \Omega_f$. Let $\delta_1 \cup \delta_2$ be the graph obtained by taking the union of their edges. It is easy to check that $\delta_1 \cup \delta_2$ is a disjoint union of cycles of even lengths. Let $\Lambda(\delta_1, \delta_2)$ be the partition of f whose parts are half the cycle length of $\delta_1 \cup \delta_2$.

Define a Markov chain M with states Ω_f in the following way. Given a 1-factor δ, the probability $R_f(1)_{\delta, \tau}$ of moving to another 1-factor τ is given by

$$R_f(1)_{\delta, \tau} = \begin{cases} \frac{1}{2f - 1} & \text{if } \delta = \tau \\ \frac{2}{f(2f - 1)} & \text{if } \tau \text{ can be obtained from } \delta \text{ by} \\ & \quad \text{switching a pair of non-adjacent points} \\ 0 & \text{otherwise.} \end{cases}$$

For example, with $f = 2$ we have

$$R_2(1) \quad = \quad \begin{pmatrix} 1/3 & 1/3 & 1/3 \\ 1/3 & 1/3 & 1/3 \\ 1/3 & 1/3 & 1/3 \end{pmatrix}.$$

Let δ be in Ω_f. The *type of* δ is the partition $\Lambda(\delta, \Delta_0)$. Type for one-factors is analogous to cycle type for permutations. Let $c(\delta)$ denote the number of parts of $\Lambda(\delta, \Delta_0)$.

Define a probability distribution Γ on Ω_f by

$$\Gamma(\delta) = \alpha^{f-c(\delta)}/G$$

where α is a real number greater than 1 and where G is the appropriate normalizing constant $G = \prod_{i=0}^{f-1} (1 + i(2\alpha))$. We now have an $M = M(1)$ and a probability distribution Γ so we are in a position to apply the Metropolis algorithm to obtain a twisted Markov chain $M(\alpha)$ whose stable distribution is Γ. Let $R_f(\alpha)$ denote the transition matrix of $M(\alpha)$. For example,

$$R_2(\alpha) \quad = \quad \frac{1}{3\alpha} \begin{pmatrix} \alpha & \alpha & \alpha \\ 1 & 3\alpha - 2 & 1 \\ 1 & 1 & 3\alpha - 2 \end{pmatrix}.$$

The next proposition we state without proof because we will prove something more general later in this section. The proposition says that the Markov chain $R_f(\alpha)$ can be reduced (or lumped) to a Markov chain on the types.

PROPOSITION. *For each partition λ of f let v_λ be defined by*

$$v_\lambda = 2^{-\ell(\lambda)} \left\{ \sum_{\substack{\delta \\ \Lambda(\delta, \Delta_n) = \lambda}} \delta \right\},$$

and let V denote the span of the v_λ. Then

$$R_f(\alpha) v_\lambda \subseteq V.$$

Let $R_f^{(f,\phi)}(\alpha)$ denote the restriction of $R_f(\alpha)$ to V written with respect to the basis $\{v_\lambda\}$. Recall from Section 2 that $J_\lambda(\mathbf{x}; \alpha)$ denotes the Jack symmetric function indexed by λ and that $c_{\lambda,\mu}^\alpha$ is the coefficient of $p_\mu(x)$ when $J_\lambda(\mathbf{x}; \alpha)$ expanded in terms of power sums. The main result of this section can now be stated.

THEOREM 1. *For each λ, the vector $\sum_{\mu \vdash f} c_{\lambda\mu}^{2\alpha} v_\mu$ is a left eigenvector of $R_f^{(f,\phi)}(\alpha)$ with corresponding eigenvalue $(f\alpha + 2\{2\alpha n(\lambda') - n(\lambda)\})/\alpha\binom{2f}{2}$.*

PROOF. To begin, we will take a closer look at the entries of the matrix $R_f^{(f,\phi)}(\alpha)$. Fix a partition $\lambda = (\lambda_1, \ldots, \lambda_\ell)$, let π_λ be the permutation in S_f given by

$$\pi = (1, 2, \ldots, \lambda_1)(\lambda_1 + 1, \ldots, \lambda_1 + \lambda_2) \cdots (\lambda_1 + \cdots + \lambda_{\ell-1} - 1, \ldots, f)$$

(so π_λ has cycle type λ) and let δ_λ be the 1-factor with an edge from i to $f+\pi(i)$ for $i = 1, 2, \ldots, f$. So δ_λ looks like

$$\delta_\lambda \quad = \quad$$

It is easy to check that δ_λ has type λ (so δ_λ is amongst the one-factors appearing in the sum v_λ).

We can make use of the one-factors δ_λ to compute the entries of $R_f^{(f,\phi)}(\alpha)$. The following lemma is derived easily from the definition of $R_f^{(f,\phi)}(\alpha)$.

LEMMA 1. *Let λ and μ be partitions of f. The λ, μ entry of $\widehat{R}_f^{(f,\phi)}(\alpha) := \alpha\binom{2f}{2}R_f^{(f,\phi)}(\alpha)$ is given by:*

(a) *If $\ell(\mu) \geq \ell(\lambda)$ then $(\widehat{R}_f^{(f,\phi)}(\alpha))_{\lambda\mu}$ is $\alpha 2^{\ell(\mu)-\ell(\lambda)}$ times the number of δ of type λ which can be obtained from δ_μ by switching a pair of points.*

(b) *If $\ell(\mu) < \ell(\lambda)$ then $(\widehat{R}_f^{(f,\phi)}(\alpha))_{\lambda\mu}$ is $2^{\ell(\mu)-\ell(\lambda)}$ times the number of δ of type λ which can be obtained from δ_μ by switching a pair of points (you count δ more than once if it can be obtained more than once).*

Return now to the proof of Theorem 1. We will fix a partition $\mu = (\mu_1, \ldots, \mu_\ell)$ and compute the λ, μ entries of $(\widehat{R}_f^{(f,\phi)}(\alpha))_{\lambda\mu}$ for all λ. According to the lemma we need to count how many δ' of each type λ arise by switching a pair of points (x, y) in δ_μ.

CASE 1. x and y come from different cycles of $\delta_\mu \cup \Delta_0$, say x comes from the cycle of length μ_i and y comes from the cycle of length μ_j (where $i \neq j$).

In $((x, y) \cdot \delta_\mu) \cup \Delta_0$ the points that were in these cycles of lengths μ_i and μ_j now form a cycle of length $2(\mu_i + \mu_j)$. So the partition "corresponding" to partition λ is

$$\lambda = \mu[\mu_i + \mu_j \longleftarrow \mu_i, \mu_j]$$

(this notation means that μ_i and μ_j are removed and replaced by $\mu_i + \mu_j$).

For fixed μ_i and μ_j we had $(2\mu_i)(2\mu_j)$ choices of the pair (x, y). In this case, $\ell(u) = \ell(\lambda) + 1$ so the λ, μ entry of $(\widehat{R}_f^{(f,\phi)}(\alpha))_{\lambda\mu}$ contains a factor of 2α.

CASE 2. x and y come from the same cycle $\mathcal{C}$ of $\delta_\mu \cup \Delta_0$. Let μ_i be the part of μ corresponding to $\mathcal{C}$ so $\mathcal{C}$ consists of $2\mu_i$ points, μ_i from each row. It is straightforward to check the following fact:

(5.3)   A) Suppose x and y are in the same row and are separated (cyclically) by gaps of b and μ_i. Then $((x, y)\delta_\mu) \cup \Delta_0$ corresponds to the partition

$$\lambda = \mu[b, \mu_i - b \longleftarrow \mu_i].$$

B) Suppose x and y are in different rows. Then $((x, y)\delta_\mu) \cup \Delta_0$ corresponds to the partition μ.

Note that in case (A), $(\widehat{R}_f^{(f,\phi)}(\alpha))_{\lambda\mu} = \frac{1}{2}$; whereas in case (B), $(\widehat{R}_f^{(f,\phi)}(\alpha))_{\lambda\mu} = \alpha$.

Applying the above facts we see that the μth column of $(\widehat{R}_f^{(f,\phi)}(\alpha))_{\lambda\mu}$ is given by the above transformation:

$$\boxed{\mu^{\text{th}} \text{ column of } \widehat{R}_f^{(f,\phi)}(\alpha)}$$

(5.4)
$$v_\mu \longrightarrow \sum_{i<j}(2\mu_i\mu_j)v_{\mu[\mu_i+\mu_j\leftarrow\mu_i,\mu_j]}(f\alpha + 2(2\alpha-1)(\sum_i\binom{\mu_i}{2}))v_\mu$$

$$+ (2\alpha)\left(\sum_i \mu_i \sum_{b=1}^{\mu_i-1} v_{\mu[b,\mu_i-b\leftarrow\mu_i]}\right).$$

It is worth commenting on how we computed the coefficient of v_μ on the right-hand side. For starters there are $\sum_i \mu_i^2$ pairs (x,y) that yield the same μ. Of these, $\sum_i \mu_i(\mu_i-1)$ satisfy $(x,y)\delta_\mu \neq \delta_\mu$ and f satisfy $(x,y)\delta_\mu = \delta_\mu$. The former pairs each contribute 1 to the v_μ, v_μ entry. The latter contribute a more complicated factor, namely, the δ_μ, δ_μ entry of $R_f(\alpha)$. We need to compute the entry. First observe that the δ_μ, δ_μ entry in $R_f(1)$ is f (pairs (x,y) which fix δ_μ must be endpoints of an edge of δ_μ). The off-diagonal entries of $\widehat{R}_f(\alpha)$ are either $0, 1$, or α. It follows that the δ_μ, δ_μ entry of $\widehat{R}_f(\alpha)$ is $f\alpha + H(\alpha-1)$ where H is the number of off-diagonal entries equal to 1 in the δ_μ^{th} row. This is exactly the number of pairs (x,y) with $c((x,y)\delta_\mu) > c(\delta_\mu)$. As seen above, this number is $2\sum_i\binom{\mu_i}{2}$. So, the δ_μ, δ_μ entry in $\widehat{R}_f(\alpha)$ is

$$2\alpha\sum_i\binom{\mu_i}{2} + f\alpha + 2(\alpha-1)(\sum_i\binom{\mu_i}{2}) = 2\{\frac{f}{2}\alpha + (2\alpha-1)(\sum_i\binom{\mu_i}{2})\}.$$

Let Λ_n be the ring of symmetric functions in $x_1,\ldots,x_n$ and let Λ_n^f be the subspace spanned by polynomials that are homogeneous of degree f. Define $\psi: V \to \Lambda_n^f$ to be the linear map satisfying

$$\psi(v_\lambda) = p_\lambda(\mathbf{x}).$$

Now let ρ be the linear transformation on Λ_n^f given by

$$\rho = \psi \circ \widehat{R}_f(\alpha) \circ \psi^{-1}.$$

The computation we've just done tells us how to compute the μth column of $\widehat{R}_f(\alpha)$, i.e., how to compute the linear map given by $\widehat{R}_f(\alpha)^t$. The matrix for ρ is the dual $\widehat{R}_f(\alpha)^t$ with respect to the inner product $\langle p_\lambda(\mathbf{x}), p_\mu(\mathbf{x})\rangle = z_\lambda^{-1}\delta_{\lambda\mu}$. A straightforward computation shows that

$$\rho(p_\lambda(\mathbf{x})) = \{(f\alpha) + 2(2\alpha-1)n(\lambda')\}p_\lambda(\mathbf{x})$$

$$+ 2(2\alpha)\sum_{u\neq v}\lambda_u\lambda_v p_{\lambda[\lambda_u,\lambda_v\leftarrow\lambda_u+\lambda_v]}(\mathbf{x})$$

(5.5)

$$+ 2(\frac{1}{2}\sum_k\lambda_k\sum_{j=1}^{\lambda_k-1}p_{\lambda[\lambda_k\leftarrow j,\lambda_k-j]}(\mathbf{x})).$$

The result now follows by comparing (5.5) with either formula (3.6) in Hanlon (1988) or the first formula in the proof of Theorem 3.1 in Stanley (1989). $\square$

We refrain from carrying out further computations for this example. Preliminary analysis shows order $a\,n\log n + cn$ steps are needed with a depending on α.

5.2 Signed one-factors.

We are going to modify the situation in the previous section so that each one-factor comes with an "orientation." Changing the orientation alters the one-factor by a sign. This change in orientation will sometimes occur when we switch points and the result will be that the transition matrix $S_f(1)$ for the untwisted Markov chain will have some negative entries. So $S_f(1)$ will not really represent a Markov chain but we can still apply the Metropolis algorithm to it as given in (2.1).

To begin, we will generalize the notion of one-factors. Let B_f be the automorphism group of Δ_0. B_f is the hyperoctahedral group of order $f!2^f$ and can be thought of as the set of signed $f \times f$ permutation matrices in the following way. An element $\sigma \in B_f$ must permute the f edges of Δ_0 which gives us the underlying $f \times f$ permutation matrix $\hat{\sigma}$. If σ maps the ith edge of Δ_0 to the jth (so there is a 1 in the i,j entry of $\hat{\sigma}$). Then $\sigma_{ij} = 1$, if σ maps the point in the top row of column j and $\sigma_{ij} = -1$ if the point in the top row of column i goes to the point in the bottom row of column j.

The group B_f has four linear characters $\delta_0, \delta_1, \delta_2, \delta_3$. To describe them, let $\sigma \in B_f$ (think of σ as an $f \times f$ signed permutation matrix). Then

$$\delta_0(\sigma) = 1$$
$$\delta_1(\sigma) = \text{sign}(\hat{\sigma})$$
$$\delta_2(\sigma) = \det(\sigma)$$
$$\delta_3(\sigma) = \delta_1(\sigma)\delta_2(\sigma).$$

There is an alternative description of the character δ_3. If we think of B_f as a subgroup of S_{2f} then δ_3 is the restriction of the sign character of S_{2f} to B_f.

We will now follow a construction given in Stembridge (1992). For each i, let e_i denote the idempotent given by δ_i,

$$e_i = \frac{1}{|B_f|} \sum_{\sigma \in B_f} \delta_i(\sigma)\sigma.$$

Let X_i denote the left deal $e_i \mathbb{C}S_{2f}$ and let V_i denote the two-sided ideal $e_i \mathbb{C}S_{2f} e_i$. Note that X_0 and V_0 are isomorphic to the spaces of left B_f-cosets and double B_f-cosets in S_{2f}.

There is a combinatorial method for identifying a basis of the X_i and V_i. Let τ be a permutation in S_{2f} written in 1-line form,

$$\tau = a_1 a_2 \cdots a_{2f}.$$

Assign to τ a one-factor $\delta(\tau)$ by putting an edge between i and j in $\delta(\tau)$ iff $|a_i - a_j| = f$.

The following result is well known for $i = 0, 3$ and can be found in Stembridge (1992) for $i = 1, 2$.

THEOREM 5.

(A) *For $i = 0, 1, 2, 3$ we have $e_i \tau = \pm e_i \pi$ iff $\delta(\tau) = \delta(\pi)$.*

(B) *For $i = 0, 3$ we have $e_i \tau e_i = \pm e_i \pi e_i$ iff $\Lambda(\delta(\tau), \Delta_0) = \Lambda(\delta(\pi), \Delta_0)$.*

(C) *For $i = 1, 2$ we have*

    (i) *$e_i \tau c_i \neq 0$ iff $\Lambda(\delta(\tau), \Delta_0)$ has all odd parts.*

    (ii) *Suppose $e_i \tau e_i$ and $e_i \pi e_i$ are nonzero. Then $e_i \tau e_i = e_i \pi e_i$ iff $\Lambda(\delta(\tau), \Delta_0) = \Delta(\delta(\pi), \Delta_0)$.*

For the moment we will construct only cases $i = 0, 3$. Theorem 5 tells us that the spaces Ω_f and V from Section 4.1 can be identified with the cosets X_0 and double cosets V_0 of B_f.

The stable distribution Γ is a simple function that is constant on double cosets. It remains to understand the Markov chain $R_f(1)$ in this context.

LEMMA 2. *Let Ω_f be identified with X_0 as above. Then $R_f(1)$ is the matrix for right multiplication by $\frac{1}{|2f|} \sum_{1 \leq i \leq j \leq 2f} (i, j)$.*

We will now follow the same construction as in Section 5.1 but with X_0 replaced by X_3. The stable distribution Γ will be the same and the untwisted Markov chain will again be multiplication by $\binom{2f}{2}^{-1} \sum (i, j)$. To make this precise, we need to specify bases for X_3 and V_3 (because the natural bases are determined only up to sign).

DEFINITION.

(A) *For each one-factor, δ, let $\tau(\delta)$ be the lexicographically minimal element α of S_{2f} with $\delta(\alpha) = \delta$.*

(B) *For each partition λ of f, let $\pi(\lambda)$ be the lexicographically minimal element β of S_{2f} with*

$$\Lambda(\delta(\beta), \Delta_0) = \lambda.$$

Let B be the basis for X_3 given by

$$B = \{e_3 \tau(\delta) : \delta \in \Omega_f\}$$

and let $\mathcal{C}$ be the vector space for V_3 given by

$$\mathcal{C} = \{2^{\ell(\lambda)} e_3 \pi(\lambda) e_3 : \lambda \leftarrow f\}.$$

We henceforth let $v_\lambda^{(3)}$ denote $2^{\ell(\lambda)} e_3$. Let $Q_f(1)$ be the matrix for right multiplication by $\binom{2f}{2}^{-1} \sum (i, j)$ with respect to the basis B. Let $Q_f(\alpha)$ be the matrix obtained by applying the Metropolis algorithm to $Q_f(1)$ and Γ. Since $Q_f(1)$ corresponds to right multiplication by a conjugacy class in S_{2f}, it commutes with right multiplication by e_3. It follows that $Q_f(\alpha)$ restricts to the space V_3. Let $Q_f^{(\phi, 1^f)}(\alpha)$ be this restriction with respect to the basic $\mathcal{C}$. We can now state the analogue of Theorem 4.

THEOREM 2. *For each λ, the vector $\sum_{\mu \vdash f} c_{\lambda\mu}^{(\alpha/2)} v_{\mu}^{(3)}$ is a left eigenvector of* $Q_f^{(\phi,f)}(\alpha)$ *with corresponding eigenvalue* $(-f\alpha + 2\{\frac{\alpha}{2}n(\lambda') - n(\lambda)\})/\alpha(\binom{2f}{2i}))$.

To prove Theorem 2 one follows much the same procedure as in the proof of Theorem 1 (but care must be taken with signs).

One reason for the discussion of this signed case is to bring up the question of what happens for the characters δ_1 and δ_2. A simple computation shows that the restriction of right multiplication by $\binom{2f}{2}^{-1} \sum(i,j)$ to V_1 (or V_2) is the identity map. So this is the wrong (untwisted) Markov chain to begin with. Another possibility is to let the untwisted Markov chain be right multiplication by $\left(2\binom{2f}{3}^{-1} \sum(i,j,k)\right)$. A recent result of Stembridge tells us that the restriction of this Markov chain to V_1 has Schur's Q-functions as eigenvectors. So applying the Metropolis algorithm to this restriction will produce a Markov chain with eigenvectors that are perhaps interesting deformations of the Q-functions.

ACKNOWLEDGEMENT. We thank Jim Fill for help with Section 3.

REFERENCES

1. D. Critchlow, *Metric Methods for Analyzing Partially Ranked Data*, Lecture Notes in Statistics, No. 34, Springer-Verlag, Berlin, 1985.
2. P. Diaconis, *A random walk on partitions and zonal polynomials*, Unpublished manuscript, 1986.
3. _____, *Group Representations in Probability and Statistics*, Institute for Mathematical Statistics, Hayward, CA, 1988.
4. P. Diaconis and L. Saloff-Coste, *Comparison theorems for reversible Markov chain*, Technical report, Dept. of Mathematics, Harvard University, 1992.
5. P. Diaconis and M. Shahshahani, *Generating a random permutation with random transpositions*, Z. Wahrscheinlichkeitstheorie Verw. Gebiete **57** (1981), 183-195.
6. P. Diaconis and S. Zabell, *Updating subjective probability*, J. Amer. Statist. Assoc. **77** (1982), 822-830.
7. J. Hammersley and D. Handscomb, *Monte Carlo Methods*, Chapman and Hall, London, 1964.
8. P. Hanlon, *Jack symmetric functions and some combinatorial properties of Young symmetrizers*, J. Combin. Th. **A 47** (1988), 37-70.
9. _____, *A Markov chain on the symmetric group and Jack's symmetric functions*, to appear, Discrete Math. (1992).
10. M. Kac, Amer. Math. Monthly **54** (1947), 369-391.
11. M. Krawtchouk, *Sur une généralisation des polynomes d'Hermite*, Comptes Rendus Acad. Sci. Paris **189** (1929), 620-622.
12. G. Letac and L. Takacs, *Random walks on the m-dimensional cube*, J. reine angew. math. **310** (1979), 187-195.
13. I. Macdonald, *Symmetric Functions and Hall Polynomials*, Clarendon Press, Oxford, 1979.
14. _____, *Orthogonal polynomials associated with root systems*, unpublished manuscript, 1989.
15. F. MacWilliams and N. Sloane, *The Theory of Error Correcting Codes*, North-Holland, Amsterdam, 1977.
16. R. Stanley, *Some combinatorial properties of the Jack polynomial*, Adv. Math. **77** (1989), 76-115.
17. J. Stembridge, *On Schur's Q-functions and the primitive idempotents of a commutative Hecke Algebra*, technical report, Department of Mathematics, University of Michigan, 1992.

DEPARTMENT OF MATHEMATICS, HARVARD UNIVERSITY, CAMBRIDGE, MA 02138
DEPARTMENT OF MATHEMATICS, UNIVERSITY OF MICHIGAN, ANN ARBOR, MICHIGAN 48109

Contemporary Mathematics
Volume **138**, 1992

Hilbert Spaces of Vector-Valued Holomorphic Functions and Irreducibility of Multiplier Representations

HONGMING DING

Introduction and Main Result

Let M be a complex manifold and G be a transitive group of holomorphic automorphisms of M. R. Kunze [4] proved that every unitary multiplier representation of G on a Hilbert space of *complex-valued* holomorphic functions on M, in which point evaluations are nonzero continuous functionals, is irreducible.

Certain multiplier representations on Hilbert spaces H of *vector-valued* holomorphic functions on a complex manifold M arise in harmonic analysis on Hermitian symmetric spaces [1], [2], [5], [6]. It is the purpose of this note to generalize Kunze's theorem to the vector–valued context, and to provide an easily verified criterion for the irreducibility of these more general multiplier representations.

Let V be any Hilbert space, $\mathcal{L}(V)$ be the space of bounded linear transformations of V and H be a Hilbert space of V–valued holomorphic functions on a complex manifold M. Assume that for each $z \in H$ the point evaluation $E_z : f \mapsto f(z)$ is a continuous linear transformation from H onto V. Then H is characterized by its reproducing kernel

$$(1) \qquad\qquad Q(z,w) = E_z E_w^*$$

in the sense that

$$(2) \qquad\qquad \langle f(w),\ v \rangle = \langle f \mid Q(\cdot, w)v \rangle$$

for all $f \in H$ and $v \in V$, where the left and right sides are inner products on V and H, respectively (cf. [3]).

1991 *Mathematics Subject Classification*. Primary 22E30, Secondary 22E45.

This paper is in final form and no version of it will be submitted for publication elsewhere .

Let $T : g \mapsto Tg$ be a given representation of G on H. We shall say that T is a *multiplier representation* if there is a map $m : M \times G \to \mathcal{L}(V)$ such that

$$(3) \qquad ((Tg)f)(z) = m(z, g)\ f(z \circ g)$$

for all $f \in H$, where $z \circ g$ denotes the image of z under the transformation of g. Since T is a representation, m satisfies the multiplier identity

$$(4) \qquad m(z, g_1 g_2) = m(z, g_1)\ m(z \circ g_1, g_2)$$

and for that reason m is called a *multiplier*.

We now give a condition that implies the irreducibility of T. If there is a fixed point $z_0 \in M$ such that for its stability subgroup K

$$(5) \qquad \{ l \in \mathcal{L}(V) : m(z_0, k)\ l\ m(z_0, k)^* = l \ \text{ for } \ k \in K \} \subset \{ cI : c \in \mathbb{C} \},$$

then we will call m a *scalarizing multiplier*.

Note by (4) that the mapping $k \mapsto m(z_0, k)$ defines a representation of K on the space $\mathcal{L}(V)$. If this representation is irreducible and unitary, then m is a scalarizing multiplier (cf. the example after the theorem).

THEOREM. *Let M be a (connected) complex manifold, G a transitive group of holomorphic automorphisms of M, and V a Hilbert space. Suppose that H is a Hilbert space of V-valued holomorphic functions on M in which point evaluations are nonzero continuous linear transformations from H onto V. Then every unitary multiplier representation of G on H defined by (3) with a scalarizing multiplier m is irreducible.*

PROOF. Rewrite (3) as

$$(6) \qquad E_z\, Tg = m(z, g)\, E_{z \circ g}$$

for $(g, z) \in G \times M$. Since T is unitary, taking adjoints in (6) we see that

$$(Tg)^{-1}\, E_z^* = E_{z \circ g}^*\ m(z, g)^*$$

for $(g, z) \in G \times M$; or equivalently,

$$(7) \qquad E_w\, E_z^* = m(w, g)\, E_{w \circ g}\, E_{z \circ g}^*\, m(z, g)^*$$

for $(g, z, w) \in G \times M \times M$. In (7) we substitute $z = w = z_0$ and $g = k \in K$ to obtain

$$(8) \qquad E_{z_0}\, E_{z_0}^* = m(z_0, k)\, E_{z_0}\, E_{z_0}^*\, m(z_0, k)^*.$$

By (5), $E_{z_0}\, E_{z_0}^*$ is a scalar. In (7) let $z = w \in M$ and choose $g \in G$ such that $z_0 = z \circ g$. Then there is a constant c such that

$$(9) \qquad E_z\, E_z^* = c\, m(z, g)\, m(z, g)^*$$

for all $z \in M$.

Now let H' be any nonzero closed subspace of H that is invariant under T, and let $E_z^{*\prime}$ be the adjoint of E_z in H'. Then the subrepresentation of T on H'

is also given by (3) with precisely the same multiplier m. Formulas (6) through (9) hold with E_z^* replaced by $E_z^{*\prime}$ and we conclude that there exists a constant c' such that

$$(10) \qquad E_z\, E_z^* = c'\, E_z\, E_z^{*\prime}$$

for all $z \in M$. As a function of z and w, $E_w\, E_z^* - c'\, E_w\, E_z^{*\prime}$ is holomorphic in z and conjugate holomorphic in w, and vanishes for all points of the form (z, z). We now invoke the lemma in [4] to conclude that

$$(11) \qquad E_w E_z^* = c'\, E_w\, E_z^{*\prime}$$

for all $(z, w) \in M \times M$. Thus, $H' = H$ and T is irreducible.

As an example (see [2]), let $\mathcal{F}$ be a real finite–dimensional division algebra, and M be the generalized Siegel upper half–plane in the complexification $(\mathcal{F}^{\mathbb{C}})^{n \times n}$ of $\mathcal{F}^{n \times n}$. That is,

$$(12) \qquad M = \{x + iy \in (\mathcal{F}^{\mathbb{C}})^{n \times n} : x, y \in \mathcal{F}^{n \times n}, x = x^*, y = y^* > 0\},$$

where $x^* = \overline{x}^t, y^* = \overline{y}^t$ are the adjoint matrices of $x, y \in \mathcal{F}^{n \times n}$. Let G be the group of biholomorphic automorphisms of M, realized as a group of 2×2 matrices

$$g = \begin{pmatrix} g_{11} & g_{12} \\ g_{21} & g_{22} \end{pmatrix},$$

with $g_{ij} \in \mathcal{F}^{n \times n}$. In customary notation, G is $Sp(n, \mathfrak{R}), U(n, n)$ or $O_*(4n)$ accordingly as $\mathcal{F}$ is real, complex, or quaternionic. The action of the group G on M is defined by the linear fractional transformation

$$z \circ g = (zg_{12} + g_{22})^{-1}(zg_{11} + g_{21}).$$

The space M is an unbounded realization of the Hermitian symmetric space G/K, where $K = \{k \in G : kk^* = I\}$ is the maximal compact subgroup of G. The group K is the stability group of $i1 \in M$ and consists of all matrices

$$k = \begin{pmatrix} a & b \\ -b & a \end{pmatrix}$$

with $a, b \in \mathcal{F}^{n \times n}$ such that $aa^* + bb^* = I_n$ and $ab^* = ba^*$. The mapping $k \mapsto u = a - ib$ is an isomorphism of K with the unitary subgroup K_1 of $GL(n, \mathcal{F}^{\mathbb{C}})$.

For each irreducible unitary finite–dimensional representation π of K_1 extended (uniquely) to a holomorphic representation of $GL(n, \mathcal{F}^{\mathbb{C}})$, we define a multiplier representation of G as follows. Let H consist of all holomorphic functions f from M to the space V_π of π, such that

$$(13) \qquad \|f\|_\pi^2 = \int_M \|\pi(y)^{\frac{1}{2}} f(x + iy)\|^2 d_* z < \infty$$

where $z = x + iy \in M$ and d_*z is a G–invariant measure on M. Of course, H may be zero. But whenever it is nonzero, H is a Hilbert space and the equation

$$(14) \qquad T(g,\pi)f)(z) = \pi(g_{12}^* z + g_{22}^*)^{-1} f(z \circ g)$$

defines a multiplier representation $T(\cdot, \pi)$ of G in H with the multiplier m given by

$$(15) \qquad m(z, g) = \pi(g_{12}^* z + g_{22}^*)^{-1}.$$

Since π is an *irreducible unitary* representation of K_1, m satisfies (5) for $z_0 = i1$. Our theorem implies the irreducibility of the representation $T(\cdot, \pi)$. By [2], $T(\cdot, \pi)$ is also square–integrable and hence $T(\cdot, \pi)$ is in the discrete series of the group G.

REFERENCES

1. H. Ding, *Operator-valued Bessel functions on Schrödinger-Fock spaces and Siegel domains of type II*, preprint.
2. K. I. Gross and R. A. Kunze, *Bessel functions and representation theory II: holomorphic discrete series and metaplectic representations*, J. Functional Analysis **25** (1977), 1-49.
3. R. A. Kunze, *Positive definite operator valued kernels and unitary representations*, in "Proceedings of the Conference on Functional Analysis at Irvine, California," Thomson Book Company (1966).
4. R. A. Kunze, *On the irreducibility of certain multiplier representations*, Bull. Amer. Math. Soc. **68** (1962), 93-94.
5. H. Rossi and M. Vergne, *Representations of certain solvable Lie groups on Hilbert spaces of holomorphic functions and application to the holomorphic discrete series of a semisimple Lie group*, J. Functional Analysis **13** (1973), 324-389.
6. H. Rossi and M. Vergne, *Analytic continuation of the holomorphic discrete series of a semi-simple Lie group*, Acta Math. **136** (1976), 1-59.

DEPARTMENT OF MATHEMATICS, UNIVERSITY OF VERMONT, BURLINGTON, VT 05405

Contemporary Mathematics
Volume **138**, 1992

Hankel Transforms Associated
to Finite Reflection Groups

CHARLES F. DUNKL

Introduction

Root systems provide a rich framework for the study of special functions and integral transforms in several variables. Finite reflection groups are subgroups of the orthogonal group which are generated by reflections. The set of suitably normalized vectors normal to the hyperplanes corresponding to the reflections in the group is called a root system. This paper concerns weight functions which are products of powers of the linear functions vanishing on the reflecting hyperplanes and which are invariant under the group. These functions are specified by as many parameters as there are conjugacy classes of reflections. There is an analogue of the exponential function $\exp(\sum_{j=1}^{n} x_j y_j)$, $x, y \in \mathbb{C}^N$, for the weight functions, which allows the definition of a transform generalizing the Fourier transform. It is called "Hankel" because the one-dimensional specialization is the classical Hankel transform. There is a Plancherel-type result. Its proof depends on an orthogonal basis of rapidly decreasing functions involving Laguerre polynomials and polynomials harmonic for the differential-difference Laplacian associated to the group.

The author has previously developed the theory of a commutative algebra of differential-difference operators which generalize partial differentiation. The Hankel transform maps these onto multiplication operators.

The weight functions under discussion are related to the Macdonald-Mehta integrals. Important work in the analysis of these integrals and related special functions, such as hypergeometric functions and Jacobi polynomials, has been done by Heckman and Opdam [6, 7, 8, 12, 13, 14].

The outline of the paper is as follows:

1991 *Mathematics Subject Classification.* 33C80, 44A15, 42B10.

During the preparation of this paper the author was partially supported by NSF Grant DMS-8802400.

This paper is in final form and no version of it will be submitted for publication elsewhere .

(1) Background: definitions and theorems from previous work;
(2) The Hankel transform: orthogonal bases involving Laguerre polynomials, the isometry theorem, action on differential-difference operators;
(3) Special harmonic polynomials: the analogues of $x \mapsto (\sum_{j=1}^{N} x_j y_j)^n$, $y \in \mathbb{C}^N$ with $\sum_{j=1}^{N} y_j^2 = 0$, formulas for integrals;
(4) Classical examples: specialization to Bessel functions, Gegenbauer and Jacobi polynomials.

1. Background

Suppose G is a finite reflection group on $\mathbb{R}^N$ with the set $\{v_i : i = 1, \ldots, m\}$ of positive roots. Let σ_i denote the reflection along v_i; thus

$$x\sigma_i := x - 2\frac{\langle x, v_i \rangle}{|v_i|^2} v_i.$$

The inner product is denoted $\langle x, y \rangle := \sum_{j=1}^{N} x_j y_j$ and the norm is

$$|x| := \left(\sum_{j=1}^{N} |x_j|^2 \right)^{1/2},$$

$x, y \in \mathbb{C}^N$. We will use $\nu(y) := \sum_{j=1}^{N} y_j^2$, especially for the complexification of $\mathbb{R}^N$. Choose positive parameters $\alpha_i, 1 \leq i \leq m$, such that $\alpha_i = \alpha_j$ whenever σ_i is conjugate to σ_j in G (assume also that $|v_i| = |v_j|$ in this case, so that $v_j = v_i w$ for some $w \in G$). The parameters α_i are denoted k_α (where α is a root label) and called a multiplicity function in Heckman's and Opdam's papers [6, 7, 8]. We will discuss the possibility of non-positive values of α_i later.

Let $h(x) := \prod_{j=1}^{m} |\langle x, v_j \rangle|^{\alpha_j}$, a positively homogeneous G-invariant function on $\mathbb{R}^N$ of degree $\gamma := \sum_{j=1}^{m} \alpha_j$. We will be concerned with the measures $h^2 d\omega$ on the unit sphere

$$S := \{x \in \mathbb{R}^N : |x| = 1\}$$

(where $d\omega$ is the normalized rotation-invariant surface measure), $h^2 dm_N$ and $h^2 d\mu$ on $\mathbb{R}^N$, where dm_N is the N-dimensional Lebesgue measure and

$$d\mu(x) = (2\pi)^{-N/2} e^{-|x|^2/2} dm_N(x),$$

the Gaussian measure. One important example is the group of type B_N for which

$$h(x) = \prod_{i=1}^{N} |x_i|^\alpha \prod_{1 \leq i < j \leq N} |x_i^2 - x_j^2|^\beta \qquad (\alpha, \beta > 0).$$

The algebra of differential-difference operators is generated by $\{T_i : i = 1, \ldots, N\}$ where

$$T_i f(x) := \frac{\partial f}{\partial x_i} + \sum_{j=1}^{m} \alpha_j \frac{f(x) - f(x\sigma_j)}{\langle x, v_j \rangle} (v_j)_i,$$

where $f \in C^\infty(\mathbb{R}^N)$ and $(v_j)_i$ is the i-th component of v_j.

Let $\mathcal{P}_n$ denote the space of polynomial functions on $\mathbb{R}^N$ homogeneous of degree n. It was shown in [2] that $T_i\mathcal{P}_n \subset \mathcal{P}_{n-1}$, $n = 1, 2, 3, \ldots$, each i. There are harmonic functions associated to h. Define the h-Laplacian $\Delta_h := \sum_{i=1}^N T_i^2$; indeed

$$\Delta_h f(x) = \Delta f(x) + \sum_{j=1}^m \alpha_j \left(\frac{2\langle \nabla f, v_j \rangle}{\langle x, v_j \rangle} - \frac{f(x) - f(x\sigma_j)}{\langle x, v_j \rangle^2} |v_j|^2 \right)$$

for $f \in C^\infty(\mathbb{R}^N)$, where Δ is the ordinary Laplacian and ∇ is the gradient. Further $\mathcal{H}_n^h := \mathcal{P}_n \cap (\ker \Delta_h)$, the space of h-harmonic polynomials of degree n. The orthogonality theorem [1, Section 1] asserts for $p \in \mathcal{P}_n$ that $\int_S pqh^2 d\omega = 0$ for all $q \in \sum_{i=0}^{n-1} \mathcal{P}_i$ if and only if $\Delta_h p = 0$. Also $L^2(S, h^2 d\omega) = \sum_{n=0}^\infty \oplus \mathcal{H}_n^h$.

We use the normalization constants

$$c_h := \left(\int_{\mathbb{R}^N} h^2 d\mu \right)^{-1}$$

and

$$c_h' := \left(\int_S h^2 d\omega \right)^{-1}$$

(called c_N and c_N' in [4]), then by use of spherical polar coordinates

$$c_h' = 2^\gamma (\Gamma(N/2 + \gamma)/\Gamma(N/2)) c_h$$

and

$$(1.1) \qquad c_h \int_{\mathbb{R}^N} fh^2 d\mu = 2^n (N/2 + \gamma)_n c_h' \int_S fh^2 d\omega$$

when f is a smooth function on $\mathbb{R}^N$ positively homogeneous of degree $2n$. The values of c_h are known (the "Macdonald-Mehta" integrals). Opdam [13, 14] has a proof that works for all finite reflection (Coxeter) groups (see also Macdonald [11]).

There is an algebraically defined pairing on polynomials closely related to the $L^2(\mathbb{R}^N; h^2 d\mu)$ inner product. For $p, q \in \mathcal{P}_n$ define

$$[p, q]_h := p(T^x)q(x),$$

(this means that the variable x_i is replaced by T_i in p, each $i = 1, \ldots, N$). By the homogeneity properties of T_i this form is scalar-valued. Extend the pairing to all polynomials by setting $[p, q]_h = 0$ when $p \in \mathcal{P}_n$, $q \in \mathcal{P}_m$, and $n \neq m$. It was shown in Th. 3.10 [4] that for any polynomials p, q,

$$(1.2) \qquad [p, q]_h = c_h \int_{\mathbb{R}^N} (e^{-\Delta_h/2} p)(e^{-\Delta_h/2} q) h^2 d\mu.$$

Here $e^{-\Delta_h/2} p$ denotes the terminating series $\sum_{j \geq 0} (-1/2)^j (\Delta_h^j p)/j!$. This formula is fundamental for our subsequent development of the Hankel transform and its isometry properties. (For ordinary differentiation, all $\alpha_i = 0$, this formula is due to Macdonald [10, 11].) The proof depends on the existence and properties of a reproducing kernel called $K_n(x,y)$ for the form $[p,q]_h$ on $\mathcal{P}_n$. Some of the properties are (where $x,y \in \mathbb{C}^N$) (Prop. 3.2[4]):

(1.3) (i) $K_n(x,y)$ is a homogeneous polynomial in x of degree n,

(ii) $K_n(y,x) = K_n(x,y)$,

(iii) $K_n(xw, yw) = K_n(x,y)$ for all $w \in G$,

(iv) $T_i^x K_n(x,y) = y_i K_{n-1}(x,y)$ where T_i^x denotes operation on the x variable),

(v) $|K_n(x,y)| \leq \max_{w \in G} |\langle xw, y \rangle|^n / n!$,

(vi) $K_n(T^x, y)p(x) = p(y)$ for $p \in \mathcal{P}_n$ (T^x means x_i is replaced by T_i^x), that is, $[K_n(\cdot, y), p]_h = p(y)$.

Note that when all $\alpha_i = 0$, $K_n(x,y) = \langle x,y \rangle^n / n!$.

There is an inductive construction of K_n (Prop. 3.2 [4]). Here is a restatement. Suppose there are s classes of reflections, with class #i consisting of $\sigma_{i,1}, \sigma_{i,2}, \ldots,$ σ_{i,m_i} and let β_i denote the common value of α_j on this class (so that $\sum_{i=1}^s m_i = m$ and $\sum_{i=1}^s m_i \beta_i = \gamma$). Further, let $\hat{G}$ denote the set of complex irreducible characters of G (in fact, all characters of Coxeter groups are real). For $\tau \in \hat{G}$, let

$$\lambda(\tau) = \sum_{i=1}^s \beta_i m_i (1 - \tau(\sigma_{i,1})/\tau(1))$$

(the coefficients of β_i are integers $n_i(\tau)$ satisfying $0 \leq n_i(\tau) \leq 2m_i$). For any value λ and each $w \in G$ the sum $\sum\{\tau(1)\tau(w) : \lambda(\tau) = \lambda, \tau \in \hat{G}\} \in Z$ (or is empty) (Th. 3.1 [3]). Then K_n is defined by

(i) $K_0(x,y) = 1$.

(ii) for $n = 0, 1, 2, \ldots,$

$$(1.4) \qquad K_{n+1}(x,y) = \frac{1}{|G|} \sum_{w \in G} \langle xw, y \rangle K_n(xw, y) \cdot \sum_{\tau \in \hat{G}} \frac{\tau(1)\tau(w)}{\lambda(\tau) + n + 1}.$$

We see that this formula defines K_n for any complex values of β_i except those for which $\sum_{i=1}^s \beta_i n_i(\tau) = -1, -2, -3, \ldots$. Not all of these values are excluded but a more precise list requires more information about the representations of G. One more condition must be imposed to obtain a nondegenerate pairing $[\cdot, \cdot]_h$ and the validity of the decomposition

$$\mathcal{P}_n = \sum_{j=0}^{[n/2]} |x|^2 \mathcal{H}_{n-2j}^h$$

(that is, each $p \in \mathcal{P}_n$ can be written uniquely as

$$p(x) = \sum_{j \leq n/2} |x|^{2j} p_{n-2j}(x)$$

with $p_{n-2j} \in \mathcal{H}_{n-2j}^h$, see Th. 1.7, 1.11, [1]), namely $N/2 + \gamma \neq 0, -1, -2, \ldots$.

From formula (1.2) we can prove that for a fixed permissible value of $(\beta_1, \beta_2, \ldots, \beta_s)$ there is a finite constant B (depending on $|G|, \beta_1, \beta_2, \ldots$) such that

$$|K_n(x,y)| \leq B|x|^n|y|^n/n!, n = 0, 1, 2, \ldots$$

and $x, y \in \mathbb{C}^N$. If each $\beta_i \geq 0$, then the stronger bound (1.3) (v) holds.

Now define the h-analogue of the exponential function (and of the Bessel function!) by $K(x,y) := \sum_{n=0}^{\infty} K_n(x,y)$. For fixed y this is entire in x and $|K(x,y)| \leq e^{B|x||y|}$ for $B < \infty, x, y \in \mathbb{C}^N$. This function is denoted Exp_G by Opdam [14] who gives an existence proof based on differential equations techniques applied to G-invariant functions.

Note that $T_i^x K(x,y) = y_i K(x,y)$. Our Hankel transform is based on the kernel $K(x, -iy)$ for $x, y \in \mathbb{R}^n$, the analogue of $e^{-i\langle x,y \rangle}$ for the Fourier transform on $\mathbb{R}^N$.

2. The Hankel Transform

We present the definition and isometry theorem for the Hankel transform by use of an explicit orthogonal basis, and the adjoint of T_i, in the Hilbert space $L^2(\mathbb{R}^N, h^2 dm_N)$. The parameters must satisfy $\alpha_i \geq 0$. Recall for $y \in \mathbb{C}^N$ that $|y|^2 = \sum_{j=1}^{N} |y_j|^2$ while $\nu(y) = \sum_{j=1}^{N} y_j^2$. We begin with an integral formula, a slight generalization of Prop. 3.12 [4].

2.1 PROPOSITION. *Let p be a polynomial on $\mathbb{R}^N$ and $y \in \mathbb{C}^N$, then*

$$c_h \int_{\mathbb{R}^N} (e^{-\Delta h/2} p(x)) K(x,y) h(x)^2 d\mu(x) = e^{\nu(y)/2} p(y).$$

PROOF. Let m be an integer larger than the degree of p, fix $y \in \mathbb{C}^N$, and let $q_m(x) = \sum_{j=0}^{m} K_j(x,y)$. Thus $[q_m, p]_h = p(y)$ (breaking p up into homogeneous components; this is a polynomial identity valid for complex y). By formula (1.2),

$$[q_m, p]_h = c_h \int_{\mathbb{R}^N} (e^{-\Delta_h/2} p)(e^{-\Delta_h/2} q_m) h^2 d\mu.$$

But $\Delta_h^x K_n(x,y) = \nu(y) K_{n-2}(x,y)$ and so

$$e^{-\Delta_h/2} q_m(x) = \sum_{j=0}^{m} \sum_{l \leq j/2} ((-\nu(y)/2)^l/l!) K_{j-2l}(x,y)$$

$$= \sum_{l \leq m/2} ((-\nu(y)/2)^l/l!) \sum_{s=0}^{m-2l} K_s(x,y).$$

Now let $m \to \infty$. The double sum converges to $e^{-\nu(y)/2} K(x,y)$ since it is dominated termwise by

$$\sum_{l=0}^{\infty} (|y|^2/l! 2^l) \sum_{s=0}^{\infty} (|x|^s|y|^s/s!) = e^{(|y|^2/2)+|x||y|},$$

which is integrable with respect to $e^{-|x|^2/2}dm_N(x)$. By the dominated convergence theorem

$$p(y) = e^{-\nu(y)/2}c_h \int_{\mathbb{R}^N} (e^{-\Delta_h/2}p(x))K(x,y)h(x)^2 d\mu(x).$$

□

We will define the Hankel transform explicitly on a dense subspace of $L^2(\mathbb{R}^N, h^2 dm_N)$ and prove its isometry properties on an orthogonal basis.

Let

$$E(\mathbb{R}^N) = \{f \in C^\infty(\mathbb{R}^N) : \int_{\mathbb{R}^N} |p(\frac{\partial}{\partial x_1}, \dots, \frac{\partial}{\partial x_N})f(x)|e^{B|x|}dm_N(x) < \infty$$
$$\text{for each polynomial } p \text{ and each } B < \infty\}.$$

(It is conjectured that $|K(x, -iy)| \le 1$ for $x, y \in \mathbb{R}^N$, in which case these boundedness conditions are excessively restrictive; on the other hand our orthogonal basis functions decrease like $e^{-|x|^2/2}$ at infinity so the proofs would not change.)

2.2. DEFINITION. *For $f \in E(\mathbb{R}^N), y \in \mathbb{R}^N$, let*

$$\hat{f}(y) := (2\pi)^{-N/2}c_h \int_{\mathbb{R}^N} f(x)K(x, -iy)h(x)^2 dm_N(x),$$

the Hankel transform of f at y.

By the dominated convergence theorem $\hat{f}$ is continuous on $\mathbb{R}^N$ (use a sequence $y_n \to y$ as $n \to \infty$, then $K(x, -iy_n) \to K(x, -iy)$).

We produce orthogonal bases for $L^2(\mathbb{R}^N, h^2 dm_N)$ in $E(\mathbb{R}^N)$ by using Laguerre polynomials and h-harmonic polynomials.

2.3. DEFINITION. *For $m, n = 0, 1, 2, \dots, p \in \mathcal{H}_n^h$ let*

$$\phi_m(p; x) := p(x)L_m^{(n+\gamma+N/2-1)}(|x|^2)e^{-|x|^2/2}(x \in \mathbb{R}^N).$$

The Laguerre polynomials $L_m^{(A)}(t)$ are defined by

$$L_m^{(A)}(t) = \frac{(A+1)_m}{m!} \sum_{j=0}^{m} \frac{(-m)_j}{(A+1)_j} \frac{t^j}{j!}$$

and satisfy the orthogonality relation

$$\Gamma(A+1)^{-1} \int_0^\infty L_n^{(A)}(t)L_m^{(A)}(t)t^A e^{-t}dt = \delta_{mn}(A+1)_m/m!$$

(see Szegö [15, Ch. V]).

2.4. PROPOSITION. *For* $k, l, m, n = 0, 1, 2, \ldots, p \in \mathcal{H}_n^h, q \in \mathcal{H}_l^h$ *the integral*

$$(2\pi)^{-N/2} c_h \int_{\mathbb{R}^N} \phi_m(p; x) \phi_k(q; x) h(x)^2 dx$$

$$= \delta_{mk} \delta_{nl} 2^{-\gamma - N/2} \frac{(N/2 + \gamma)_{n+m}}{m} c_h' \int_S pqh^2 d\omega.$$

PROOF. By spherical polar coordinates the first integral equals

$$c_h \int_0^\infty L_m^{(n+\gamma+N/2-1)}(r^2) L_k^{(l+\gamma+N/2-1)}(r^2) \cdot e^{-r^2} r^{n+l+2\gamma+N-1} dr$$

$$\cdot \frac{2^{1-N/2}}{\Gamma(N/2)} \int_S pqh^2 d\omega.$$

The inner integral is zero if $n \neq l$. Assume $n = l$, make the change of variable $r^2 = t$. Then the outer integral equals $(1/2)\delta_{mk}\Gamma(n + \gamma + N/2 + m)/m!$. $\square$

The following is a several-variable restatement of a classical result.

2.5. THEOREM. *The linear space* span$\{\phi_m(p): m, n = 0, 1, 2, \ldots, p \in \mathcal{H}_n^h\}$ *is dense in* $L^2(\mathbb{R}^N, h^2 dm_N)$.

PROOF. The space is exactly

$$\{q(x)e^{-|x|^2/2} : q \text{ is a polynomial}\}$$

because every polynomial can be written as $q(x) = \sum_{j=0}^M \sum_{k=0}^M |x|^{2k} q_{kj}(x)$ where $q_{kj} \in \mathcal{H}_j^h$ (some integer M) (by Th. 1.7 [1]). It suffices to show that if $f \in L^2(\mathbb{R}^N, h^2 dm_N)$ and

$$\int_{\mathbb{R}^N} f(x) q(x) e^{-|x|^2/2} h(x)^2 dm_N(x) = 0$$

for each polynomial q then $f(x) = 0$ a.e. Rewrite this equation as

$$\int_{\mathbb{R}^N} (f(x)e^{-|x|^2/4}) q(x) d\mu_1(x) = 0,$$

where

$$d\mu_1(x) = e^{-|x|^2/4} h(x)^2 dm_N(x).$$

Now $f(x)e^{-|x|^2/4} \in L^2(d\mu_1)$ because

$$\int_{\mathbb{R}^N} |f(x)e^{-|x|^2/4}|^2 d\mu_1(x) \leq \int_{\mathbb{R}^N} |f(x)|^2 h(x)^2 e^{-3|x|^2/4} dm_N(x)$$

$$\leq \int_{\mathbb{R}^N} |f(x)|^2 h(x)^2 dm_N(x) < \infty.$$

Further $f(x)e^{-|x|^2/4}$ is orthogonal to all polynomials in $L^2(d\mu_1)$ and μ_1 satisfies the hypotheses of (the N-dimensional version of) Hamburger's theorem [5]: if $\int_{\mathbb{R}^N} e^{c|x|} d\mu_1(x) < \infty$ for some $c > 0$, then the polynomials are dense in $L^2(\mu_1)$. Thus $f(x)e^{-|x|^2/4} = 0$ for almost all x.

We can explicitly determine a basis of eigenfunctions of the Hankel transform. The eigenvalues are powers of i and this proves the isometry properties.

2.6. THEOREM. *For* $m, n = 0, 1, 2, \ldots$, $p \in \mathcal{H}_n^h$, $y \in \mathbb{R}^N$, $\phi_m(p)\check{\ }(y) = (-i)^{n+2m}\phi_m(p; y)$.

PROOF. For brevity, let $A := n + N/2 + \gamma - 1$. By Proposition 2.1, and the fact that

$$e^{-\Delta_h/2}(|x|^{2j}p(x)) = (-1)^j j! 2^j L_j^{(A)}(|x|^2/2)p(x)$$

for any $j = 0, 1, 2, \ldots$ (Prop. 3.9 [4]), we have

$$(2\pi)^{-N/2}c_h \int_{\mathbb{R}^N} L_j^{(A)}(|x|^2/2)p(x)K(x,y)h(x)^2 e^{-|x|^2/2} dm_N(x)$$

$$= (-1)^j (j! 2^j)^{-1} e^{\nu(y)/2}\nu(y)^j p(y), \quad y \in \mathbb{C}^N.$$

We change the argument in the Laguerre polynomial by using the identity

$$L_m^{(A)}(t) = \sum_{j=0}^m 2^j \frac{(A+1)_m}{(A+1)_j} \frac{(-1)^{m-j}}{(m-j)!} L_j^{(A)}(t/2), \qquad t \in \mathbb{R}$$

(a special case of problem 67, p. 387, Szegö [15], prove with the generating function). Use this expansion together with the above integral to obtain

$$(2\pi)^{-N/2}c_h \int_{\mathbb{R}^N} L_m^{(A)}(|x|^2)p(x)e^{-|x|^2/2}K(x,y)h(x)^2 dm_N(x)$$

$$= e^{\nu(y)/2}p(y)(-1)^m \frac{(A+1)_m}{m!} \sum_{j=0}^m \frac{(-m)_j}{(A+1)_j} \frac{(-\nu(y))^j}{j!}.$$

Now replace y by $-iy$ with $y \in \mathbb{R}^N$, then $\nu(y)$ becomes $-|y|^2$ and $p(y)$ becomes $(-i)^n p(y)$, and the sum yields a Laguerre polynomial. The integral equals $(-1)^m(-i)^n p(y) L_m^{(A)}(|y|^2)e^{-|y|^2/2}$.

2.7. COROLLARY. *The Hankel transform has period 4 and extends to an isometry of* $L^2(\mathbb{R}^N, h^2 dm_N)$ *onto itself. The square of the transform is the central involution, that is, if* $f \in L^2(\mathbb{R}^N, h^2 dm_N)$, $\hat{f} = g$, *then* $\hat{g}(x) = f(-x)$ *(almost all* $x \in \mathbb{R}^N$*).*

By restricting the Hankel transform to smooth functions on S (producing a type of Hankel-Stieltjes transform), we obtain another interesting integral formula, as well as a method for finding eigenfunctions of Δ_h (for the classical result see Helgason [9, p. 25]).

2.8. PROPOSITION. *Let* $f \in \mathcal{H}_n^h$, $n = 0, 1, 2, \ldots$, $y \in \mathbb{C}^N$, *then*

$$c_h' \int_S f(x)K(x,y)h(x)^2 d\omega(x) = f(y) \sum_{m=0}^\infty \frac{\nu(y)^m}{2^{2m+n}m!(\frac{N}{2}+\gamma)_{m+n}}.$$

Further if $y \in \mathbb{R}^N$, $\lambda > 0$, then the function

$$g(y) := c_h' \int_S f(x) K(x, -iy\lambda) h(x)^2 d\omega(x)$$

satisfies $\Delta_h g = -\lambda^2 g$ and

$$g(y) = (-i)^n (\lambda|y|/2)^{-N/2-\gamma+1} f(y/|y|) \Gamma(\frac{N}{2} + \gamma) J_{n+N/2+\gamma-1}(\lambda|y|)$$

(the Bessel function).

PROOF. In the formula (Proposition 2.1)

$$c_h \int_{\mathbb{R}^N} f(x) K(x, y) h(x)^2 d\mu(x) = e^{\nu(y)/2} f(y)$$

the part homogeneous of degree $n + 2m$ in y $(m = 0, 1, 2, \ldots)$ yields the equation

$$c_h \int_{\mathbb{R}^N} f(x) K_{n+2m}(x, y) h(x)^2 d\mu(x) = \frac{\nu(y)^m}{2^m m!} f(y).$$

Then

$$c_h' \int_S f(x) K(x, y) h(x)^2 d\omega(x)$$

$$= \sum_{m=0}^{\infty} \frac{1}{2^{n+m}(N/2 + \gamma)_{n+m}} c_h \int_{\mathbb{R}^N} f(x) K_{n+2m}(x, y) h(x)^2 d\mu(x)$$

(identity (1.1)), and this gives the stated formula (clearly,

$$\int_S f(x) K_j(x, y) h(x)^2 d\omega(x) = 0$$

if $j < n$ or $j \not\equiv n$ modulo 2).

Replace y by $-i\lambda y$ for $\lambda > 0$, $y \in \mathbb{R}^N$. Let $A = n + N/2 + \gamma - 1$ and recall the Bessel function

$$J_A(t) = \Gamma(A + 1)^{-1}(t/2)^A \sum_{m=0}^{\infty} \frac{1}{m!(A + 1)_m}(-t^2/4)^m$$

for $t > 0$. This leads to the expression for g in terms of J_A. To find $\Delta_h g$ we can interchange the integral and Δ_h^y (because the resulting integral of a series $\sum_{n=0}^{\infty} \Delta_h^y K_n(x, -iy)$ converges absolutely). Indeed

$$\Delta_h^y K(x, -i\lambda y) = \Delta_h^y K(-i\lambda x, y)$$

$$= \sum_{j=1}^{N}(-i\lambda x_j)^2 K(-i\lambda x, y)$$

$$= -\lambda^2 |x|^2 K(x, -i\lambda y).$$

But $|x|^2 = 1$ on S and so $\Delta_h g = -\lambda^2 g$.

The Hankel transform diagonalizes each T_i (as the Fourier-Plancherel transform does to $\partial/\partial x_i$). First we prove a symmetry relation with some technical restrictions.

2.9 LEMMA. *Let $f \in E(\mathbb{R}^N)$ and let $g \in C^\infty(\mathbb{R}^N)$ such that g and all its partial derivatives are $O(e^{B|x|})$ for some $B < \infty$ (this includes $g(x) = K(x, y)$ for fixed y). Then*

$$\int_{\mathbb{R}^N} (T_j f) g h^2 dm_N = - \int_{\mathbb{R}^N} f(T_j g) h^2 dm_N, \quad j = 1, \ldots, N.$$

PROOF. The following integration by parts is justified by the rapid decrease of f at infinity. We also require $\alpha_i \geq 1$ for each i so that $1/\langle x, v_i \rangle$ is integrable for $h^2 dm_N$. After the formula is established the restriction can be dropped (back to $\alpha_i \geq 0$) by analytic continuation. Now

$$\int_{\mathbb{R}^N} (T_j f) g h^2 dm_N$$

$$= - \int_{\mathbb{R}^N} f(x) \frac{\partial}{\partial x_j} (g(x) h(x)^2) dm_N(x)$$

$$+ \sum_{\ell=1}^m \alpha_\ell (v_\ell)_j \int_{\mathbb{R}^N} \frac{f(x) - f(x\sigma_\ell)}{\langle x, v_\ell \rangle} g(x) h(x)^2 dm_N(x)$$

$$= - \int_{\mathbb{R}^N} \left[f(x) \frac{\partial}{\partial x_j} g(x) + 2f(x) \sum_{\ell=1}^m \alpha_\ell \frac{(v_\ell)_j}{\langle x, v_\ell \rangle} g(x) \right] h(x)^2 dm_N(x)$$

$$+ \sum_{\ell=1}^m \alpha_\ell (v_\ell)_j \int_{\mathbb{R}^N} f(x) \frac{g(x) + g(x\sigma_\ell)}{\langle x, v_\ell \rangle} h(x)^2 dm_N(x)$$

$$= - \int_{\mathbb{R}^N} f(T_j g) h^2 dm_N.$$

The substitution $x \mapsto x\sigma_\ell$ for which $\langle x, v_\ell \rangle$ becomes $\langle x\sigma_\ell, v_\ell \rangle = \langle x, v_\ell \sigma_\ell \rangle = -\langle x, v_\ell \rangle$ was used to show

$$\int_{\mathbb{R}^N} \frac{f(x\sigma_\ell) g(x)}{\langle x, v_\ell \rangle} h(x)^2 dm_N(x) = - \int_{\mathbb{R}^N} \frac{f(x) g(x\sigma_\ell)}{\langle x, v_\ell \rangle} h(x)^2 dm_N(x)$$

($h^2 dm_N$ is G-invariant). $\square$

2.10 THEOREM. *For $f \in E(\mathbb{R}^n)$, $y \in \mathbb{R}^N$, $j = 1, \ldots, N$, $(T_j f)^\smile(y) = iy_j \hat{f}(y)$. The operator $-iT_j$ is densely defined on $L^2(\mathbb{R}^N, h^2 dm_N)$ and is self-adjoint.*

PROOF. For fixed $y \in \mathbb{R}^N$ put $g(x) = K(x, -iy)$ in the lemma. Then $T_j g(x) = -iy_j K(x, -iy)$ and $(T_j f)^\smile(y) = (-1)(-iy_j)\hat{f}(y)$. The multiplication operator defined by $M_j f(y) = y_j f(y)$ $(j = 1, \ldots, N)$ is densely defined and self-adjoint on $L^2(\mathbb{R}^N, h^2 dm_N)$. Further $-iT_j$ is the inverse image of M_j under the Hankel transform, an isometric isomorphism. $\square$

2.11. COROLLARY. *For $f \in E(\mathbb{R}^N)$, $j = 1, \ldots, N, g_j(x) := x_j f(x)$, the transform $(g_j)\check{\ } (y) = iT_j \hat{f}(y)(y \in \mathbb{R}^N)$.*

The self-adjointness of $-iT_j$ is the "infinitesimal" analogue of the same property of Heckman's [7] "global" differential-difference operators acting on the group algebra of the root lattice of a Weyl group.

The action T_j on the functions $\phi_m(p)$, $p \in \mathcal{H}_n^h$ can be explicitly found (but it is not a three-term recurrence).

First we recall the formula Th. 2.1 [2] for the adjoint of T_j on $L^2(S, h^2 d\omega)$ (denoted here by $\tilde{T}_j$ to avoid confusion with the $L^2(\mathbb{R}^N, h^2 dm_N)$ adjoint): if $n = 0, 1, 2, \ldots, f \in \mathcal{H}_n^h$, then

$$\tilde{T}_j f(x) := (N + 2n + 2\gamma)(x_j f(x) - (N + 2n + 2\gamma - 2)^{-1}|x|^2 T_j f(x))$$

is in $\mathcal{H}_{n+1}^h$.

2.12. PROPOSITION. *For $m, n = 0, 1, 2, \ldots, p \in \mathcal{H}_n^h$ $(j = 1, \ldots, n)$*

$$T_j \phi_m(p) = \frac{m+1}{2A} \phi_{m+1}(T_j p) + \frac{A+m}{2A} \phi_m(T_j p)$$
$$- \frac{1}{2(A+1)} \phi_m(\tilde{T}_j p) - \frac{1}{2(A+1)} \phi_{m-1}(\tilde{T}_j p),$$

and

$$x_j \phi_m(p; x) = -\frac{m+1}{2A} \phi_{m+1}(T_j p; x) + \frac{A+m}{2A} \phi_m(T_j p; x)$$
$$+ \frac{1}{2(A+1)} \phi_m(\tilde{T}_j p; x) - \frac{1}{2(A+1)} \phi_{m-1}(\tilde{T}_j p; x)$$

(where $A = n + N/2 + \gamma - 1$).

PROOF. Let $t := |x|^2$, then by the product rule

$$T_j \phi_m(p; x) = (T_j p(x)) L_m^{(A)}(t) e^{-t/2} + 2x_j p(x) \frac{d}{dt}(L_m^{(A)}(t) e^{-t/2}).$$

Use the formula

$$x_j p(x) = \frac{1}{2(A+1)} \tilde{T}_j p(x) + \frac{t}{2A} T_j p(x).$$

In $T_j \phi_m(p; x)$ the coefficient of $(\tilde{T}_j p(x)) e^{-t/2}$ is

$$\frac{1}{(A+1)}(\frac{d}{dt} L_m^{(A)}(t) - \frac{1}{2} L_m^{(A)}(t)) = \frac{-1}{2(A+1)}(L_m^{(A+1)}(t) + L_{m-1}^{(A+1)}(t)).$$

The coefficient of $(T_j p(x)) e^{-t/2}$ is

$$L_m^{(A)}(t) + \frac{t}{A} \frac{d}{dt} L_m^{(A)}(t) - \frac{t}{2A} L_m^{(A)}(t) = \frac{m+1}{2A} L_{m+1}^{(A-1)}(t) + \frac{A+m}{2A} L_m^{(A-1)}(t).$$

This used the identities $\frac{d}{dt} L_m^{(A)}(t) = -L_{m-1}^{(A+1)}(t), tL_M^{(A)}(t) = -(m+1)L_{m+1}^{(A-1)}(t) + (A+m)L_m^{(A-1)}(t)$, and $L_m^{(A)}(t) = L_m^{(A+1)}(t) - L_{m-1}^{(A+1)}(t)$ (which can all be proved directly from the definition of $L_m^{(A)}(t)$). The expansion of $x_j \phi_m(p; x)$ can be proven similarly, or by use of the Hankel transform and Theorems 2.6 and 2.10. $\square$

With the Hankel transform we can easily exhibit the effect of Δ_h on $\phi_m(p)$.

2.13. PROPOSITION. *For* $m, n = 0, 1, 2, \ldots, p \in \mathcal{H}_n^h$,

$$\Delta_h \phi_m(p) = -(m+1)\phi_{m+1}(p) - (n + N/2 + \gamma + 2m)\phi_m(p)$$
$$- (n + N/2 + \gamma + m - 1)\phi_{m-1}(p).$$

PROOF. The recurrence for $L_m^{(A)}(t)$ is

$$t L_m^{(A)}(t) = -(m+1)L_{m+1}^{(A)}(t) + (A + 2m + 1)L_m^{(A)}(t) - (A + m)L_{m-1}^{(A)}(t)$$

(with $A = n + N/2 + \gamma - 1$). This shows that

$$|y|^2 \phi_m(p; y) = -(m+1)\phi_{m+1}(p; y) + (A + 2m + 1)\phi_m(p; y) - (A + m)\phi_{m-1}(p; y).$$

Take the Hankel transform and use Theorems 2.6 and 2.10. $\square$

The various formulas and transforms can be specialised to the G-invariant functions. The kernel K be replaced by

$$K_G(x, y) := \frac{1}{|G|} \sum_{w \in G} K(x, yw)$$

(thus $K_G(xw, y) = K_G(x, y)$ for all $x, y \in \mathbb{C}^N, w \in G$). For example, if p is a G-invariant polynomial then

$$p(y)e^{\nu(y)/2} = c_h \int_{\mathbb{R}^N} e^{-\Delta_h/2} p(x) K_G(x, y) h(x)^2 d\mu(x).$$

The function $K_G(x, -iy)$ is called the Bessel function associated to the root system, by Opdam [14], who denotes it by J_G. In Section 4 we show that it is the classical Bessel function when $G = Z_2$.

3. Special harmonic polynomials

Observe that $\sum_{i=1}^{N}(\partial/\partial x_i)^2 (\sum_{j=1}^{N} x_j y_j)^n = 0$ when $\nu(y) (= \sum_{j=1}^{N} y_j^2) = 0$, $y \in \mathbb{C}^N$ for any $n = 0, 1, 2, \ldots$, and the set of such functions spans the space of harmonic polynomials. The analogue for $\mathcal{H}_n^h$ is $K_n(x, y)$ with $\nu(y) = 0$, and similar results hold.

3.1. PROPOSITION. *For* $n = 0, 1, 2, \ldots$ *the set of polynomials* $\{x \mapsto K_n(x, y) : y \in \mathbb{C}^N, \nu(y) = 0\}$ *spans* $\mathcal{H}_n^h$.

PROOF. Since $\Delta_h^x K_n(x, y) = \nu(y) K_{n-2}(x, y)$ for any $y \in \mathbb{C}^N$, the specified set is indeed a subset of $\mathcal{H}_n^h$. The set $\{x \mapsto \langle x, y \rangle^n : \nu(y) = 0\}$ spans the space of ordinary harmonic polynomials $\mathcal{H}_n$ because it is invariant under the orthogonal group, and H_n provides an irreducible representation of this group (also see Helgason [9, p. 17]). It was shown in [3] that there exists an isomorphism V of polynomials satisfying $V\mathcal{P}_n = \mathcal{P}_n$, $V\frac{\partial}{\partial x_j} = T_j V$ for each j, in particular, $V\mathcal{H}_n = \mathcal{H}_n^h$. Further $K_n(x,y) = V(\langle x,y\rangle^n/n!)$ for fixed $y \in \mathbb{C}^N$ (in fact, for any $p \in \mathcal{P}_n$, $Vp(x) = K_n(x, \nabla_y)p(y)$, where ∇_y is the formal variable $(\partial/\partial y_1, \partial/\partial y_2, \ldots, \partial/\partial y_N)$. $\square$

The utility of these harmonic polynomials comes from the fact that their inner products in $L^2(S, h^2 d\omega)$ can be computed in a closed form involving K_n.

3.2. THEOREM. *For $y, z \in \mathbb{C}^N$,*

$$c_h \int_{\mathbb{R}^N} K(x,y)K(x,z)h(x)^2 d\mu(x) = e^{(\nu(y)+\nu(z))/2}K(y,z).$$

PROOF. The formula

$$p(y)e^{\nu(y)/2} = c_h \int_{\mathbb{R}^N} K(x,y)e^{-\Delta_h/2}p(x)h(x)^2 d\mu(x)$$

was established in Proposition 2.1 for polynomials p. Fix $z \in \mathbb{C}^n$, and let $p_m(x) = \sum_{j=0}^m K_j(x,z)$. As in 2.1, $p_m(y) \to K(y,z)$ and $e^{-\Delta_h/2}p_m(x) \to e^{-\nu(z)/2}K(x,z)$ as $m \to \infty$. The dominated convergence and Fubini theorems apply to prove the desired result. $\square$

3.3 COROLLARY. *For $y, z \in \mathbb{C}^N$, $m, n = 0, 1, 2, \ldots$,*

$$c_h \int_{\mathbb{R}^N} K_n(x,y)K_m(x,z)h(x)^2 d\mu(x)$$

$$= \begin{cases} 0, & \text{if } m \not\equiv n \text{ modulo } 2 \\ \sum_{\ell=0}^{[n/2]} \frac{\nu(y)^\ell \nu(z)^{\ell+k}}{2^{2\ell+k}\ell!(\ell+k)!} K_{n-2\ell}(y,z), & \text{if } m = n + 2k \ (k = 0, 1, 2, \ldots). \end{cases}$$

PROOF. Extract the part homogeneous of degree m in y, n in z both sides of the formula in the Theorem. $\square$

3.4. COROLLARY. *For $n = 0, 1, 2, \ldots$, $y, z \in \mathbb{C}^N$, $\nu(y) = 0 = \nu(z)$, $K_n(x,y)$ and $K_n(x,z)$ are in $\mathcal{H}_n^h$ and*

$$c_h' \int_S K_n(x,y)K_n(x,z)h(x)^2 d\mu(x) = \frac{1}{2^n(\frac{N}{2}+\gamma)_n}K_n(y,z).$$

PROOF. Use formula (1.1) in Corollary 3.3. $\square$

We will discuss specializations of this formula to the classical Gegenbauer and Jacobi polynomials which are associated to the groups Z_2 and $Z_2 \times Z_2$ respectively.

4. Classical examples

The kernel $K(x,z)$ is easy to determine when the group is Z_2. The various formulas involve Bessel functions and Gegenbauer and Jacobi polynomials, and are mostly well known. By discussing this situation we are motivating further study of the kernel for general finite reflection groups. The goal is to get more explicit knowledge about K. Already we are able to put some classical results about Hankel transform into a unifying and conceptual framework.

Although for $N = 1, G = Z_2, K(x,y)$ can be found directly from first principles (solve $T_1^x K(x,y) = yK(x,y)$ with a power series in x), it is illuminating to use formula (1.4). Consider G as $\{1, \sigma\}$ with $x\sigma = -x, \hat{G} = \{\tau_0, \tau_1\}$ with $\tau_0(1) = \tau_0(\sigma) = \tau_1(1) = 1$ and $\tau_1(\sigma) = -1$. Thus $\lambda(\tau_0) = 0, \lambda(\tau_1) = 2\alpha, K_0 = 1$ and

$$
\begin{aligned}
K_{n+1}(x,y) &= \frac{1}{2}\left[xyK_n(x,y)\left(\frac{1}{n+1} + \frac{1}{2\alpha + n + 1}\right) \right. \\
&\quad \left. + (-xy)K_n(-x,y)\left(\frac{1}{n+1} - \frac{1}{2\alpha + n + 1}\right)\right] \\
&= \frac{1}{2}xyK_n(x,y)\left(\frac{1 - (-1)^n}{n+1} + \frac{1 + (-1)^n}{2\alpha + n + 1}\right)
\end{aligned}
$$

(since $K_n(-x,y) = (-1)^n K_n(x,y)$), for $x, y \in \mathbb{C}$. Set $K(x,y) = \sum_{n=0}^{\infty} a_n x^n y^n$. The above formula shows

$$
a_{2n+1} = \frac{1}{2\alpha + 2n + 1}a_{2n} \quad \text{and} \quad a_{2n} = \frac{1}{2n}a_{2n-1},
$$

which implies

$$
a_{2n} = \frac{2^{-2n}}{n!(\alpha + 1/2)_n} \quad \text{and} \quad a_{2n+1} = \frac{2^{-2n-1}}{n!(\alpha + 1/2)_{n+1}}.
$$

Thus

$$
K(x,y) = {}_0F_1\left(\begin{array}{c} - \\ \alpha + 1/2 \end{array}; \left(\frac{xy}{2}\right)^2\right) + \frac{xy}{2\alpha + 1}{}_0F_1\left(\begin{array}{c} - \\ \alpha + 3/2 \end{array}; \left(\frac{xy}{2}\right)^2\right).
$$

Let $x, y \in \mathbb{R}$, then $K(x, -iy) = E_\alpha(xy)$ where

$$
E_\alpha(t) := \Gamma(\alpha + 1/2)(|t|/2)^{-\alpha+1/2}(J_{\alpha-1/2}(|t|) - i(\mathrm{sgn}\ t)J_{\alpha+1/2}(|t|))
$$

for $t \in \mathbb{R}$ (here E_α is a purely local definition allowing some concise formulas). The G-invariant part is

$$
\begin{aligned}
K_G(x, -iy) &= \frac{1}{2}(K(x, -iy) + K(x, iy)) \\
&= \Gamma(\alpha + 1/2)(|t|/2)^{-\alpha+1/2}J_{\alpha-1/2}(|t|),
\end{aligned}
$$

so the transform $\hat{f}$ specializes to the classical Hankel transform on $\mathbb{R}_+$.

4.1. EXAMPLE. (Gegenbauer Polynomials). Using the group Z_2 on $\mathbb{R}^2$ with $h(x) = |x_2|^\alpha$ we obtain Gegenbauer polynomials as spherical harmonics. In polar coordinates,

$$
f_n(r,\theta) := r^n C_n^\alpha(\cos\theta), \quad g_n(r,\theta) := r^n \sin\theta\ C_{n-1}^{\alpha+1}(\cos\theta)
$$

provide a basis for $\mathcal{H}_n^h$. Note f_n is G-invariant (the involution maps $x_2 \mapsto -x_2, \theta \mapsto -\theta$) and g_n is skew-invariant (the character τ_1). Because G leaves x_1

invariant the kernel $K(x, -iy) = e^{-ix_1 y_1} E_\alpha(x_2 y_2)$ for $x, y \in \mathbb{R}^2$. We write out the G-invariant case for Proposition 2.8:

$$c_\alpha \int_0^\pi C_n^\alpha(\cos\theta) \exp(-i\rho\cos\theta\cos\phi)\Gamma(\alpha + 1/2)$$

$$\cdot ((\rho\sin\theta\sin\phi)/2)^{-\alpha+1/2} J_{\alpha-1/2}(\rho\sin\theta\sin\phi)(\sin\theta)^{2\alpha} d\theta$$

$$= (-i)^n (\rho/2)^{-\alpha} C_n^\alpha(\cos\phi)\Gamma(\alpha+1)J_{n+\alpha}(\rho),$$

where $y = (\rho\cos\phi, \rho\sin\phi)$, $0 < \phi < \pi$ and

$$c_\alpha = \left(\int_0^\pi (\sin\theta)^{2\alpha} d\theta \right)^{-1} = \frac{\Gamma(\alpha+1)}{\Gamma(\alpha+1/2)\Gamma(1/2)},$$

$n = 0, 1, 2, \ldots$.

4.2. EXAMPLE. (Jacobi Polynomials). For the group $Z_2 \times Z_2$ on $\mathbb{R}^2$ with $h(x) = |x_1|^\beta |x_2|^\alpha$ the harmonic polynomials involve four families of Jacobi polynomials (with indices $\alpha \pm 1/2, \beta \pm 1/2$), see §3.15 [2]. Here we will mention only the G-invariants $f_{2n}(r, \theta) = r^{2n} P_n^{(\alpha-1/2, \beta-1/2)}(\cos 2\theta)$ (note the reflections $x_1 \mapsto -x_1, x_2 \mapsto -x_2$ are equivalent to $\theta \mapsto \pi - \theta$ and $\theta \mapsto -\theta$ respectively). Because G is decomposable we have $K(x, -iy) = E_\beta(x_1 y_1)E_\alpha(x_2 y_2)$. The formula from 2.8 becomes

$$c_{\alpha,\beta} \int_0^{\pi/2} P_n^{(\alpha-1/2, \beta-1/2)}(\cos 2\theta)\Gamma(\alpha+1)\Gamma(\beta+1)$$

$$\cdot ((\rho\cos\theta\cos\phi)/2)^{1/2-\beta}((\rho\sin\theta\sin\phi)/2)^{1/2-\alpha}$$

$$\cdot J_{\alpha-1/2}(\rho\sin\theta\sin\phi)J_{\beta-1/2}(\rho\cos\theta\cos\phi)(\sin\theta)^{2\alpha}(\cos\theta)^{2\beta} d\theta$$

$$= (-i)^{2n}(\rho/2)^{-\alpha-\beta} P_n^{(\alpha-1/2, \beta-1/2)}(\cos 2\phi)\Gamma(\alpha+\beta+1)J_{2n+\alpha+\beta}(\rho),$$

for $\rho > 0$, $y = (\rho\cos\phi, \rho\sin\phi)$ with $0 < \phi < \pi/2$, and

$$c_{\alpha,\beta} = \frac{2\Gamma(\alpha+\beta+1)}{\Gamma(\alpha+1/2)\Gamma(\beta+1/2)}.$$

Here is a quick look at the meaning of $K_n(x, y)$ with $\nu(y) = 0$ for $N = 2$ (G is a dihedral group). This is essentially the situation presented in §3 [2] dealing with analogues of the Cauchy-Riemann equations. In the notation of [2], let

$$T := \frac{1}{2}(T_1 - iT_2) \quad \text{and} \quad \bar{T} := \frac{1}{2}(T_1 + iT_2)$$

(when $\alpha, \beta = 0$, T and $\bar{T}$ specialize to $\partial/\partial z$ and $\partial/\partial \bar{z}$ respectively, $z := x_1 + ix_2$). Choose $\xi \in \mathbb{C}$, $\xi \neq 0$, and let $y_\epsilon = \xi(1, \epsilon i) \in \mathbb{C}^2$ with $\epsilon = \pm 1$. Thus $\langle x, y \rangle^n = \xi^n(x_1 + i\epsilon x_2)^n$, and $\epsilon = 1$ gives analytic polynomials in $z = x_1 + ix_2$, annihilated by $\partial/\partial \bar{z}$. Analogously we want polynomials annihilated by $\bar{T}$. It was shown in [2] that for each $n = 0, 1, 2, \ldots$, $\dim(\mathcal{H}_n^h \cap (\ker \bar{T})) = 1$ and these polynomials were explicitly determined. Now $\bar{T}K_n(x, y_\epsilon) = \frac{1}{2}\xi(1 - \epsilon)K_{n-1}(x, y_\epsilon) = 0$ for $\epsilon = 1$. This determines $K_n(x, y_1)$ up to a factor depending only on y_1. The factor is found by using a recurrence based on $TK_n(x, y_1) = \xi K_{n-1}(x, y_1)$ and the formula from §3.20 [2]. In particular, for $G = Z_2$, $h(x) = |x_2|^\alpha$,

$$K_n(x, \xi(1, i)) = \frac{\xi^n}{(2\alpha + 1)_n} C_n^{(\alpha, \alpha+1)}(x_1 + ix_2)$$

$$= \frac{\xi^n}{(2\alpha + 1)_n} \sum_{j=0}^{n} \frac{(\alpha)_j (\alpha + 1)_{n-j}}{j!(n-j)!} (x_1 - ix_2)^j (x_1 + ix_2)^{n-j},$$

$x \in \mathbb{C}^2$ (a Heisenberg polynomial).

4.3. CLOSING REMARKS. We have found a generalization of the Fourier or Hankel transforms to measures coming from root systems. It remains to find more explicit formulas (fractional integrals, perhaps) for $K(x, y)$. It is plausible that $|K(x, -iy)| \leq 1$ for all $x, y \in \mathbb{R}^N$, which would allow an L^1-transform. There may be asymptotics comparable to ordinary Bessel functions - a speculation: for fixed y, $|K(x, -iy)|h(x)^{-1/2}$ as $x \mapsto \infty$ on a ray for which $h(x) \neq 0$.

REFERENCES

1. C. F. Dunkl, *Reflection groups and orthogonal polynomials on the sphere*, Math. Z. **197** (1988), 33-60.
2. _____, *Differential-difference operators associated to reflection groups*, Trans. Amer. Math. Soc **311** (1989), 167-183.
3. _____, *Operators commuting with Coxeter group actions on polynomials*, In: Invariant Theory and Tableaux, (D. Stanton, ed.) pp. 107-117, Springer-Verlag, New York, 1990.
4. _____, *Integral kernels with reflection group invariance,*, Canadian J. Math. **43** (1991), 1213-1227.
5. H. Hamburger, *Zur Konvergenztheorie der Stieltjesschen Kettenbrüche*, Math. Z. **4** (1919), 186-222.
6. G. J. Heckman, *Root systems and hypergeometric functions II*, Compositio Math **64** (1987), 353-373.
7. _____, *An elementary approach to the hypergeometric shift operators of Opdam*, Invent. Math **103** (1991), 341-350.
8. G. J. Heckman and E. M. Opdam, *Root systems and hypergeometric functions, I*, Compositio Math. **64** (1987), 329-352.
9. S. Helgason, *Groups and Geometric Analysis*, Academic Press, New York, 1984.
10. I. G. Macdonald, *The volume of a compact Lie group*, Invent. Math. **56** (1980), 93-95.
11. _____, *Some conjectures for root systems*, SIAM J. Math. Anal. **13** (1982), 988-1007.
12. E. M. Opdam, *Root systems and hypergeometric functions III, IV*, Compositio Math. **67** (1988), 21-49, 191-209.
13. _____, *Some applications of hypergeometric shift operators*, Invent. Math. **98** (1989), 1-18.
14. _____, *Dunkl operators, Bessel functions, and the discriminant of a finite Coxeter group*, Compositio Math., to appear.
15. G. Szegö, *Orthogonal Polynomials,*, 4th ed., Colloquium Pub. 23, Amer. Math. Soc., Providence, 1975.

DEPARTMENT OF MATHEMATICS, UNIVERSITY OF VIRGINIA, CHARLOTTESVILLE, VIRGINIA 22903

Contemporary Mathematics
Volume **138**, 1992

Prolongement Analytique des Series de Taylor Spheriques

JACQUES FARAUT

1. Introduction

Considérons la série de Taylor

$$F(z) = \sum_{m=0}^{\infty} a(m)z^m,$$

dont les coefficients sont les valeurs aux entiers positifs ou nuls d'une fonction analytique a. Le théorème suivant, qui est une variante d'un résultat démontré par Leroy en 1900 [8], établit une relation entre les propriétés analytiques de la fonction a et le prolongement analytique de la fonction F.

THÉORÈME 1. *Si la fonction a est holomorphe dans le demi-plan $\Re s > -\frac{1}{2}$, et s'il existe des constantes A et N telles que*

$$|a(s)| \le A(1+|s|)^N, \quad \Re s > -\frac{1}{2},$$

alors la série de Taylor $\sum a(m)z^m$ converge dans le disque unité $D = \{z \mid |z| < 1\}$, et la fonction F admet un prolongement analytique dans le plan coupé $\mathbb{C} \setminus [1, \infty[$.

Par exemple ce théorème montre que la fonction hypergéométrique,

$$_2F_1(\alpha, \beta; \gamma; z) = \sum_{m=0}^{\infty} \frac{(\alpha)_m (\beta)_m}{(\gamma)_m} \frac{z^m}{m!},$$

admet un prolongement analytique dans le plan coupé $\mathbb{C} \setminus [1, \infty[$, en considérant la fonction

$$a(s) = \frac{\Gamma(\gamma)}{\Gamma(\alpha)\Gamma(\beta)} \frac{\Gamma(\alpha+s)\Gamma(\beta+s)}{\Gamma(\gamma+s)\Gamma(1+s)}.$$

1991 *Mathematics Subject Classification.* 22E30, 22E45, 33C80.
This paper is in final form and no version of it will be submitted for publication elsewhere .

La démonstration originale de Leroy utilise une représentation intégrale de F obtenue à l'aide du théorème des résidus et d'une déformation du contour d'intégration. Nous en donnons ci-dessous une démonstration différente qui repose sur un théorème du type Paley-Wiener pour la transformation de Mellin.

Remarquons d'abord que nous pouvons supposer $N < -1$. Choisissons en effet un entier $M > N + 1$, et posons

$$a_1(s) = \frac{a(s)}{(1+s)^M},$$

$$F_1(z) = \sum_{m=0}^{\infty} a_1(m) z^m,$$

alors

$$|a_1(s)| \leq A_1 (1 + |s|)^{N-M},$$

et

$$F(z) = (I + z\frac{d}{dz})^M F_1(z).$$

Si $N < -1$, la fonction a est la transformée de Mellin d'une fonction f dont le support est contenu dans $[1, \infty[$,

$$a(s) = \int_1^{\infty} f(t) t^{-s} \frac{dt}{t},$$

et f vérifie

$$\int_1^{\infty} |f(t)|^2 dt < \infty.$$

Par suite,

$$F(z) = \sum_{m=0}^{\infty} \left(\int_1^{\infty} t^{-m} \frac{dt}{t} \right) z^m$$

$$= \int_1^{\infty} f(t) \sum_{m=0}^{\infty} \left(\frac{z}{t} \right)^m \frac{dt}{t}$$

$$= \int_1^{\infty} f(t) \frac{1}{t - z} dt.$$

De cette représentation intégrale on déduit que F admet un prolongement analytique au plan coupé $\mathbb{C} \setminus [1, \infty[$. $\square$

Si le support de f est compact, $\text{supp}(f) \subset [1, R]$, alors la fonctio a est entière, la fonction F est holomorphe pour $|z| > R$, et admet un développement de Laurent à l'infini qui s'écrit

$$F(z) = - \sum_{m=0}^{\infty} a(-m-1) z^{-m-1}.$$

Lindelöf étudie dans son livre [9] des questions semblables. Des résultats du même type ont été obtenus par Stein et Wainger pour des séries de polynômes de Legendre [10], et également par Avanissian dans le cas de plusieurs variables [1].

Dans cet article nous démontrons un analogue du théorème 1 pour des séries de polynômes sphériques sur une algèbre de Jordan. Nous utilisons une propriété du type Paley-Wiener pour la transformation de Fourier sphérique sur un cône symétrique, c'est la proposition 1. Nous l'avons démontrée grâce à des indications qui nous ont été fournies par Jean-Philippe Anker. C'est Jacques Bros qui a porté à notre connaissance le résultat du type "edge of the wedge" (proposition 3) qui nous permet d'achever la démonstration du théorème 2.

2. Algèbres de Jordan et polynômes sphériques

Une algèbre de Jordan réelle V est un espace vectoriel réel muni d'un produit vérifiant

$$(J1) \quad xy = yx,$$

$$(J2) \quad x^2(xy) = x(x^2y).$$

Si on note $L(x)$ la transformation de V définie par

$$L(x) : y \mapsto xy,$$

la propriété (J2) signifie que $L(x)$ et $L(x^2)$ commutent. L'algèbre de Jordan V est dite euclidienne s'il existe sur V un produit scalaire euclidien associatif, c'est à dire

$$(xy|z) = (x|yz).$$

Par exemple l'espace $V = \mathrm{Sym}(m, \mathbb{R})$ des matrices symétriques réelles $m \times m$ muni du produit de Jordan

$$x \circ y = \frac{1}{2}(xy + yx),$$

et du produit scalaire

$$(x|y) = \mathrm{tr}(xy) = \mathrm{tr}(x \circ y),$$

est une algèbre de Jordan euclidienne. Nous supposons dans la suite que V est une algèbre de Jordan euclidienne de dimension finie. Une telle algèbre admet un élément neutre que nous notons e. Nous supposons également que V est simple, c'est à dire qu'il n'existe pas dans V d'ideal non trivial. Soit K la composante connexe de l'identité dans le groupe $\mathrm{Aut}(V)$ des automorphismes de V. Le groupe K est compact. Fixons un système maximal d'idempotents deux à deux orthogonaux $c_1, \ldots, c_r$,

$$c_i^2 = c_i, \ c_ic_j = 0 \ (i \neq j),$$

$$c_1 + \cdots + c_r = e.$$

Le nombre r d'éléments d'un tel système est appelé le rang de V. Tout élément x de V peut s'écrire

$$x = k \sum_{i=1}^{r} \lambda_i c_i, \ \lambda_i \in \mathbb{R}, \ k \in K,$$

c'est la décomposition spectrale de x. Les nombres λ_i sont appelés les valeurs propres de x, $\mathrm{tr}(x) = \sum \lambda_i$, $\Delta(x) = \prod \lambda_i$ sont la trace et le déterminant de x.

Le déterminant Δ est un polynôme homogène de degré r, et on peut supposer que le produit scalaire est donné par

$$(x|y) = \operatorname{tr}(xy).$$

Les valeurs propres de l'endomorphisme $L(c_i)$ sont $1, \frac{1}{2}$ et 0. On pose, pour $i \neq j$,

$$V_{ij} = \{x \in V \mid L(c_i)x = \frac{1}{2}x, \ L(c_j)x = \frac{1}{2}x\}.$$

La dimension d de V_{ij} est indépendente de i et j. La dimension n de V est donnée par

$$n = r + \frac{d}{2}r(r-1).$$

Posons $e_j = c_1 + \cdots + c_j$, et

$$V_j = \{x \mid e_j x = x\}.$$

Alors V_j est une sous-algèbre de V de rang j. Le polynôme Δ_j défini par

$$\Delta_j(x) = \det_{V_j}(\operatorname{pr}_{V_j} x)$$

(déterminant relativement à la sous-algèbre V_j de la projection orthogonale de x sur V_j), est appelé le j-ième déterminant mineur principal. Pour $\mathbf{m} = (m_1, \ldots, m_r)$, où $m_1 \geq \ldots \geq m_r \geq 0$ sont des entiers (on notera $\mathbf{m} \geq 0$), on pose

$$\Delta_{\mathbf{m}}(x) = \Delta_1(x)^{m_1 - m_2} \Delta_2(x)^{m_2 - m_3} \ldots \Delta_r(x)^{m_r}.$$

Alors $\Delta_{\mathbf{m}}$ est un polynôme homogène de degré $|\mathbf{m}| = m_1 + \cdots + m_r$. Soit $\mathfrak{g}$ la sous-algèbre de Lie de $\mathfrak{gl}(V)$ engendrée par les transformations $L(a)$ ($a \in V$), et soit G le sous-groupe analytique de $GL(V)$ d'algèbre de Lie $\mathfrak{g}$. Alors K est un sous-groupe compact maximal de G. Soit $\mathcal{P}_{\mathbf{m}}$ l'espace des fonctions polynômes engendrées par $\Delta_{\mathbf{m}}(gx)$ ($g \in G$). L'espace $\mathcal{P}_{\mathbf{m}}$ est irréductible sous l'action de G et l'espace $\mathcal{P}$ de toutes les fonctions polynômes sur V est la somme directe des sous-espaces $\mathcal{P}_{\mathbf{m}}$,

$$\mathcal{P} = \bigoplus_{\mathbf{m} \geq 0} \mathcal{P}_{\mathbf{m}}.$$

Le polynôme sphérique $\Phi_{\mathbf{m}}$ défini par

$$\Phi_{\mathbf{m}}(x) = \int_K \Delta_{\mathbf{m}}(kx)\,dk$$

(dk désigne la mesure de Haar normalisée de K), est invariant par K, et tout polynôme K-invariant de $\mathcal{P}_{\mathbf{m}}$ est proportionnel à $\Phi_{\mathbf{m}}$. Le polynôme $\Phi_{\mathbf{m}}(x)$ ne dépend que des valeurs propres $\lambda_1, \ldots, \lambda_r$ de x, et les polynômes sphériques s'expriment à l'aide des polynomes de Jack,

$$\Phi_{\mathbf{m}}(x) = \frac{J_{\mathbf{m}}^{(\alpha)}(\lambda_1, \ldots, \lambda_r)}{J_{\mathbf{m}}^{(\alpha)}(1, \ldots, 1)},$$

où $J_{\mathbf{m}}^{(\alpha)}$ est le polynôme de Jack associé à la partition $\mathbf{m} = (m_1, \ldots, m_r)$ et $\alpha = \frac{2}{d}$.

3. Séries de Taylor sphériques

Une série de Taylor sphérique est une série de la forme

$$F(x) = \sum_{\mathbf{m} \geq 0} a_{\mathbf{m}} \Phi_{\mathbf{m}}(x).$$

Si cette série converge unifromément sur tout compact d'un ouvert de $V^{\mathbb{C}}$, sa somme est une fonction holomorphe qui est invariante par le groupe $K^{\mathbb{C}}$, complexifié de K. Pour généraliser le théormème 1 à de telles séries nous devons définir l'analogue du disque unité, de la demi-droite $]0, \infty[$, et du plan coupé $\mathbb{C} \setminus [1, \infty[$.

a) Soit U le groupe $G^{\mathbb{C}} \cap U(V^{\mathbb{C}})$, intersection du groupe complexifié $G^{\mathbb{C}}$ de G et du groupe unitaire de l'espace hermitien $V^{\mathbb{C}}$. Tout élément z de $V^{\mathbb{C}}$ peut s'écrire

$$z = u \sum_{j=1}^{r} \lambda_j c_j, \ \lambda_j \geq 0, \ u \in U,$$

et alors $|z| = \sup \lambda_j$ est une norme sur $V^{\mathbb{C}}$. Le disque généralisé D est défini par

$$D = \{ z \in V^{\mathbb{C}} : |z| < 1 \}.$$

Par exemple, si $V = \mathrm{Sym}(r, \mathbb{R})$, alors $V^{\mathbb{C}} = \mathrm{Sym}(r, \mathbb{C})$, et D est le disque de Siegel

$$D = \{ z \in \mathrm{Sym}(r, \mathbb{C}) \mid I - z\bar{z} >> 0 \}.$$

b) L'ensemble des carrés de l'algèbre de Jordan V est un cône convexe fermé. Son intérieur Ω est un cône symétrique, c'est à dire homogène et autodual. Par exemple, si $V = \mathrm{Sym}(r, \mathbb{R})$, alors Ω est le cône des matrices symétriques définies positives. Un élément x de V appartient à Ω si et seulement si

$$\Delta_j(x) > 0, \ j = 1, \ldots, r.$$

Par suite, si x appartient à Ω, on peut définir

$$\Phi_{\mathbf{s}}(x) = \int_K \Delta_{\mathbf{s}}(kx) dk, \ \mathbf{s} \in \mathbb{C}^r.$$

Le cône Ω est un espace riemannien symétrique, $\Omega \simeq G/K$, et les fonctions $\Phi_{\mathbf{s}}$ en sont les fonctions sphériques.

c) Soit $T_\Omega = V + i\Omega$ le tube dans $V^{\mathbb{C}}$ de base Ω, et posons

$$\mathcal{H} = \bigcup_{0 \leq \theta \leq \pi} e^{-i\theta} T_\Omega.$$

Remarquons que si $r = 1$ ($V^{\mathbb{C}} = \mathbb{C}$), alors $\mathcal{H}$ est un plan coupé, $\mathcal{H} = \mathbb{C} \setminus] -\infty, 0[$.

Notons $\rho_j = \frac{d}{4}(2j - r - 1)$ $(j = 1, \ldots, r)$, $\rho = (\rho_1, \ldots, \rho_r)$, et soit $d_{\mathbf{m}}$ la dimension de l'espace $\mathcal{P}_{\mathbf{m}}$.

THÉORÈME 2. *Soit F une série de Taylor sphérique de la forme*

(S) $$F(z) = \sum_{\mathbf{m} \geq 0} d_{\mathbf{m}} a(\mathbf{m}) \Phi_{\mathbf{m}}(z),$$

où a est une fonction holomorphe pour $\Re s_j > \rho_j - \frac{n}{2r}$, vérifiant

(i) $|a(\mathbf{s})| \leq A(1 + ||\mathbf{s}||)^N$,

(ii) *la fonction $\mu \mapsto a(\mu + \rho)$ est invariante par le groupe des permutations $W = \mathfrak{S}_r$.*

Alors la série (S) converge dans D et admet un prolongement analytique dans le domaine $e - \mathcal{H}$.

Ce théorème s'applique aux fonctions hypergéométriques $_2F_1$ associées au cônes symétriques [4]. La fonction gamma du cône Ω est définie par

$$\Gamma_\Omega(\mathbf{s}) = \int_\Omega e^{-\operatorname{tr}(x)} \Delta_{\mathbf{s}}(x) \Delta(x)^{-\frac{n}{r}} dx, \ (\mathbf{s} \in \mathbb{C}^r),$$

les coefficients binomiaux par

$$(\alpha)_{\mathbf{m}} = \frac{\Gamma_\Omega(\alpha + \mathbf{m})}{\Gamma_\Omega(\alpha)}, \ (\alpha \in \mathbb{C}),$$

où $\alpha + \mathbf{m} = (\alpha + m_1, \ldots, \alpha + m_r)$, et la fonction hypergéométrique par

$$_2F_1(\alpha, \beta; \gamma; z) = \sum_{\mathbf{m} \geq 0} \frac{(\alpha)_{\mathbf{m}} (\beta)_{\mathbf{m}}}{(\gamma)_{\mathbf{m}}} \frac{1}{(\frac{n}{r})_{\mathbf{m}}} d_{\mathbf{m}} \Phi_{\mathbf{m}}(z).$$

4. Démonstration du théorème 2

Effectuons la substitution

$$\mathbf{s} \mapsto i\lambda + \rho,$$

et posons

$$\varphi_\lambda = \Phi_{i\lambda + \rho}.$$

La transformée de Fourier sphérique $\tilde{f}$ d'une fonction f sur Ω qui est K-invariante est définie par

$$\tilde{f}(\lambda) = \int_\Omega \varphi_\lambda(\omega^{-1}) f(\omega) \Delta(\omega)^{-\frac{n}{r}} d\omega.$$

PROPOSITION 1. *Soit ψ une fonction holomorphe de $\lambda = (\lambda_1, \ldots, \lambda_r)$ pour*

$$\Im \lambda_j < \frac{n}{2r}, \ j = 1, \ldots, r,$$

vérifiant

(i) *il existe $A > 0$ et $N > n$ tels que*

$$|\psi(\lambda)| \leq A(1 + ||\lambda||)^{-N},$$

(ii) *pour toute permutation w de $W = \mathfrak{S}_r$,*

$$\psi(w\lambda) = \psi(\lambda).$$

Alors il existe une fonction f sur Ω, dont le support est contenu dans $e + \overline{\Omega}$, vérifiant

$$\int_{e+\Omega} |f(\omega)|^2 d\omega < \infty,$$

telle que

$$\psi(\lambda) = \int_{e+\Omega} \varphi_\lambda(\omega^{-1}) f(\omega) \Delta(\omega)^{-\frac{n}{r}} d\omega.$$

DÉMONSTRATION. Posons

$$f(\omega) = \frac{1}{|W|} \int_{\mathbb{R}^r} \varphi_\lambda(\omega) \psi(\lambda) \frac{d\lambda}{|c(\lambda)|^2},$$

où c est la fonction de Harish-Chandra de l'espace riemannien symétrique $\Omega = G/K$,

$$c(\lambda) = \frac{I(i\lambda)}{I(\rho)},$$

$$I(\mu) = \prod_{j<k} B(\mu_k - \mu_j, \frac{d}{2}).$$

Alors f est une fonction continue sur Ω qui est invariante par K.

(a) Montrons d'abord que le support de f est contenu dans $e + \overline{\Omega}$. Pour $t = \mathrm{diag}(t_1, \ldots, t_r)$, $t_1 < \cdots < t_r$,

$$\varphi_\lambda(\exp t) = \sum_{w \in W} c(w\lambda) \Phi(t, w\lambda),$$

avec

$$\Phi(t, \lambda) = e^{(i\lambda - \rho | t)} \sum_{\mu \in \Lambda} \Gamma_\mu(\lambda) e^{-(\mu | t)}.$$

(Voir [6], Chapter IV, Theorem 5.5). Il existe $\varepsilon > 0$ tel que, pour t fixé, la série converge uniformément en λ pour

$$\Im(\lambda_j - \lambda_k) \leq \varepsilon, \ j < k.$$

En utilisant l'invariance par W de la fonction ψ on peut écrire

$$f(\exp t) = \int_{\mathbb{R}^r} \Phi(t, -\lambda) \psi(-\lambda) \frac{d\lambda}{c(\lambda)}.$$

Si $\Im(\lambda_k - \lambda j) \leq 0$, $j < k$, alors

$$\frac{1}{|c(\lambda)|} \leq B(1 + ||\lambda||)^{n-r},$$

et, si $\Im\lambda_j \leq 0$,

$$|\psi(\lambda)| \leq A(1 + ||\lambda||)^N.$$

En déplaçant le domaine d'intégration dans $\mathbb{C}^r$ grâce au théorème de Cauchy on obtient

$$f(\exp t) = \int_{\mathbb{R}^r} \Phi(t, -\lambda - i\tau\eta) \psi(-\lambda - i\tau\eta) \frac{d\lambda}{c(\lambda + i\tau\eta)},$$

pout $\tau > 0$ et $\eta_1 > \eta_2 > \cdots > \eta_r > 0$. On en déduit la majoration

$$|f(\exp t)| \leq M e^{\tau(\eta|t)},$$

où $M = M(t, \eta)$ est une constante qui ne dépend pas de τ. Si $\exp t$ n'appartient pas à $e + \overline{\Omega}$, l'un des nombres t_j est négatif, et on peut choisir η tel que $(\eta|t) < 0$. En faisant tendre τ vers $+\infty$ on en déduit que $f(\exp t) = 0$.

(b) Montrons maintenant que

$$\int_{e+\Omega} |f(\omega)|^2 d\omega < \infty.$$

D'après le théorème de Plancherel pour la transformation de Fourier sphérique ([6], Chapter IV, Theorem 7.5), pour $\alpha < \frac{n}{r}$,

$$\int_{e+\Omega} |f(\omega)\Delta(\omega)^{\frac{\alpha}{2}}|^2 \Delta(\omega)^{-\frac{n}{r}} d\omega = \frac{1}{|W|} \int_{\mathbb{R}^r} |\psi(\lambda + i\alpha)|^2 \frac{d\lambda}{|c(\lambda)|^2}$$

$$\leq C \int_{\mathbb{R}^r} (1 + \|\lambda\|)^{n-r-2N} d\lambda.$$

En faisant tendre α vers $\frac{n}{r}$, et en utilisant le lemme de Fatou, on en déduit que

$$\int_{e+\Omega} |f(w)|^2 < \infty.$$

□

PROPOSITION 2. *Si z appartient à D et ω à $e + \Omega$,*

$$\sum_{\mathbf{m} \geq 0} d_{\mathbf{m}} \Phi_{\mathbf{m}}(z) \Phi_{\mathbf{m}}(\omega^{-1}) = \Delta(\omega)^{\frac{n}{r}} \int_K \Delta(k\omega - z)^{-\frac{n}{r}} dk.$$

DÉMONSTRATION. D'après [5] (Remark 2 p.79), et aussi [7], si x appartient à D,

$$\Delta(e - x)^{-\frac{n}{r}} = \sum_{\mathbf{m} \geq 0} d_{\mathbf{m}} \Phi_{\mathbf{m}}(x).$$

En posant $x = P(\omega)^{-\frac{1}{2}} k^{-1} z$, où $k \in K$, $\omega \in e + \Omega$, et où $P(a)$ désigne la représentation quadratique de l'algèbre de Jordan V,

$$P(a) = 2L(a)^2 - L(a^2),$$

nous obtenons

$$\Delta(e - x) = \Delta(\omega)^{-1} \Delta(k\omega - z),$$

et par suite

$$\Delta(\omega)^{\frac{n}{r}} \int_K \Delta(k\omega - z)^{-\frac{n}{r}} dk = \sum_{\mathbf{m} \geq 0} d_{\mathbf{m}} \int_K \Phi_{\mathbf{m}}\big(P(\omega^{-\frac{1}{2}}) k^{-1} z\big) dk.$$

Pour finir on utilise la relation fonctionnelle des fonctions sphériques,

$$\int_K \Phi_{\mathbf{m}}\big(P(\omega^{-\frac{1}{2}}) k^{-1} z\big) dk = \Phi_{\mathbf{m}}(\omega^{-1}) \Phi_{\mathbf{m}}(z).$$

□

Pour achever la démonstration du théorème 2 nous aurons besoin d'un résultat du type "edge of the wedge" qui est démontré dans [2].

PROPOSITION 3. *Soit Ω un cône convexe ouvert dans un espace vectoriel réel V de dimension finie. Toute fonction F qui est holomorphe dans l'ouvert $T_+ \cup T_- \cup \Omega \subset V^{\mathbb{C}}$, où $T_\pm = V \pm i\Omega$, admet un prolongement analytique dans le domaine $\mathcal{H}$, et $\mathcal{H}$ est l'enveloppe d'holomorphie de $T_+ \cup T_- \cup \Omega$.*

DÉMONTRONS MAINTENANT LE THÉORÈME 2. Il existe des constantes C_1 et N_1 telles que
$$d_{\mathbf{m}} \le C_1(1 + |\mathbf{m}|)^{N_1}.$$
De plus
$$|\Phi_{\mathbf{m}}(z)| \le |z|^{|\mathbf{m}|}$$
Par suite la série (S) converge uniformémént sur tout compact de D.

Les fonctions sphériques sont fonctions propres des opérateurs diffé rentiels invariants de l'espace riemannien symétrique $\Omega = G/K$. En particulier, si $\mathcal{D} = \Delta(x)\Delta(\frac{\partial}{\partial x})$, alors
$$\mathcal{D}\Phi_{\mathbf{s}} = \gamma(\mathbf{s})\Phi_{\mathbf{s}},$$
avec
$$\gamma(\mathbf{s}) = \prod_{j=1}^{r}\left(s_j + \frac{d}{2}(r - j)\right).$$
Choisissons $M > N + n$ et posons
$$a_1(\mathbf{s}) = \frac{a(\mathbf{s})}{\left(1 + \gamma(\mathbf{s})\right)^M},$$
$$F_1(z) = \sum_{\mathbf{m} \ge 0} a_1(\mathbf{m})d_{\mathbf{m}}\Phi_{\mathbf{m}}(z),$$
alors
$$|a_1(\mathbf{s}| \le A_1(1 + ||\mathbf{s}||)^{N-M},$$
et
$$F(z) = (I + \mathcal{D})^M F_1(z).$$
Quitte à remplacer F par F_1 ceci permet de supposer $N < -n$. Alors, d'après la proposition 1, il existe une fonction f K-invariante de support contenu dans $e + \overline{\Omega}$ vérifiant
$$\int_{e+\Omega} |f(\omega)|^2 d\omega < \infty,$$
telle que
$$a(\mathbf{s}) = \int_{e+\Omega} \Phi_{\mathbf{s}}(\omega^{-1})f(\omega)\Delta(\omega)^{-\frac{n}{r}} d\omega.$$
En utilisant la proposition 2 on en déduit que F admet la représentation intégrale suivante
$$F(z) = \int_{e+\Omega} \Delta(\omega - z)^{-\frac{n}{r}} f(\omega) d\omega.$$

La fonction $\Delta(z)^{-\frac{n}{r}}$ est bien définie dans le tube $\Omega + iV$, notamment par la formule

$$\Delta(z)^{-\frac{n}{r}} = \frac{1}{c} \int_\Omega e^{-(z|u)} du,$$

où $c = \int_\Omega e^{-\mathrm{tr}(u)} du$. Dans $T_\pm = V \pm i\Omega$ posons

$$F_\pm(z) = i^{\pm n} \int_{e+\Omega} \Delta\big(\pm i(\omega - z)\big)^{-\frac{n}{r}} f(\omega) d\omega.$$

Les fonctions F_+ et F_- sont holomorphes et ont même valeur au bord sur $e - \Omega$. Le théorème 2 se déduit alors de la proposition 3. $\square$

Supposons que $\frac{n}{r}$ soit un entier ($d = 1$ et r est impair, ou d est pair). Si le support de f est compact,

$$\mathrm{supp}(f) \subset \{x \in e + \overline{\Omega} \mid |x| \le R\}, \ (R > 1),$$

alors a est une fonction entière, F est holomorphe au voisinage de l'infini et admet un développement de Laurent généralisé pour $|z^{-1}| < R^{-1}$,

$$F(z) = (-1)^{\frac{n}{r}} \sum_{\mathbf{m} \ge 0} d_{\mathbf{m}} a(-\frac{n}{r} - \mathbf{m}^*) \Phi_{\mathbf{m}}(z^{-1}) \Delta(z)^{-\frac{n}{r}},$$

où $\mathbf{m}^* = (m_r, m_{r-1}, \ldots, m_1)$. On démontre ce développement à l'aide de la relation

$$\Phi_{\mathbf{m}}(x^{-1}) = \Phi_{-\mathbf{m}^*}(x).$$

Remarquons que la démonstration du théorème 2 utilise l'analyse harmonique de l'espace symétrique U/K (développements en séries de polynômes sphériques) et celle de $\Omega \simeq G/K$ (transformation de Fourier sphérique), et que l'espace symétrique U/K est le dual compact de l'espace symétrique G/K. Une situation analogue se présente dans le travail récent de Bros et Viano [3].

REFERENCES

1. Avanissian V., *Quelques applications des fonctionnelles analytiques*, prépublication.
2. Bros J., Epstein H., Glaser V., Stora R., *Quelques aspects globaux des problèmes d'Edge of the Wedge, Hyperfonctions and theoretical physics*, Springer, New York, Lecture Notes in Math. 449 (1975), 185–218.
3. Bros J., Viano G., *On the connection between harmonic analysis on the sphere and on the one-sheeted hyperboloid : an analytic continuation viewpoint*, prépublication.
4. Faraut J., Koranyi A., *Fonctions hypergéométriques associées aux cônes symétriques*, C.R. Acad. Sci. Paris **307** (1988), 555–558.
5. Faraut J., Koranyi A., *Function spaces and reproducing kernels on bounded symmetric domains*, J. Functional Analysis **88** (1990), 64–89.
6. Helgason S., *Groups and Geometric Analysis*, Academic Press, 1984.
7. Lassalle M., *Noyau de Szegő, K-types et algèbres de Jordan*, C. R. Acad. Sci. Paris **310** (1990), 253–256.
8. Le Roy E., *Sur les séries divergentes et les fonctions définies par un développement de Taylor*, Ann. Faculté des Sci. Univ. de Toulouse **2** (1900), 317–430.
9. Lindelöf E., *Le Calcul des Résidus et ses Applications à la Théorie des Fonctions*, Gauthier-Villars, Paris, 1905.

10. Stein E.M., Wainger S., *Analytic properties of expansions, and some variants of Parseval-Plancherel formulas*, Arkiv för Mat. 5 (1963), 553–567.

Département de Mathématiques, Analyse complexe et géométrie, Université Pierre et Marie Curie, 4 place Jussieu, 75 252 Paris Cedex 05

Contemporary Mathematics
Volume **138**, 1992

Some Combinatorial Aspects of the Spectra
of Normally Distributed Random Matrices

PHILIP J. HANLON, RICHARD P. STANLEY
AND JOHN R. STEMBRIDGE

Introduction

Let U be a real $n \times n$ matrix whose entries u_{ij} are random variables, and let A and B be fixed $n \times n$ real symmetric matrices. Statisticians (e.g., see [**OU**]) have been interested in the distribution of the eigenvalues $\theta_1, \ldots, \theta_n$ of the matrix $AUBU^t$, where ^{t} denotes transpose. Of particular interest are the quantities $\text{tr}((AUBU^t)^k) = \sum \theta_i^k$ for $k = 1, 2, \ldots$, since these determine the eigenvalues.

More generally, one may consider the distribution of arbitrary symmetric functions of the eigenvalues of $AUBU^t$. Thus let us regard any symmetric polynomial f (say with real coefficients) in the variables $x_1, \ldots, x_n$ as a function on $n \times n$ matrices by defining $f(U)$ to be the value of f at the eigenvalues of the matrix U. In these terms, we have

$$\text{tr}((AUBU^t)^k) = p_k(AUBU^t),$$

where p_k denotes the kth power-sum symmetric function.

We will follow the symmetric function notation and terminology of [**M1**]. In particular, if $\lambda = (\lambda_1, \lambda_2, \ldots)$ is any partition of n (denoted $\lambda \vdash n$ or $|\lambda| = n$), $p_\lambda = p_{\lambda_1} p_{\lambda_2} \cdots$ will denote the product of power sums indexed by the parts of λ. The p_λ's are a linear basis for the symmetric polynomials f.

Let $\mathcal{E}_U$ denote expectation with respect to a random $n \times n$ matrix U of independent standard normal variables. Given that A and B are fixed real symmetric matrices, it is easy to show that $\mathcal{E}_U(f(AUBU^t))$ is a symmetric polynomial function of the eigenvalues of A and B separately, and thus is a linear combination of $p_\mu(A) p_\nu(B)$. The main result of this paper (Theorem 3.5) is an explicit expansion for $\mathcal{E}_U(p_\lambda(AUBU^t))$ in terms of $p_\mu(A) p_\nu(B)$, together with a combinatorial

1991 *Mathematics Subject Classification.* 05E05, 15A52.

Research supported by grants from the NSF.

This paper is in final form and no version of it will be submitted for publication elsewhere.

interpretation of the coefficients involving perfect matchings on $2k$ points, where $|\lambda| = k$. We will also give analogous results for complex and quaternionic matrices (Theorems 2.3 and 4.1). Tables of these polynomials are provided in the Appendix.

Let I_n and O_n denote the $n \times n$ identity and zero matrices, respectively. Of particular interest is the special case in which $A = I_m \oplus O_{n-m}$ and $B = I_n$. Under these circumstances, one has

$$AUBU^t = \begin{bmatrix} \tilde{U}\tilde{U}^t & * \\ 0 & O_{n-m} \end{bmatrix},$$

where $\tilde{U}$ consists of the first m rows of U. Thus the eigenvalues of $AUBU^t$ are just the squares of the singular values of the random $m \times n$ matrix $\tilde{U}$, together with $n - m$ irrelevant zeroes. In this special case, the quantity $\mathcal{E}_U(p_\lambda(AUBU^t))$ turns out to be a polynomial $Q_\lambda(m, n)$ with nonnegative integral coefficients which sum to $1 \cdot 3 \cdot 5 \cdots (2k - 1)$ (where $|\lambda| = k$), and our combinatorial interpretation of $\mathcal{E}_U(p_\lambda(AUBU^t))$ simplifies considerably (see Corollary 3.6). Again there are similar results for complex and quaternionic matrices (Corollaries 2.4 and 4.2). We will also derive a more explicit formula (Theorem 5.4) for $Q_\lambda(m, n)$ in the case $\lambda = (k)$ that is more efficient for computational purposes.

The paper is organized as follows. In the first section, we review the theory of zonal spherical functions of Hermitian matrices over finite-dimensional real division algebras. The spherical functions are essentially Jack symmetric functions for special choices of the parameter α. The crucial point of this section is Theorem 1.1, which reduces the problem of computing the expectation of $p_\lambda(AUBU^t)$ to the theory of (Jack) symmetric functions. In the next three sections, we analyze the individual details of the complex, real and quaternionic cases. We do the complex case first, primarily because the spherical functions for complex Hermitian matrices (i.e., Schur functions) are easier to work with than the spherical functions for real symmetric matrices (i.e., zonal polynomials). Although the quaternionic case must be formulated more carefully than the others, the corresponding results can be deduced easily from the real case, since the spherical functions for quaternionic Hermitian matrices are closely related to the real symmetric case. In the final section, we present an (incomplete) combinatorial approach to computing the expectation of $p_\lambda(AUBU^t)$. If successful, this would yield a completely elementary method of deriving the main results of this paper without the theory of spherical functions. We will also point out how some recent work of Goulden and Jackson [GJ] provides some partial success for this approach in the complex case.

1. Zonal Spherical Functions

Let F be a finite dimensional real division algebra; i.e., $F = \mathbf{R}$, $\mathbf{C}$, or the quaternions $\mathbf{H}$, and let $x \mapsto \bar{x}$ denote the usual conjugation on F. (In particular, $\bar{x} = x$ when $F = \mathbf{R}$.) Let $\Re(x) = (x + \bar{x})/2$ denote the real part of x. An F-valued random variable u will be said to be *standard normal* if $\mathcal{E}(u) = 0$, $\mathcal{E}(u\bar{u}) = 1$, and the distribution is Gaussian.

Let $\mathcal{S}_n(F)$ denote the real vector space of $n \times n$ "Hermitian" matrices A satisfying $A = A^*$, where A^* denotes the conjugated transpose. The general linear group $GL_n(F)$ acts on $\mathcal{S}_n(F)$ via $A \mapsto XAX^*$. Since any Hermitian A can be diagonalized by a member of the "unitary" group $K_n = \{X \in GL_n(F) : XX^* = I_n\}$, it follows that the K_n-invariant polynomials on $\mathcal{S}_n(F)$ (i.e., the polynomials f satisfying $f(XAX^*) = f(A)$ for all $X \in K_n$, $A \in \mathcal{S}_n(F)$) can be identified with the symmetric polynomial functions of n (real) variables. For example, the polynomial function corresponding to the kth power sum p_k is

$$(1.1) \qquad p_k(A) = \text{tr}(A^k) \quad (A \in \mathcal{S}_n(F)),$$

and these particular functions generate the entire algebra of K_n-invariants.

Since $(GL_n(F), K_n)$ is a Gelfand pair, it follows that each irreducible $GL_n(F)$-invariant subspace of polynomial functions on $\mathcal{S}_n(F)$ contains a unique (up to scalar multiples) K_n-invariant polynomial, known as a *(zonal) spherical function* for the pair $(GL_n(F), K_n)$. Furthermore, these spherical functions form a basis for the space of all K_n-invariant polynomials. (For further details, e.g., see [**GR**]).

Macdonald [**M2**] has observed that in each of the cases $F = \mathbf{R}$, $\mathbf{C}$ and $\mathbf{H}$, the spherical functions for the pair $(GL_n(F), K_n)$ may be uniformly described as Jack symmetric functions $J_\lambda(x; \alpha)$ for special instances of the parameter α (depending on F). Although we do not intend to define these functions here,[1] we should explain that for any positive real number α, the Jack symmetric functions $J_\lambda(x; \alpha)$ form a basis for the homogeneous symmetric polynomials of degree k in the variables $x = (x_1, x_2, \ldots)$, where λ ranges over the partitions of k. Furthermore, when J_λ is restricted to n variables (i.e., $x_m = 0$ for all $m > n$), one has $J_\lambda = 0$ if and only if λ has more than n parts, and the remaining nonzero Jack symmetric functions form a basis for the symmetric polynomial functions of $(x_1, \ldots, x_n)$. In these terms, the spherical functions for the pair $(GL_n(F), K_n)$ can be obtained by setting $\alpha = 2$, 1, or $1/2$, according to whether $F = \mathbf{R}$, $\mathbf{C}$, or $\mathbf{H}$.

As a second remark, we should note that although the Jack symmetric functions are well-defined only up to scalar multiplication, we will follow the usual convention and insist that J_λ be normalized so that the coefficient of $x_1 \cdots x_k$ in $J_\lambda(x; \alpha)$ is $k!$, assuming $|\lambda| = k$.

In what follows, we need to extend any K_n-invariant polynomial f on $\mathcal{S}_n(F)$ (or equivalently, any symmetric function) to the full matrix algebra $M_n(F)$, so that expressions such as $f(AUBU^*)$ are well-defined. For this it is enough to decide how to extend the power sums, since these generate all symmetric functions. In the real and complex cases, let us use the obvious extension of (1.1)

---

[1] For this and other details about Jack symmetric functions, see [**St**].

to all of $M_n(F)$; this is consistent with the conventions of the Introduction, where we defined $f(U)$ to be the value of f at the diagonal matrix of eigenvalues of U. In the quaternionic case, more delicacy is required. Indeed, if we attempted to use (1.1) for all of $M_n(\mathbf{H})$, then the value of an arbitrary symmetric function f on $M_n(\mathbf{H})$ would not be well-defined (e.g., $p_r(U)$ and $p_s(U)$ need not commute). To avoid this difficulty, we define

$$(1.2) \qquad\qquad p_k(U) = \Re(\mathrm{tr}(U^k))$$

for all $u \in M_n(\mathbf{H})$. This convention is not entirely *ad hoc*; it derives from the fact that $\Re(\mathrm{tr}(U))$ is one-half the trace of U in the standard embedding $M_n(\mathbf{H}) \hookrightarrow M_{2n}(\mathbf{C})$. In particular, since $2\Re(\mathrm{tr}(\,\cdot\,))$ is therefore a trace over a *complex* matrix algebra, it follows that $f(UV) = f(VU)$ for all symmetric functions f and all $U, V \in M_n(F)$, even in the quaternionic case.

Now fix two matrices $A, B \in \mathcal{S}_n(F)$, and let $U = [u_{ij}]$ be a random $n \times n$ matrix of independent F-valued standard normal variables. The following key property of zonal polynomials follows from the theory of spherical functions and goes back to Gelfand [**G**] and Godement [**Go**, Thm. 10] (see also James [**J**, (29)]). The earliest explicit formulation of it we have found in the form below is in [**T**, p. 31, Thm. 3] (the real case) and [**T**, p. 88, Thm. 4] (the complex case). We know of no previous formulation of this particular version of the quaternionic case, although it follows from the general theory in the same way as the real and complex cases.

1.1 THEOREM. *Let $\alpha = 2, 1$, or $1/2$, according to whether $F = \mathbf{R}, \mathbf{C}$ or $\mathbf{H}$. If λ is any partition, then*

$$\mathcal{E}_U(J_\lambda(AUBU^*;\alpha)) = J_\lambda(A;\alpha)J_\lambda(B;\alpha).$$

SKETCH OF PROOF. Let $U = RX$ denote the "polar coordinate" decomposition of U in which $X \in K_n$ and R is the unique positive semidefinite square root of UU^*. Since the distribution of U is invariant under left and right multiplication by K_n, it follows that for fixed R, X is uniformly distributed on K_n according to Haar measure. We therefore have

$$(1.3)\quad \mathcal{E}_U(f(AUBU^*)) = \mathcal{E}_R(\mathcal{E}_X(f(ARXBX^*R))) = \mathcal{E}_R(\mathcal{E}_X(f(RARXBX^*))),$$

for any K_n-invariant function f on $\mathcal{S}_n(F)$.

However, by the general theory of spherical functions (e.g., [**GR**, Prop. 5.5]), one has

$$\mathcal{E}_X(J_\lambda(AXBX^*;\alpha)) = \frac{J_\lambda(A;\alpha)J_\lambda(B;\alpha)}{J_\lambda(I_n;\alpha)},$$

so (1.3) implies

$$\mathcal{E}_U(J_\lambda(AUBU^*;\alpha)) = \frac{J_\lambda(B;\alpha)}{J_\lambda(I_n;\alpha)}\mathcal{E}_R(J_\lambda(RAR;\alpha)).$$

Now since the expectation of $f(AUBU^*)$ is invariant under interchanging A and B (recall that $f(UV) = f(VU)$), we must therefore have

$$(1.4) \qquad \mathcal{E}_U(J_\lambda(AUBU^*;\alpha)) = c_\lambda J_\lambda(A;\alpha) J_\lambda(B;\alpha)$$

for some scalar c_λ (namely, $\mathcal{E}_U(J_\lambda(UU^*;\alpha))/J_\lambda(I_n;\alpha)^2$).

To determine this scalar, we first note that by bordering A and B with zeroes, one can show that c_λ does not depend on n (cf. [T, p. 32]). We may therefore assume $n \geq k$, where $|\lambda| = k$. Now let $A = \text{diag}(a_1, \ldots, a_n)$ and $B = \text{diag}(b_1, \ldots, b_n)$, and consider the coefficient of $a_1 b_1 \cdots a_k b_k$ in (1.4). On the right side, one obtains $(k!)^2 c_\lambda$, by the normalization convention for J_λ. For the left side, note that among the power sums $p_\mu(x_1, \ldots, x_n)$ with $\mu \vdash k$, the only case for which the coefficient of $x_1 \cdots x_k$ is nonzero is the case $\mu = (1^k)$; in that case, the coefficient is $k!$. Thus when J_λ is expanded in terms of power sums, the coefficient of p_1^k is 1.

Finally, observe that the coefficient of $a_1 b_1 \cdots a_k b_k$ in

$$p_1(AUBU^*)^k = \Big(\sum_{i,j=1}^k a_i b_j u_{ij} \bar{u}_{ij}\Big)^k$$

will be a sum of $(k!)^2$ terms, each of which is of the form $(u_1 \bar{u}_1) \cdots (u_k \bar{u}_k)$, where $u_1, \ldots, u_k$ are independent standard normal variables. Since $\mathcal{E}(u_i \bar{u}_i) = 1$, it will thus follow that the coefficient of $a_1 b_1 \cdots a_k b_k$ on the left side of (1.4) is $(k!)^2$ (and hence, $c_\lambda = 1$), provided that the coefficient of $a_1 b_1 \cdots a_k b_k$ in $\mathcal{E}_U(p_\mu(AUBU^*))$ is zero whenever $\mu \neq (1^k)$. This can be established by elementary methods, as we shall see in §6. $\square$

2. The Complex Case

We first briefly review a few facts about group characters. Let G be a finite group, and let $C_1, \ldots, C_t$ denote the conjugacy classes of G. By abuse of notation, we identify each C_i with its corresponding sum in the complex group algebra $\mathbf{C}G$. These class sums form a basis for the center of $\mathbf{C}G$. If $\chi^{(1)}, \ldots, \chi^{(t)}$ are the irreducible (complex) characters of G and $\deg(\chi^{(i)}) = d_i$, then the elements

$$E_i = \frac{d_i}{|G|} \sum_{j=1}^t \bar{\chi}_j^{(i)} C_j \qquad (1 \leq i \leq t)$$

form a complete set of orthogonal idempotents for the center of $\mathbf{C}G$, where we use $\chi_j^{(i)}$ to denote the common value of $\chi^{(i)}$ at any $w \in C_j$. Inverting this yields

$$C_j = |C_j| \sum_{i=1}^t \frac{\chi_j^{(i)}}{d_i} E_i,$$

by the orthogonality properties of characters [**B**, §236]. Since the E_j's are orthogonal idempotents (i.e., $E_i E_j = \delta_{i,j} E_i$), it follows that

$$
(2.1) \qquad C_i C_j = |C_i| \cdot |C_j| \sum_{r=1}^{t} \frac{\chi_i^{(r)} \chi_j^{(r)}}{d_r^2} E_r
$$

$$
= \frac{|C_i| \cdot |C_j|}{|G|} \sum_{r=1}^{t} \frac{\chi_i^{(r)} \chi_j^{(r)}}{d_r} \sum_{k=1}^{t} \bar{\chi}_k^{(r)} C_k
$$

$$
= \frac{|C_i| \cdot |C_j|}{|G|} \sum_{k=1}^{t} C_k \sum_{r=1}^{t} \frac{1}{d_r} \chi_i^{(r)} \chi_j^{(r)} \bar{\chi}_k^{(r)}.
$$

If we specialize to the symmetric group S_k, then the characters and conjugacy classes are indexed by partitions λ of k; say χ^λ and C_λ, respectively. We will write $\chi^\lambda(\mu)$ for the value of χ^λ at any $w \in C_\mu$ (i.e., any $w \in S_k$ of cycle-type μ).

Let $\tau(w)$ denote the cycle-type of a permutation $w \in S_k$, and for each partition λ of k, choose a fixed element w_λ of C_λ, so that $\tau(w_\lambda) = \lambda$. It is easy to see that

$$
C_\mu C_\nu = \sum_{\lambda} a_{\mu,\nu}^\lambda C_\lambda,
$$

where $a_{\mu,\nu}^\lambda$ is the number of pairs $u, v \in S_k$ such that $\tau(u) = \mu$, $\tau(v) = \nu$ and $uv = w_\lambda$. The quantity $|C_\lambda| a_{\mu,\nu}^\lambda$ has a more symmetric interpretation as the number of triples $u, v, w \in S_k$ such that $uvw = \mathrm{id}$, $\tau(u) = \mu$, $\tau(v) = \nu$, and $\tau(w) = \lambda$.

Let H_λ denote the product of the hook-lengths of λ [**M1**, p.9], so that the degree of χ^λ is given by $f^\lambda = k!/H_\lambda$, and let $z_\lambda = k!/|C_\lambda|$ denote the size of the centralizer of w_λ.

2.1 LEMMA. *We have* $a_{\mu,\nu}^\lambda = z_\mu^{-1} z_\nu^{-1} \sum_\beta H_\beta \, \chi^\beta(\mu) \chi^\beta(\nu) \chi^\beta(\lambda)$.

PROOF. The characters of S_k are all real-valued, so (2.1) implies

$$
(2.2) \qquad C_\mu C_\nu = z_\mu^{-1} z_\nu^{-1} \sum_\lambda C_\lambda \sum_\beta H_\beta \, \chi^\beta(\mu) \chi^\beta(\nu) \chi^\beta(\lambda).
$$

Extract the coefficient of C_λ. $\square$

For each partition λ of k, let s_λ denote the Schur function indexed by λ [**M1**], and recall that the power-sum expansion of s_λ due to Frobenius [**M1**, (7.10)] is given by

$$
(2.3) \qquad s_\lambda = \sum_{\mu \vdash k} z_\mu^{-1} \chi^\lambda(\mu) p_\mu.
$$

Using this, it is easy to reformulate Lemma 2.1 as a symmetric function identity. In fact, the following is a slightly more general result.

2.2 PROPOSITION. *If* $x^{(1)}, \ldots, x^{(l)}$ *are disjoint sets of variables, then*

$$\sum_{\lambda \vdash k} H_\lambda^{l-2} s_\lambda(x^{(1)}) \cdots s_\lambda(x^{(l)}) = \frac{1}{k!} \sum_{\substack{w_1 \cdots w_l = \text{id} \\ \text{in } S_k}} p_{\tau(w_1)}(x^{(1)}) \cdots p_{\tau(w_l)}(x^{(l)}).$$

PROOF. Fix an l-tuple of partitions $\mu^{(1)}, \ldots, \mu^{(l)} \vdash k$, and set $\mu = \mu^{(l)}$. Now extract the coefficient of $p_{\mu^{(1)}}(x^{(1)}) \cdots p_{\mu^{(l)}}(x^{(l)})$ from the left side of the claimed identity, using (2.3). By the obvious extension of (2.2) to multiple products of class sums, one may identify the resulting sum as $1/z_\mu$ times the coefficient of C_μ in $C_{\mu^{(1)}} \cdots C_{\mu^{(l-1)}}$, or equivalently, $1/z_\mu$ times the number of solutions to the equation $w_1 \cdots w_{l-1} = w_\mu$ with $\tau(w_i) = \mu^{(i)}$. However, $|C_\mu| = k!/z_\mu$, so this is also $1/k!$ times the number of solutions to the equation $w_1 \cdots w_{l-1} = w_l$ with $\tau(w_i) = \mu^{(i)}$. $\square$

Returning to the main theme of this paper, let U be an $n \times n$ matrix whose entries are independent standard normal complex variables. For any pair of $n \times n$ Hermitian matrices A and B, define

$$P_\lambda^{\mathbf{C}}(A, B) = \mathcal{E}_U(p_\lambda(AUBU^*)).$$

The main result of this section is as follows.

2.3 THEOREM. *If* $\lambda \vdash k$, *then*

$$P_\lambda^{\mathbf{C}}(A, B) = \sum_{\mu, \nu \vdash k} a_{\mu, \nu}^\lambda p_\mu(A) p_\nu(B) = \sum_{w \in S_k} p_{\tau(w)}(A) p_{\tau(ww_\lambda)}(B).$$

PROOF. The two expressions for $P_\lambda^{\mathbf{C}}(A, B)$ are clearly equivalent, so it suffices to prove the first one. For this, we begin with the well-known fact that the Schur functions are known to be scalar multiples of the Jack symmetric functions at $\alpha = 1$; more precisely, $J_\lambda(x; 1) = H_\lambda s_\lambda$ [**St**, Prop. 1.2]. By Theorem 1.1, it follows that

$$\mathcal{E}_U(s_\lambda(AUBU^*)) = H_\lambda s_\lambda(A) s_\lambda(B).$$

Now since $p_\lambda = \sum_\beta \chi^\beta(\lambda) s_\beta$ [**M1**, (7.8)], the linearity of expectation yields

(2.4) $$\mathcal{E}_U(p_\lambda(AUBU^*)) = \sum_\beta H_\beta \chi^\beta(\lambda) s_\beta(A) s_\beta(B).$$

In view of (2.3), we therefore have

$$P_\lambda^{\mathbf{C}}(A, B) = \sum_\beta H_\beta \chi^\beta(\lambda) \sum_{\mu, \nu} z_\mu^{-1} z_\nu^{-1} \chi^\beta(\mu) \chi^\beta(\nu) p_\mu(A) p_\nu(B)$$

$$= \sum_{\mu, \nu} p_\mu(A) p_\nu(B) z_\mu^{-1} z_\nu^{-1} \sum_\beta H_\beta \chi^\beta(\lambda) \chi^\beta(\mu) \chi^\beta(\nu).$$

Apply Lemma 2.1 to complete the proof. $\square$

Now consider the consequences of setting $A = I_m \oplus O_{n-m}$ and $B = I_n$, as discussed in the Introduction. If $c(w)$ denotes the number of cycles of a permutation w (i.e., the number of parts in $\tau(w)$), then clearly $p_{\tau(w)}(I_m \oplus O_{n-m}) = m^{c(w)}$ and $p_{\tau(w)}(I_n) = n^{c(w)}$. Therefore, if we define

$$Q_\lambda^{\mathbf{C}}(m,n) = \mathcal{E}_U(p_\lambda(AUBU^*)) = \mathcal{E}_V(p_\lambda(VV^*)),$$

where V is a random $m \times n$ matrix with independent standard complex normal entries, we obtain the following.

2.4 COROLLARY. *If* $\lambda \vdash k$, *then* $Q_\lambda^{\mathbf{C}}(m,n) = \sum_{w \in S_k} m^{c(w)} n^{c(ww_\lambda)}$.

Thus $Q_\lambda^{\mathbf{C}}(m,n)$ is a polynomial function of m and n with nonnegative integer coefficients. Note that the coefficient of $m^i n^j$ can also be interpreted as the number of pairs $u, v \in S_k$ such that $c(u) = i$, $c(v) = j$, and $uv = w_\lambda$.

It is a well-known fact that $d(u,v) := k - c(u^{-1}v)$ defines a metric on S_k. Indeed, $d(u,v)$ is the distance from u to v in the Cayley graph of S_k generated by transpositions. In particular, by the triangle inequality, we have $d(\mathrm{id}, w_\lambda) \leq d(\mathrm{id}, w^{-1}) + d(w^{-1}, w_\lambda)$, so

(2.5) $c(w) + c(ww_\lambda) \leq k + c(w_\lambda) = k + \ell(\lambda)$,

where $\ell(\lambda)$ denotes the number of parts of λ. Hence, the total degree of $Q_\lambda^{\mathbf{C}}(m,n)$ is at most $k + \ell(\lambda)$. In fact, we claim that the total degree is *equal* to $k + \ell(\lambda)$. To see this, consider any shortest path from the identity to w_λ (necessarily of length $k - \ell(\lambda)$) in the Cayley graph of S_k. The permutation u at distance j along this path generates a solution to the equation with $uv = w_\lambda$ with $c(u) = k - j$ and $c(v) = j + \ell(\lambda)$, so it follows that the coefficient of $m^i n^{k+\ell(\lambda)-i}$ is nonzero whenever $\ell(\lambda) \leq i \leq k$ (and this condition is obviously necessary).

The case $\lambda = (k)$ is the one of most interest to statisticians. For example, by a result of Wigner (see [**OU**, p. 623]), it is known that

$$[m^i n^{k+1-i}] Q_k^{\mathbf{C}}(m,n) = \frac{1}{k}\binom{k}{i}\binom{k}{i-1},$$

where $[\,\cdot\,]$ denotes the operation of coefficient extraction. In fact, this result holds even for more general classes of random matrices than those we are considering here.

2.5 THEOREM. *We have*

$$Q_k^{\mathbf{C}}(m,n) = \frac{1}{k}\sum_{i=1}^k (-1)^{i-1}\frac{(m+k-i)_k(n+k-i)_k}{(k-i)!(i-1)!},$$

where $(a)_k := a(a-1)\cdots(a-k+1)$.

PROOF. Substituting $A = I_m \oplus O_{n-m}$ and $B = I_n$ into (2.4), we obtain

$$Q_\lambda^{\mathbf{C}}(m,n) = \sum_{\beta \vdash k} H_\beta \chi^\beta(\lambda) s_\beta(I_m) s_\beta(I_n).$$

δ_1 δ_2

FIGURE 1

$\delta_1 \cup \delta_2$

FIGURE 2

By the Murnaghan-Nakayama rule for $\chi^\beta(\lambda)$ [**M1**, Ex. I.7.5], it is known that

$$(2.6) \qquad \chi^\beta(k) = \begin{cases} (-1)^{i-1}, & \text{if } \beta = (k-i+1, 1^{i-1}) \\ 0, & \text{otherwise.} \end{cases}$$

When $\beta = (k - i + 1, 1^{i-1})$, it is easy to show that $H_\beta = k(k-i)!(i-1)!$ and $s_\beta(I_m) = (m+k-i)_k/H_\beta$, using [**M1**, Ex. I.3.4]. This yields the claimed identity. □

3. The Real Case

Let $\mathcal{F}_k$ denote the set of 1-factors (or perfect matchings, or 1-regular graphs) on the vertices $\{1, 2, \ldots, 2k\}$. Let $\varepsilon \in \mathcal{F}_k$ be the "identity 1-factor" in which i is adjacent to $k+i$ for $1 \leq i \leq k$.

For each $w \in S_{2k}$, define $\delta(w)$ to be the 1-factor in which i is adjacent to j if and only if $|w(i) - w(j)| = k$. Note that $\delta(\mathrm{id}) = \varepsilon$. The union $\delta_1 \cup \delta_2$ of two 1-factors is a 2-regular graph, and therefore a disjoint union of even-length cycles. Let $\Lambda(\delta_1, \delta_2)$ denote the partition of k whose parts are half the cycle-lengths of $\delta_1 \cup \delta_2$. For example, in Figure 1 are two 1-factors δ_1 and δ_2 whose union (Figure 2) consists of a 6-cycle and a 4-cycle; thus, $\Lambda(\delta_1, \delta_2) = (3, 2)$.

Let B_k denote the hyperoctahedral group, embedded in S_{2k} as the centralizer of the involution $(1, k+1)(2, k+2) \cdots (k, 2k)$. Note that B_k is the automorphism group of ε. The following well-known result shows that the right cosets of B_k in S_{2k} are indexed by 1-factors $\delta \in \mathcal{F}_k$, and that the double cosets $B_k \backslash S_{2k} / B_k$ are indexed by partitions of k.

3.1 LEMMA. *Let* $w_1, w_2 \in S_{2k}$.

(a) $B_k w_1 = B_k w_2$ *if and only if* $\delta(w_1) = \delta(w_2)$.
(b) $B_k w_1 B_k = B_k w_2 B_k$ *if and only if* $\Lambda(\varepsilon, \delta(w_1)) = \Lambda(\varepsilon, \delta(w_2))$.

PROOF. (a) Clearly, left multiplication by B_k preserves $\delta(w)$. On the other hand, the number of $w \in S_{2k}$ that fix any given 1-factor is clearly $|B_k|$, so the 1-factors must completely separate the right cosets of B_k in S_{2k}.

(b) (See also [M2], [BG], [Ste].) Right multiplication by B_k amounts to simultaneous permutation of the vertices $\{1, \ldots, k\}$ and $\{k+1, \ldots, 2k\}$ of $\delta(w)$, along with interchanging the vertices i and $k+i$. In particular, these operations preserve the isomorphism class of $\varepsilon \cup \delta(w)$ (or equivalently, preserve $\Lambda(\varepsilon, \delta(w))$). Conversely, the cycles of $\varepsilon \cup \delta(w)$ must alternate between edges of ε and $\delta(w)$, so if $\varepsilon \cup \delta(w_1) \cong \varepsilon \cup \delta(w_2)$, then there is an isomorphism that preserves ε; i.e., there exists $x \in B_k$ such that $\delta(w_1 x) = \delta(w_2)$. By (a), it follows that $B_k w_1 B_k = B_k w_2 B_k$. $\square$

For each partition λ of k, let K_λ denote the double coset consisting of all $w \in S_{2k}$ with $\Lambda(\varepsilon, \delta(w)) = \lambda$, and choose a fixed representative $\hat{w}_\lambda$ from K_λ. Define $\delta_\lambda := \delta(\hat{w}_\lambda)$. By analogy with the conventions of §2, we will identify K_λ with its sum in $\mathbf{C}S_{2k}$. These sums form a basis for a certain commutative subalgebra of $\mathbf{C}S_{2k}$; namely, the Hecke algebra of the Gelfand pair (S_{2k}, B_k) (see [BG], [Ste]). In particular, it is easy to see that

$$(3.1) \qquad\qquad K_\mu K_\nu = \sum_\lambda b_{\mu,\nu}^\lambda K_\lambda,$$

where $b_{\mu,\nu}^\lambda$ is the number of $u, v \in S_{2k}$ such that $u \in K_\mu$, $v \in K_\nu$ and $uv = \hat{w}_\lambda$.

3.2 LEMMA. *If $\lambda, \mu, \nu \vdash k$, then $|B_k|^{-1} b_{\mu,\nu}^\lambda$ is the number of 1-factors δ such that $\Lambda(\delta, \delta_\lambda) = \mu$ and $\Lambda(\varepsilon, \delta) = \nu$.*

PROOF. For each 1-factor δ, choose a representative $v_\delta \in S_{2k}$ of the right coset of B_k indexed by δ. Since $uv_\delta = (ux^{-1})(xv_\delta)$ as x varies through B_k, it follows that $|B_k|^{-1} b_{\mu,\nu}^\lambda$ is the number of 1-factors δ such that the equation $uv_\delta = \hat{w}_\lambda$ has a solution with $u \in K_\mu$ and $v_\delta \in K_\nu$. In other words, since u must be $\hat{w}_\lambda v_\delta^{-1}$, $|B_k|^{-1} b_{\mu,\nu}^\lambda$ is the number of 1-factors δ such that $\Lambda(\varepsilon, \delta(\hat{w}_\lambda v_\delta^{-1})) = \mu$ and $\Lambda(\varepsilon, \delta) = \nu$. To complete the proof, note that $\Lambda(\varepsilon, \delta(\hat{w}_\lambda v_\delta^{-1})) = \Lambda(\delta, \delta(\hat{w}_\lambda))$, since for any $w \in S_{2k}$, the operation $\delta(u) \cup \delta(v) \mapsto \delta(uw) \cup \delta(vw)$ preserves the isomorphism class of $\delta(u) \cup \delta(v)$. $\square$

We now need to review a few facts about the Hecke algebra $\mathcal{H}_k$ spanned by the K_λ's. For further details, see [BG] or [Ste]. Since left and right multiplication by the idempotent $e = |B_k|^{-1} \sum_{x \in B_k} x \in \mathbf{C}S_{2k}$ corresponds to averaging over the double cosets $B_k \backslash S_{2k} / B_k$, it follows that

$$\mathcal{H}_k = e\mathbf{C}S_{2k}e.$$

Furthermore, since $\mathcal{H}_k$ is commutative and semisimple, it has a basis of orthogonal idempotents. In fact, since the induction of the trivial character of B_k to S_{2k} is the multiplicity-free sum of the characters $\chi^{2\lambda}$ as λ varies over the partitions of k,[2] it follows that the orthogonal idempotents of $\mathcal{H}_k$ are

$$(3.2) \qquad E_\lambda := ee_{2\lambda}e = e_{2\lambda}e,$$

where λ ranges over partitions of k, and $e_{2\lambda} = H_{2\lambda}^{-1} \sum_{\mu \vdash 2k} \chi^{2\lambda}(\mu) C_\mu$ denotes the primitive central idempotent of CS_{2k} indexed by 2λ.

The following result is analogous to Lemma 2.1.

3.3 LEMMA. *We have*

$$b^\lambda_{\mu,\nu} = \frac{1}{|K_\lambda|} \sum_{\beta \vdash k} \frac{1}{H_{2\beta}} \varphi^\beta(\lambda) \varphi^\beta(\mu) \varphi^\beta(\nu),$$

where $\varphi^\beta(\mu) = \sum_{w \in K_\mu} \chi^{2\beta}(w)$.

PROOF. Extend the characters of S_{2k} linearly so that $\chi^{2\beta}$ is the trace of an irreducible representation of CS_{2k}. Since E_μ acts as a rank one idempotent in the representation of CS_{2k} indexed by 2β, it follows that $\chi^{2\beta}(E_\mu) = \delta_{\beta,\mu}$. We may therefore deduce that

$$(3.3) \qquad K_\mu = \sum_{\beta \vdash k} \varphi^\beta(\mu) E_\beta,$$

by applying $\chi^{2\beta}$ to both sides.

Since the E_β's are orthogonal idempotents, it follows that

$$(3.4) \qquad K_\mu K_\nu = \sum_{\beta \vdash k} \varphi^\beta(\mu) \varphi^\beta(\nu) E_\beta.$$

However, (3.2) implies

$$E_\lambda = \frac{1}{|B_k|^2} \sum_{x_1, x_2 \in B_k} \frac{1}{H_{2\lambda}} \sum_{w \in S_{2k}} \chi^{2\lambda}(w) x_1 w x_2,$$

so the coefficient of $\hat{w}_\lambda$ in E_β is

$$\frac{1}{H_{2\beta}} \cdot \frac{1}{|B_k|^2} \sum_{x_1, x_2 \in B_k} \chi^{2\beta}(x_1 \hat{w}_\lambda x_2) = \frac{1}{H_{2\beta}} \cdot \frac{1}{|K_\lambda|} \varphi^\beta(\lambda).$$

In other words,

$$(3.5) \qquad E_\beta = \frac{1}{H_{2\beta}} \sum_{\lambda \vdash k} \frac{1}{|K_\lambda|} \varphi^\beta(\lambda) K_\lambda.$$

Using this to extract the coefficient of K_λ from (3.4) yields the claimed result. $\square$

For each partition λ, let $Z_\lambda(x) = J_\lambda(x; 2)$ denote the Jack symmetric function at $\alpha = 2$; these are the classical zonal polynomials first studied by James and Hua. The power-sum expansion of Z_λ, due to James [J, Thm. 4] (cf. also [M2], [BG], and [Ste]), can be described as follows.

---

[2] This can be deduced from [M1, Ex. I.5.5] and [M1, Remark 2, p. 66], for example.

3.4 THEOREM. *For any $\lambda \vdash k$, we have*

$$Z_\lambda = \frac{1}{|B_k|} \sum_{\mu \vdash k} \varphi^\lambda(\mu) p_\mu = \frac{1}{|B_k|} \sum_{w \in S_{2k}} \chi^{2\lambda}(w) p_{\Lambda(\epsilon, \delta(w))}.$$

Now let U be a random $n \times n$ matrix whose entries are independent standard normal real variables, and for any $n \times n$ symmetric matrices A and B, define

$$P_\lambda^{\mathbf{R}}(A, B) = \mathcal{E}_U(p_\lambda(AUBU^t)).$$

The main result of this section is as follows.

3.5 THEOREM. *If $\lambda \vdash k$, then*

$$P_\lambda^{\mathbf{R}}(A, B) = \frac{1}{|B_k|} \sum_{\mu, \nu \vdash k} b_{\mu,\nu}^\lambda p_\mu(A) p_\nu(B) = \sum_{\delta \in \mathcal{F}_k} p_{\Lambda(\epsilon, \delta)}(A) p_{\Lambda(\delta, \delta_\lambda)}(B).$$

PROOF. Lemma 3.2 implies that the two expressions for $P_\lambda^{\mathbf{R}}(A, B)$ are equivalent, so it suffices to prove the first one.

If we define a linear transformation from $\mathcal{H}_k$ to symmetric functions of degree k by setting $K_\nu \mapsto |K_\nu| p_\nu$, then (3.5) and Theorem 3.4 imply $E_\beta \mapsto |B_k| H_{2\beta}^{-1} Z_\beta$. Therefore by (3.3), we have

$$(3.6) \qquad\qquad p_\lambda = \frac{1}{|K_\lambda|} \sum_{\beta \vdash k} \frac{|B_k|}{H_{2\beta}} \varphi^\beta(\lambda) Z_\beta.$$

Since expectation is linear, this identity and Theorem 1.1 imply

$$P_\lambda^{\mathbf{R}}(A, B) = \frac{1}{|K_\lambda|} \sum_{\beta \vdash k} \frac{|B_k|}{H_{2\beta}} \varphi^\beta(\lambda) Z_\beta(A) Z_\beta(B).$$

Using Theorem 3.4 to expand $Z_\beta(A)$ and $Z_\beta(B)$ therefore yields

$$P_\lambda^{\mathbf{R}}(A, B) = \frac{1}{|K_\lambda|} \sum_{\beta \vdash k} \frac{1}{H_{2\beta}} \cdot \frac{\varphi^\beta(\lambda)}{|B_k|} \sum_{\mu, \nu \vdash k} \varphi^\beta(\mu) \varphi^\beta(\nu) p_\mu(A) p_\nu(B).$$

Lemma 3.3 can now be used to extract the coefficient of $p_\mu(A) p_\nu(B)$. □

Now consider setting $A = I_m \oplus O_{n-m}$ and $B = I_n$, as discussed in the Introduction. By analogy with the notation of §2, let $c(\delta_1 \cup \delta_2)$ denote the number of cycles in the graph $\delta_1 \cup \delta_2$ (i.e., the number of parts in $\Lambda(\delta_1, \delta_2)$), and define

$$Q_\lambda^{\mathbf{R}}(m, n) = \mathcal{E}_U(p_\lambda(AUBU^t)) = \mathcal{E}_V(p_\lambda(VV^t)),$$

where V is a random $m \times n$ matrix with independent standard real normal entries. The counterpart of Corollary 2.4 is as follows.

3.6 COROLLARY. *If* $\lambda \vdash k$, *then* $Q_\lambda^{\mathbf{R}}(m,n) = \sum_{\delta \in \mathcal{F}_k} m^{c(\varepsilon \cup \delta)} n^{c(\delta \cup \delta_\lambda)}$.

Thus $Q_\lambda^{\mathbf{R}}(m,n)$ is a polynomial function of m and n with nonnegative integer coefficients. We claim that the total degree of $Q_\lambda^{\mathbf{R}}(m,n)$ is $k + \ell(\lambda)$ (the same as $Q_\lambda^{\mathbf{C}}(m,n)$). To see this, define a graph structure on $\mathcal{F}_k$ by declaring δ_1 adjacent to δ_2 if $c(\delta_1 \cup \delta_2) = k - 1$, or equivalently, $\Lambda(\delta_1, \delta_2) = (2, 1^{k-2})$. Note that $k - c(\delta_1 \cup \delta_2)$ is a metric on $\mathcal{F}_k$, since it is the distance from δ_1 to δ_2 in this graph. In particular, the triangle inequality implies

$$c(\varepsilon \cup \delta) + c(\delta \cup \delta_\lambda) \leq k + c(\varepsilon \cup \delta_\lambda) = k + \ell(\lambda),$$

so the total degree of $Q_\lambda^{\mathbf{R}}(m,n)$ is at most $k + \ell(\lambda)$. Conversely, one can show that the coefficient of $m^i n^{k+\ell(\lambda)-i}$ is nonzero for $\ell(\lambda) \leq i \leq k$ by an argument similar to the one we used for $Q_\lambda^{\mathbf{C}}(m,n)$ in §2. In fact, one can also prove that the subgraph of shortest paths from ε to δ_λ is isomorphic to the subgraph of shortest paths from the identity to w_λ in the Cayley graph of S_k (cf. §2), so it follows that the terms of highest total degree in $Q_\lambda^{\mathbf{R}}(m,n)$ and $Q_\lambda^{\mathbf{C}}(m,n)$ are identical.

4. The Quaternionic Case

Let u be a standard normal quaternionic variable, as in §1. By a simple calculation, one can show that $\mathcal{E}((u\bar{u})^k) = (k+1)!/2^k$ for any nonnegative integer k. It follows that

(4.1) $$\mathcal{E}((v\bar{v})^k) = (k+1)!$$

for the random variable $v = \sqrt{2}u$. In order to to avoid non-integral coefficients in what follows, we will therefore assume that $U = [u_{ij}]$ is an $n \times n$ matrix of random independent quaternionic variables, each distributed identically to v, rather than u.

Given this modified distribution for U, let us define

$$P_\lambda^{\mathbf{H}}(A, B) = \mathcal{E}_U(p_\lambda(AUBU^*)),$$

for any $n \times n$ Hermitian matrices A and B (over $\mathbf{H}$).

In order to use Theorem 1.1 to give an explicit formula for $P_\lambda^{\mathbf{H}}(A, B)$, we will need the power-sum expansion for the Jack symmetric functions at $\alpha = 1/2$. For this there is a dual relationship, due to Ian Macdonald [St, Thm. 3.3], between the Jack symmetric functions for the parameters α and $1/\alpha$. To describe this relationship, let ω_α denote the unique automorphism of the ring of symmetric functions satisfying $\omega_\alpha(p_k) = \alpha p_k$ for $k \geq 1$, and let λ' denote the conjugate of λ [M1, (1.3)]. In these terms, we have

(4.2) $$J_\lambda(x; 1/\alpha) = (-1/\alpha)^k \omega_{-\alpha} J_{\lambda'}(x; \alpha),$$

if λ is a partition of k. This dual relationship will allow us to express $P_\lambda^{\mathbf{H}}(A, B)$ in terms of its real counterpart $P_\lambda^{\mathbf{R}}(A, B)$.

4.1 THEOREM. *If $\lambda \vdash k$, then $P_\lambda^{\mathbf{H}}(A, B) = |B_k|^{-1} \sum_{\mu, \nu \vdash k} c_{\mu, \nu}^\lambda p_\mu(A) p_\nu(B)$, where*

$$c_{\mu, \nu}^\lambda = (-1)^k (-2)^{\ell(\mu) + \ell(\nu) - \ell(\lambda)} b_{\mu, \nu}^\lambda,$$

and $b_{\mu, \nu}^\lambda$ is defined as in (3.1).

PROOF. Taking into account the modified distribution for U, Theorem 1.1 implies

$$\mathcal{E}_U(J_\lambda(AUBU^*; 1/2)) = 2^k J_\lambda(A; 1/2) J_\lambda(B; 1/2),$$

for any partition λ of k. However, if we use (4.2) to apply ω_{-2} to (3.6), we obtain

$$p_\lambda(x) = \frac{(-2)^{k - \ell(\lambda)}}{|K_\lambda|} \sum_{\beta \vdash k} \frac{|B_k|}{H_{2\beta}} \varphi^\beta(\lambda) J_{\beta'}(x; 1/2),$$

and therefore

$$P_\lambda^{\mathbf{H}}(A, B) = (-1)^k \frac{(-2)^{2k - \ell(\lambda)}}{|K_\lambda|} \sum_{\beta \vdash k} \frac{|B_k|}{H_{2\beta}} \varphi^\beta(\lambda) J_{\beta'}(A; 1/2) J_{\beta'}(B; 1/2).$$

Now by Theorem 3.4 and a second application of (4.2), we have

$$J_{\beta'}(x; 1/2) = (-1/2)^k \omega_{-2} Z_\beta(x) = \frac{1}{|B_k|} \sum_{\mu \vdash k} (-1/2)^{k - \ell(\mu)} \varphi^\beta(\mu) p_\mu(x).$$

Using this to expand $J_{\beta'}(A; 1/2) J_{\beta'}(B; 1/2)$, we obtain

$$P_\lambda^{\mathbf{H}}(A, B) = \frac{(-1)^k}{|K_\lambda|} \sum_{\beta \vdash k} \frac{1}{H_{2\beta}} \cdot \frac{\varphi^\beta(\lambda)}{|B_k|}$$

$$\times \sum_{\mu, \nu \vdash k} (-2)^{\ell(\mu) + \ell(\nu) - \ell(\lambda)} \varphi^\beta(\mu) \varphi^\beta(\nu) p_\mu(A) p_\nu(B).$$

By Lemma 3.3, the coefficient of $p_\mu(A) p_\nu(B)$ in this expression is $|B_k|^{-1} c_{\mu, \nu}^\lambda$. $\square$

Now consider the special case $A = I_m \oplus O_{n-m}$, $B = I_n$, and define

$$Q_\lambda^{\mathbf{H}}(m, n) = \mathcal{E}_U(p_\lambda(AUBU^*)) = \mathcal{E}_V(p_\lambda(VV^*)),$$

where V is an $m \times n$ matrix of independent quaternionic random variables, each distributed identically to v. An immediate consequence of Theorem 4.1 is the fact that $Q_\lambda^{\mathbf{H}}(m, n)$ can be easily expressed in terms of its real counterpart $Q_\lambda^{\mathbf{R}}(m, n)$.

4.2 COROLLARY. *If $\lambda \vdash k$, then $Q_\lambda^{\mathbf{H}}(m, n) = (-1)^{k + \ell(\lambda)} 2^{-\ell(\lambda)} Q_\lambda^{\mathbf{R}}(-2m, -2n)$.*

Since the terms of highest total degree in $Q_\lambda^{\mathbf{R}}(m, n)$ are contributed by the terms $p_\mu(A) p_\nu(B)$ with $\ell(\mu) + \ell(\nu) = k + \ell(\lambda)$, it follows that $c_{\mu, \nu}^\lambda = 2^k b_{\mu, \nu}^\lambda$ in such cases. In other words, the terms of highest total degree in $Q_\lambda^{\mathbf{H}}(m, n)$ are 2^k times the corresponding terms in $Q_\lambda^{\mathbf{R}}(m, n)$.

The following result shows that $|B_k|^{-1} c_{\mu, \nu}^\lambda$ is an integer divisible by $2^{\ell(\mu)}$; in particular, it follows that $Q_\lambda^{\mathbf{H}}(m, n)$ has integer coefficients.

4.3 PROPOSITION. *If* $\lambda, \mu, \nu \vdash k$, *then* $|B_k|^{-1} 2^{\ell(\nu) - \ell(\lambda)} b^\lambda_{\mu, \nu}$ *is an integer.*

PROOF. Let $\mathcal{B}^\lambda_{\mu, \nu}$ be the set of 1-factors δ with $\Lambda(\delta, \delta_\lambda) = \mu$ and $\Lambda(\delta, \varepsilon) = \nu$. We know that $|\mathcal{B}^\lambda_{\mu, \nu}| = |B_k|^{-1} b^\lambda_{\mu, \nu}$, by Lemma 3.2. Now if $w \in S_{2k}$ is any permutation that preserves both ε and δ_λ, then $\mathcal{B}^\lambda_{\mu, \nu}$ will be stable under the action of w. Thus it suffices to find a subgroup G of $\operatorname{Aut}(\varepsilon) \cap \operatorname{Aut}(\delta_\lambda)$ with the property that the orbit of any 1-factor δ is divisible by $2^{\ell(\lambda) - c(\varepsilon \cup \delta)}$.

For this, consider the cycles $C_1, \ldots, C_l$ of $\varepsilon \cup \delta_\lambda$ (where $l = \ell(\lambda)$), labeled so that C_i is a cycle of length $2\lambda_i$. For each cycle C_i, let $w_i \in S_{2k}$ be the permutation that interchanges each pair of vertices in C_i at distance λ_i, and fixes the remaining vertices. It is easy to see that w_i preserves both ε and δ_λ. Now let G be the group (isomorphic to the direct product of l copies of $\mathbf{Z}_2$) generated by $w_1, \ldots, w_l$.

To prove that this group satisfies the necessary properties, choose a 1-factor δ, and let c denote the number of connected components of $\varepsilon \cup \delta \cup \delta_\lambda$. The G-stabilizer of δ must be a divisor of 2^c, so the G-orbit of δ will be a multiple of 2^{l-c}. However, the number of connected components of $\varepsilon \cup \delta$ is $c(\varepsilon \cup \delta)$, so we have $l - c \geq l - c(\varepsilon \cup \delta)$. $\square$

As a final remark, note that (4.1) implies $Q^{\mathbf{H}}_\lambda(1, 1) = (k+1)!$, or equivalently,

$$Q^{\mathbf{R}}_\lambda(-2, -2) = (-1)^{k + \ell(\lambda)} 2^{\ell(\lambda)} (k+1)!.$$

It would be interesting to give a combinatorial proof of this based on Corollary 3.6.

5. An Explicit Formula for $Q^{\mathbf{R}}_k(m, n)$

In this section, we will derive an explicit Jack symmetric function expansion for p_k; this will lead to an explicit formula for $Q^{\mathbf{R}}_k(m, n)$ analogous to the formula for $Q^{\mathbf{C}}_k(m, n)$ we gave in Theorem 2.5. In view of Corollary 4.2, this also yields a formula for $Q^{\mathbf{H}}_k(m, n)$. Alternatively, one could use this technique to give a uniform treatment of all three cases.

To begin, let us introduce a scalar product $\langle \cdot, \cdot \rangle_\alpha$ on the ring of symmetric functions (say with real coefficients) by setting

$$(5.1) \qquad \langle p_\mu, p_\nu \rangle_\alpha = z_\mu \alpha^{\ell(\mu)} \delta_{\mu, \nu}.$$

It is known that the Jack symmetric functions $J_\lambda = J_\lambda(x; \alpha)$ are orthogonal with respect to this inner product; in fact, by Theorem 5.8 of [St], we have

$$\langle J_\lambda, J_\mu \rangle_\alpha = j_\lambda(\alpha) \delta_{\lambda, \mu},$$

where

$$(5.2) \qquad j_\lambda(\alpha) = \prod_{(i,j) \in \lambda} (\lambda'_j - i + \alpha(\lambda_i - j + 1))(\lambda'_j - i + 1 + \alpha(\lambda_i - j)).$$

Here we identify λ with its diagram $\{(i, j) \in \mathbf{Z}^2 : 1 \leq i \leq \ell(\lambda), 1 \leq j \leq \lambda_i\}$.

By Theorem 5.4 of [St], there is an explicit formula for the value of J_λ at the identity matrix; namely,

$$(5.3) \qquad J_\lambda(I_n; \alpha) = \prod_{(i,j) \in \lambda} (n - (i-1) + \alpha(j-1)).$$

5.1 LEMMA. *We have*

$$p_k(x) = k\alpha \sum_{\lambda \vdash k} \frac{1}{j_\lambda(\alpha)} J_\lambda(x; \alpha) \prod_{\substack{(i,j) \in \lambda \\ (i,j) \neq (1,1)}} (\alpha(j-1) - (i-1)).$$

PROOF. Let $\psi_\mu^\lambda(\alpha)$ denote the coefficient of p_μ in J_λ, so that $J_\lambda = \sum \psi_\mu^\lambda(\alpha) p_\mu$. In view of (5.3), we have

$$\prod_{(i,j) \in \lambda} (n - (i-1) + \alpha(j-1)) = \sum_{\mu \vdash k} \psi_\mu^\lambda(\alpha) n^{\ell(\mu)}.$$

Note that both sides of this identity are polynomials in n. Since the term on the left side indexed by $(i, j) = (1, 1)$ is n, it follows that if we extract the coefficient of n from both sides, we obtain

$$(5.4) \qquad \psi_k^\lambda(\alpha) = \prod_{\substack{(i,j) \in \lambda \\ (i,j) \neq (1,1)}} (\alpha(j-1) - (i-1)).$$

Now let $c_\lambda(\alpha)$ denote the coefficient of J_λ in p_k, so that

$$(5.5) \qquad p_k(x) = \sum_{\lambda \vdash k} c_\lambda(\alpha) J_\lambda(x; \alpha).$$

Taking the scalar product of both sides with J_λ yields

$$\langle J_\lambda, p_k \rangle_\alpha = c_\lambda(\alpha) \langle J_\lambda, J_\lambda \rangle_\alpha = c_\lambda(\alpha) j_\lambda(\alpha).$$

On the other hand, (5.1) implies

$$\langle J_\lambda, p_k \rangle_\alpha = k\alpha \psi_k^\lambda(\alpha),$$

so we have

$$c_\lambda(\alpha) = k\alpha j_\lambda(\alpha)^{-1} \psi_k^\lambda(\alpha).$$

The result is now an immediate consequence of (5.2) and (5.4). $\square$

5.2 COROLLARY. *If $\lambda \vdash k$ and α is a positive integer, then $c_\lambda(\alpha)$ (the coefficient of J_λ in p_k) is zero if $\lambda_{\alpha+1} \geq 2$. In other words, if $c_\lambda(\alpha) \neq 0$, then $(\alpha + 1, 2) \notin \lambda$.*

PROOF. If $(\alpha + 1, 2) \in \lambda$, then (5.4) vanishes. $\square$

Note that when $\alpha = 1$, we have $c_\lambda(1) \neq 0$ only if λ is of the form $(i, 1^{k-i})$ (i.e., a hook). In this case, it is easy to show that Lemma 5.1 is equivalent to (2.6).

It would be interesting to generalize Lemma 5.1 to arbitrary power sums p_μ. To do so, it would suffice to give a rule for the J_λ-expansion of $p_k J_\mu$; this would amount to a generalization of the Murnaghan-Nakayama rule [M1, Ex. I.3.11] to

Jack symmetric functions. Although this is still an open problem, one can show that the coefficient of J_λ in $p_k J_\mu$ is nonzero only if $\lambda \supseteq \mu$ (diagram inclusion) and the skew diagram $\lambda - \mu$ contains no $(\alpha + 1) \times 2$ rectangle (assuming α is a positive integer).

Returning to the problem of determining a formula for $Q_k^{\mathbf{R}}(m, n)$, note that by Corollary 5.2, the partitions $\lambda \vdash k$ for which $c_\lambda(2) \neq 0$ are of the form $(a, b, 1^{k-a-b})$, where $a \geq b \geq 1$ or $a = k$, $b = 0$. The following formulas for $c_\lambda(2)$ are easily obtained by specializing Lemma 5.1; we leave the details to the reader.

5.3 LEMMA.

(a) *If* $\lambda = (a, b, 1^{k-a-b})$ *and* $a \geq b \geq 1$, *then*[3]

$$c_\lambda(2) = (-1)^k \frac{(-2)^{a-b+1} k(2a - 2b + 1)(a - 1)!}{(k + a - b + 1)_2 (k - a + b)_2 (k - a - b)!(2a - 1)!(b - 1)!}.$$

(b) *If* $\lambda = (k)$, *then* $c_\lambda(2) = 2^k k!/(2k)! = 1/(1 \cdot 3 \cdots (2k - 1))$. $\square$

Now if $\lambda = (a, b, 1^{k-a-b})$ (where $a \geq b \geq 1$ or $a = k$, $b = 0$), let us define

$$F_\lambda(x) = 2^{a-b}(\frac{x}{2} + a - 1)_{a-b}(x + 2b - 2)_{k-a+b}.$$

5.4 THEOREM. *We have* $Q_k^{\mathbf{R}}(m, n) = \sum_\lambda c_\lambda(2) F_\lambda(m) F_\lambda(n)$, *where* λ *ranges over all partitions of the form* $(a, b, 1^{k-a-b})$, *and* $c_\lambda(2)$ *is given by Lemma 5.3.*

PROOF. Proceed by analogy with the proof of Theorem 2.5. If we apply (5.5) and Theorem 1.1 in the case $\alpha = 2$, we obtain

$$Q_k^{\mathbf{R}}(m, n) = \sum_{\lambda \vdash k} c_\lambda(2) J_\lambda(I_m; 2) J_\lambda(I_n; 2),$$

by the linearity of expectation. However, if $\lambda = (a, b, 1^{k-a-b})$, then we have $J_\lambda(I_n; 2) = F_\lambda(n)$, by (5.3). Since $c_\lambda(2) = 0$ unless λ is of the form $(a, b, 1^{k-a-b})$ (Corollary 5.2), the result follows. $\square$

This formula for $Q_k^{\mathbf{R}}(m, n)$ is rather unsightly, and it is by no means obvious that it has integer coefficients. But the number of terms is only a quadratic function of k, and each term is a simple product, so it is more efficient for computational purposes than Corollary 3.6, which has exponential complexity.

In some unpublished work on Wishart eigenvalues, Mallows and Wachter have considered some combinatorial aspects of the polynomials $Q_k^{\mathbf{R}}(m, n)$ (see the related paper [**MW**], although it does not explicitly mention $Q_k^{\mathbf{R}}(m, n)$). In particular, Mallows has raised the question of computing the coefficient t_k of mn in $Q_k^{\mathbf{R}}(m, n)$; indeed, the sequence

$$(t_1, t_2, \dots) = (1, 1, 4, 20, 148, 1348, 15104, \dots)$$

---

[3]Recall that $(a)_i$ denotes the falling factorial $a(a - 1) \cdots (a - i + 1)$.

is number 1447 in [Sl]. Our work yields two expressions for t_k. By Corollary 3.6, it follows that t_k is the number of 1-factors $\delta \in \mathcal{F}_k$ with $c(\varepsilon \cup \delta) = c(\delta \cup \delta_k) = 1$, whereas by Theorem 5.4, we have

$$t_k = \sum_\lambda c_\lambda(2) \left[\frac{2^{a-b-1}(2b)!(a-1)!(k-a-b+1)!}{(2b-1)!\, b!} \right]^2,$$

where λ is restricted as in Theorem 5.4, and $c_\lambda(2)$ is given by Lemma 5.3.

6. An Alternative Combinatorial Approach

Returning to the general setting of §1, let F be a finite-dimensional real division algebra, and let $U = [u_{ij}]$ be an $n \times n$ matrix of independent F-valued standard normal variables. As usual, let U^* denote the conjugated transpose of U (so that $U^* = U^t$ in the real case). Recall that if A and B are fixed $n \times n$ Hermitian matrices and f is a symmetric function, then $\mathcal{E}_U(f(AUBU^*))$ is invariant under unitary transformations $A \mapsto XAX^*$ and $B \mapsto YBY^*$. We will therefore assume in what follows that A and B are (real) diagonal matrices; say $A = \mathrm{diag}(a_1, \ldots, a_n)$, $B = \mathrm{diag}(b_1, \ldots, b_n)$.

In the real and complex cases, it is possible to give a direct combinatorial interpretation of $\mathcal{E}_U(p_\lambda(AUBU^*))$ (as is apparent from [OU, §3]), which raises the possibility that a more combinatorial proof of Theorems 2.3 and 3.5 could be given. In the quaternionic case, the noncommutativity causes certain complications which make this approach more difficult, as we shall see below. To describe the combinatorial interpretation, first note that the (i,j)-entry of $AUBU^*$ is $\sum_k a_i b_k u_{ik} \bar{u}_{jk}$, so it follows that

$$(6.1) \quad p_k(AUBU^*) = \sum (a_{i_1} b_{j_1} \cdots a_{i_k} b_{j_k}) \cdot (u_{i_1 j_1} \bar{u}_{i_2 j_1} u_{i_2 j_2} \bar{u}_{i_3 j_2} \cdots u_{i_k j_k} \bar{u}_{i_1 j_k}),$$

where the indices $i_1, j_1, \ldots, i_k, j_k$ all range from 1 to n. Since conjugation is an anti-automorphism of $\mathbf{H}$, it follows that the above expression is real (regardless of F), and therefore consistent with the conventions of §1 (cf. (1.2)).

Now let $X = \{x_1, \ldots, x_n\}$ and $Y = \{y_1, \ldots, y_n\}$ be two n-element alphabets, and let $\mathcal{W}_k$ be the set of all words w of length $2k$ in the elements of $X \cup Y$, starting with a letter of X, and alternating thereafter between Y and X; say,

$$(6.2) \qquad\qquad w = x_{i_1} y_{j_1} x_{i_2} y_{j_2} \cdots x_{i_k} y_{j_k}.$$

There is an obvious one-to-one correspondence between $\mathcal{W}_k$ and the terms of $p_k(AUBU^*)$ (identify the indices $i_1, j_1, \ldots, i_k, j_k$ in (6.1) and (6.2)). More generally, for each partition $\lambda = (\lambda_1, \ldots, \lambda_l)$, there is a one-to-one correspondence between the terms of $p_\lambda(AUBU^*)$ and the Cartesian product

$$\mathcal{W}_\lambda := \mathcal{W}_{\lambda_1} \times \cdots \times \mathcal{W}_{\lambda_l}.$$

In order to determine the amount contributed to $\mathcal{E}_U(p_\lambda(AUBU^*))$ by a term indexed by a typical member of $\mathcal{W}_\lambda$, it suffices to determine the expectation of an arbitrary monomial of independent standard normal variables and their conjugates. In the commutative cases (i.e., $F = \mathbf{R}$ or $\mathbf{C}$), this is easy to do, but in the quaternionic case, we do not know the answer in general. In spite of this, we can still prove the following result for all three cases, thereby completing our proof of Theorem 1.1.

6.1 PROPOSITION. *If $\lambda \vdash k$ and $\lambda \neq (1^k)$, then the coefficient of $a_1 b_1 \cdots a_k b_k$ in $\mathcal{E}_U(p_\lambda(AUBU^*))$ is zero.*

PROOF. Let $w = (w_1, \ldots, w_l)$ be a typical member of $\mathcal{W}_\lambda$, and assume $\lambda_1 \geq 2$. Let $x_i y_j$ be the first two letters of w_1. In order for the term indexed by w to contribute to the coefficient of $a_1 b_1 \cdots a_k b_k$, there can be no other occurrence of x_i or y_j among $w_1, \ldots, w_l$. In particular, the term indexed by w will have at most one occurrence of u_{ij}, and since $\lambda_1 \geq 2$, there cannot be any occurrence of $\bar{u}_{ij}$. (In graph-theoretic terms, this is equivalent to the fact that a closed path of length $2k$ with no repeated vertices cannot have repeated edges unless $k = 1$.) It follows that the term indexed by w is homogeneous of degree one in the variable u_{ij}, and therefore has zero expectation. $\square$

For the remainder of this section, we assume $F = \mathbf{R}$ or $\mathbf{C}$.

Given any $w = (w_1, \ldots, w_l) \in \mathcal{W}_\lambda$, let us define $\pi(w) = \pi(w_1) \cdots \pi(w_l)$ to be the monomial obtained by substituting a_i and b_j for each occurrence of x_i and y_j. Furthermore, let us define $m_{ij}(w)$ to be the number of times $x_i y_j$ occurs consecutively as a subword in $w_1, w_2, \ldots,$ and w_l. Similarly define $m'_{ij}(w)$ to be the number of times $y_j x_i$ occurs as a subword, using the convention that the last letter of each w_r is followed by its first letter (i.e., each w_r is a "circular word").

6.2 PROPOSITION. *Assuming $F = \mathbf{R}$ or $\mathbf{C}$, we have*

$$P_\lambda^F(A, B) = \mathcal{E}_U(p_\lambda(AUBU^*)) = \sum_{w \in \mathcal{W}_\lambda} \pi(w) \prod_{1 \leq i, j \leq n} \kappa_{ij}^F(w),$$

where

$$\kappa_{ij}^{\mathbf{R}}(w) = \begin{cases} 1 \cdot 3 \cdots (2r - 1), & \text{if } m_{ij}(w) + m'_{ij}(w) = 2r \\ 0, & \text{if } m_{ij}(w) + m'_{ij}(w) \text{ is odd,} \end{cases}$$

$$\kappa_{ij}^{\mathbf{C}}(w) = \begin{cases} r!, & \text{if } m_{ij}(w) = m'_{ij}(w) = r \\ 0, & \text{if } m_{ij}(w) \neq m'_{ij}(w). \end{cases}$$

PROOF. This is an immediate consequence of the fact that

$$p_\lambda(AUBU^*) = \sum_{w \in \mathcal{W}_\lambda} \pi(w) \cdot \prod_{1 \leq i, j \leq n} u_{ij}^{m_{ij}(w)} \bar{u}_{ij}^{m'_{ij}(w)},$$

together with the fact that if u is a standard normal variable, then

(6.3) $$\mathcal{E}(u^{2r}) = 1 \cdot 3 \cdots (2r - 1)$$

in the real case, and

(6.4) $$\mathcal{E}(u^r \bar{u}^s) = r! \, \delta_{r,s}$$

in the complex case. □

Thus one could establish Theorem 2.3 by directly proving that

$$(6.5) \qquad \sum_{w \in \mathcal{W}_\lambda} \pi(w) \prod_{1 \le i,j \le n} \kappa_{ij}^{\mathbf{C}}(w) = \sum_{\mu,\nu \vdash k} a_{\mu,\nu}^\lambda p_\mu(A) p_\nu(B)$$

for all partitions λ of k, and a similar approach to Theorem 3.5 could be attempted. If successful, this would also yield proofs of the real and complex cases of Theorem 1.1 that avoid any integration beyond the simple and well-known (6.3) and (6.4).

For certain pairs $\mu, \nu \vdash k$, there is a known combinatorial proof that $a_{\mu,\nu}^\lambda$ is indeed the coefficient of $p_\mu(A) p_\nu(B)$ on the left side of (6.5). By the triangle inequality (cf. (2.5)), we have $a_{\mu,\nu}^\lambda = 0$ unless $\ell(\mu) + \ell(\nu) \le k + \ell(\lambda)$, so the terms on the right side of (6.5) with $\ell(\mu) + \ell(\nu) = k + \ell(\lambda)$ are the terms of highest total degree with respect to $p_i(A)$ and $p_j(B)$. It follows that if we expand $p_\mu(x) p_\nu(y)$ in terms of monomial symmetric functions $m_\alpha(x) m_\beta(y)$, then the coefficient of $m_\mu(x) m_\nu(y)$ is one, while any other term $m_\alpha(x) m_\beta(y)$ appearing with nonzero coefficient must satisfy $\ell(\alpha) + \ell(\beta) < \ell(\mu) + \ell(\nu)$. Hence if we let $\mathcal{W}_\lambda^*$ be the set of $w \in \mathcal{W}_\lambda$ such that $\pi(w)$ is a term of $m_\alpha(A) m_\beta(B)$ with $\ell(\alpha) + \ell(\beta) = \ell(\lambda) + k$, then (6.5) implies

$$\sum_{w \in \mathcal{W}_\lambda^*} \pi(w) \prod_{1 \le i,j \le n} \kappa_{ij}^{\mathbf{C}}(w) = \sum_{\substack{\mu,\nu \vdash k \\ \ell(\mu)+\ell(\nu)=\ell(\lambda)+k}} a_{\mu,\nu}^\lambda m_\mu(A) m_\nu(B).$$

Goulden and Jackson have given a combinatorial proof of a result easily seen to be equivalent to the case $\lambda = (k)$ of this identity [**GJ**, Thm. 2.1], and they remark that the proof generalizes to arbitrary λ [**GJ**, §4]. It would be interesting to try to extend this reasoning to give a proof of (6.5).

Finally, let us remark that when $\lambda = (k)$ and $\ell(\mu) + \ell(\nu) = \ell(\lambda) + k = k + 1$, Bédard and Goupil [**BGo**], and later Goulden and Jackson [**GJ**, Thm. 2.2], have shown that

$$a_{\mu,\nu}^{(k)} = \frac{k(\ell(\mu) - 1)! \, (\ell(\lambda) - 1)!}{m_1(\mu)! \, m_1(\nu)! \, m_2(\mu)! \, m_2(\nu)! \cdots},$$

where $m_j(\lambda)$ denotes the multiplicity of j in λ.

Appendix

The following are tables for the polynomials $P_\lambda^F(A, B) = \mathcal{E}_U(p_\lambda(AU\,BU^*))$ and $Q_\lambda^F(m, n)$ for all partitions λ of size at most 4, and each of the division algebras $F = \mathbf{R}, \mathbf{C}$ and $\mathbf{H}$. We use p_r and q_r as abbreviations for $p_r(A)$ and $p_r(B)$, respectively.

The polynomials $P_\lambda^{\mathbf{R}}(A, B)$.

2	$p_1^2 q_2 + p_2 q_1^2 + p_2 q_2$
11	$p_1^2 q_1^2 + 2 p_2 q_2$

3	$p_1^3 q_3 + 3p_1 p_2 q_1 q_2 + p_3 q_1^3 + 3p_1 p_2 q_3 + 3p_3 q_1 q_2 + 4p_3 q_3$
21	$p_1^3 q_1 q_2 + p_1 p_2 q_1^3 + p_1 p_2 q_1 q_2 + 4p_1 p_2 q_3 + 4p_3 q_1 q_2 + 4p_3 q_3$
1^3	$p_1^3 q_1^3 + 6p_1 p_2 q_1 q_2 + 8p_3 q_3$

4	$p_1^4 q_4 + 4p_1^2 p_2 q_1 q_3 + 2p_1^2 p_2 q_2^2 + 4p_1 p_3 q_1^2 q_2 + 2p_2^2 q_1^2 q_2 + p_4 q_1^4$ $+6p_1^2 p_2 q_4 + 8p_1 p_3 q_1 q_3 + 4p_1 p_3 q_2^2 + 4p_2^2 q_1 q_3 + p_2^2 q_2^2$ $+6p_4 q_1^2 q_2 + 16p_1 p_3 q_4 + 5p_2^2 q_4 + 16p_4 q_1 q_3 + 5p_4 q_2^2 + 20p_4 q_4$
31	$p_1^4 q_1 q_3 + 3p_1^2 p_2 q_1^2 q_2 + p_1 p_3 q_1^4 + 3p_1^2 p_2 q_1 q_3 + 3p_1 p_3 q_1^2 q_2$ $+6p_1^2 p_2 q_4 + 10p_1 p_3 q_1 q_3 + 6p_1 p_3 q_2^2 + 6p_2^2 q_1 q_3 + 6p_4 q_1^2 q_2$ $+12p_1 p_3 q_4 + 6p_2^2 q_4 + 12p_4 q_1 q_3 + 6p_4 q_2^2 + 24p_4 q_4$
22	$p_1^4 q_2^2 + 2p_1^2 p_2 q_1^2 q_2 + p_2^2 q_1^4 + 2p_1^2 p_2 q_2^2 + 2p_2^2 q_1^2 q_2 + 8p_1^2 p_2 q_4 + 16p_1 p_3 q_1 q_3$ $+5p_2^2 q_2^2 + 8p_4 q_1^2 q_2 + 16p_1 p_3 q_4 + 4p_2^2 q_4 + 16p_4 q_1 q_3 + 4p_4 q_2^2 + 20p_4 q_4$
211	$p_1^4 q_1^2 q_2 + p_1^2 p_2 q_1^4 + p_1^2 p_2 q_1^2 q_2 + 8p_1^2 p_2 q_1 q_3 + 2p_1^2 p_2 q_2^2 + 8p_1 p_3 q_1^2 q_2 + 2p_2^2 q_1^2 q_2$ $+8p_1 p_3 q_1 q_3 + 2p_2^2 q_2^2 + 16p_1 p_3 q_4 + 8p_2^2 q_4 + 16p_4 q_1 q_3 + 8p_4 q_2^2 + 24p_4 q_4$
1^4	$p_1^4 q_1^4 + 12p_1^2 p_2 q_1^2 q_2 + 32p_1 p_3 q_1 q_3 + 12p_2^2 q_2^2 + 48p_4 q_4$

The polynomials $Q_\lambda^{\mathbf{R}}(m, n)$.

2	$m^2 n + m n^2 + m n$
11	$m^2 n^2 + 2mn$

3	$m^3 n + 3m^2 n^2 + m n^3 + 3m^2 n + 3m n^2 + 4mn$
21	$m^3 n^2 + m^2 n^3 + m^2 n^2 + 4m^2 n + 4m n^2 + 4mn$
1^3	$m^3 n^3 + 6m^2 n^2 + 8mn$

4	$m^4 n + 6m^3 n^2 + 6m^2 n^3 + m n^4 + 6m^3 n$ $+17m^2 n^2 + 6m n^3 + 21m^2 n + 21m n^2 + 20mn$
31	$m^4 n^2 + 3m^3 n^3 + m^2 n^4 + 3m^3 n^2 + 3m^2 n^3 + 6m^3 n$ $+22m^2 n^2 + 6m n^3 + 18m^2 n + 18m n^2 + 24mn$
22	$m^4 n^2 + 2m^3 n^3 + m^2 n^4 + 2m^3 n^2 + 2m^2 n^3 + 8m^3 n$ $+21m^2 n^2 + 8m n^3 + 20m^2 n + 20m n^2 + 20mn$
211	$m^4 n^3 + m^3 n^4 + m^3 n^3 + 10m^3 n^2 + 10m^2 n^3$ $+10m^2 n^2 + 24m^2 n + 24m n^2 + 24mn$
1^4	$m^4 n^4 + 12m^3 n^3 + 44m^2 n^2 + 48mn$

The polynomials $P_\lambda^{\mathbf{C}}(A, B)$.

2	$p_1^2 q_2 + p_2 q_1^2$
11	$p_1^2 q_1^2 + p_2 q_2$

3	$p_1^3 q_3 + 3p_1 p_2 q_1 q_2 + p_3 q_1^3 + p_3 q_3$
21	$p_1^3 q_1 q_2 + p_1 p_2 q_1^3 + 2p_1 p_2 q_3 + 2p_3 q_1 q_2$
1³	$p_1^3 q_1^3 + 3p_1 p_2 q_1 q_2 + 2p_3 q_3$

4	$p_1^4 q_4 + 4p_1^2 p_2 q_1 q_3 + 2p_1^2 p_2 q_2^2 + 4p_1 p_3 q_1^2 q_2 + 2p_2^2 q_1^2 q_2$ $+ p_4 q_1^4 + 4p_1 p_3 q_4 + p_2^2 q_4 + 4p_4 q_1 q_3 + p_4 q_2^2$
31	$p_1^4 q_1 q_3 + 3p_1^2 p_2 q_1^2 q_2 + p_1 p_3 q_1^4 + 3p_1^2 p_2 q_4 + 4p_1 p_3 q_1 q_3$ $+ 3p_1 p_3 q_2^2 + 3p_2^2 q_1 q_3 + 3p_4 q_1^2 q_2 + 3p_4 q_4$
22	$p_1^4 q_2^2 + 2p_1^2 p_2 q_1^2 q_2 + p_2^2 q_1^4 + 4p_1^2 p_2 q_4$ $+ 8p_1 p_3 q_1 q_3 + 2p_2^2 q_2^2 + 4p_4 q_1^2 q_2 + 2p_4 q_4$
211	$p_1^4 q_1^2 q_2 + p_1^2 p_2 q_1^4 + 4p_1^2 p_2 q_1 q_3 + p_1^2 p_2 q_2^2 + 4p_1 p_3 q_1^2 q_2$ $+ p_2^2 q_1^2 q_2 + 4p_1 p_3 q_4 + 2p_2^2 q_4 + 4p_4 q_1 q_3 + 2p_4 q_2^2$
1⁴	$p_1^4 q_1^4 + 6p_1^2 p_2 q_1^2 q_2 + 8p_1 p_3 q_1 q_3 + 3p_2^2 q_2^2 + 6p_4 q_4$

The polynomials $Q_\lambda^{\mathbf{C}}(m, n)$.

2	$m^2 n + mn^2$
11	$m^2 n^2 + mn$

3	$m^3 n + 3m^2 n^2 + mn^3 + mn$
21	$m^3 n^2 + m^2 n^3 + 2m^2 n + 2mn^2$
1³	$m^3 n^3 + 3m^2 n^2 + 2mn$

4	$m^4 n + 6m^3 n^2 + 6m^2 n^3 + mn^4 + 5m^2 n + 5mn^2$
31	$m^4 n^2 + 3m^3 n^3 + m^2 n^4 + 3m^3 n + 10m^2 n^2 + 3mn^3 + 3mn$
22	$m^4 n^2 + 2m^3 n^3 + m^2 n^4 + 4m^3 n + 10m^2 n^2 + 4mn^3 + 2mn$
211	$m^4 n^3 + m^3 n^4 + 5m^3 n^2 + 5m^2 n^3 + 6m^2 n + 6mn^2$
1⁴	$m^4 n^4 + 6m^3 n^3 + 11m^2 n^2 + 6mn$

The polynomials $P_\lambda^{\mathbf{H}}(A, B)$.

2	$4p_1^2 q_2 + 4p_2 q_1^2 - 2p_2 q_2$
11	$4p_1^2 q_1^2 + 2p_2 q_2$

3	$8p_1^3 q_3 + 24p_1 p_2 q_1 q_2 + 8p_3 q_1^3 - 12p_1 p_2 q_3 - 12p_3 q_1 q_2 + 8p_3 q_3$
21	$8p_1^3 q_1 q_2 + 8p_1 p_2 q_1^3 - 4p_1 p_2 q_1 q_2 + 8p_1 p_2 q_3 + 8p_3 q_1 q_2 - 4p_3 q_3$
1³	$8p_1^3 q_1^3 + 12p_1 p_2 q_1 q_2 + 4p_3 q_3$

4	$16p_1^4q_4 + 64p_1^2p_2q_1q_3 + 32p_1^2p_2q_2^2 + 64p_1p_3q_1^2q_2 + 32p_2^2q_1^2q_2 + 16p_4q_1^4$
	$-48p_1^2p_2q_4 - 64p_1p_3q_1q_3 - 32p_1p_3q_2^2 - 32p_2^2q_1q_3 - 8p_2^2q_2^2$
	$-48p_4q_1^2q_2 + 64p_1p_3q_4 + 20p_2^2q_4 + 64p_4q_1q_3 + 20p_4q_2^2 - 40p_4q_4$
31	$16p_1^4q_1q_3 + 48p_1^2p_2q_1^2q_2 + 16p_1p_3q_1^4 - 24p_1^2p_2q_1q_3 - 24p_1p_3q_1^2q_2$
	$+24p_1^2p_2q_4 + 40p_1p_3q_1q_3 + 24p_1p_3q_2^2 + 24p_2^2q_1q_3 + 24p_4q_1^2q_2$
	$-24p_1p_3q_4 - 12p_2^2q_4 - 24p_4q_1q_3 - 12p_4q_2^2 + 24p_4q_4$
22	$16p_1^4q_2^2 + 32p_1^2p_2q_1^2q_2 + 16p_2^2q_1^4 - 16p_1^2p_2q_2^2$
	$-16p_2^2q_1^2q_2 + 32p_1^2p_2q_4 + 64p_1p_3q_1q_3 + 20p_2^2q_2^2 + 32p_4q_1^2q_2$
	$-32p_1p_3q_4 - 8p_2^2q_4 - 32p_4q_1q_3 - 8p_4q_2^2 + 20p_4q_4$
211	$16p_1^4q_1^2q_2 + 16p_1^2p_2q_1^4 - 8p_1^2p_2q_1^2q_2 + 32p_1^2p_2q_1q_3$
	$+8p_1^2p_2q_2^2 + 32p_1p_3q_1^2q_2 + 8p_2^2q_1^2q_2 - 16p_1p_3q_1q_3 - 4p_2^2q_2^2$
	$+16p_1p_3q_4 + 8p_2^2q_4 + 16p_4q_1q_3 + 8p_4q_2^2 - 12p_4q_4$
1^4	$16p_1^4q_1^4 + 48p_1^2p_2q_1^2q_2 + 32p_1p_3q_1q_3 + 12p_2^2q_2^2 + 12p_4q_4$

The polynomials $Q_\lambda^{\mathbf{H}}(m,n)$.

2	$4m^2n + 4mn^2 - 2mn$
11	$4m^2n^2 + 2mn$

3	$8m^3n + 24m^2n^2 + 8mn^3 - 12m^2n - 12mn^2 + 8mn$
21	$8m^3n^2 + 8m^2n^3 - 4m^2n^2 + 8m^2n + 8mn^2 - 4mn$
1^3	$8m^3n^3 + 12m^2n^2 + 4mn$

4	$16m^4n + 96m^3n^2 + 96m^2n^3 + 16mn^4 - 48m^3n$
	$-136m^2n^2 - 48mn^3 + 84m^2n + 84mn^2 - 40mn$
31	$16m^4n^2 + 48m^3n^3 + 16m^2n^4 - 24m^3n^2 - 24m^2n^3$
	$+24m^3n + 88m^2n^2 + 24mn^3 - 36m^2n - 36mn^2 + 24mn$
22	$16m^4n^2 + 32m^3n^3 + 16m^2n^4 - 16m^3n^2 - 16m^2n^3$
	$+32m^3n + 84m^2n^2 + 32mn^3 - 40m^2n - 40mn^2 + 20mn$
211	$16m^4n^3 + 16m^3n^4 - 8m^3n^3 + 40m^3n^2$
	$+40m^2n^3 - 20m^2n^2 + 24m^2n + 24mn^2 - 12mn$
1^4	$16m^4n^4 + 48m^3n^3 + 44m^2n^2 + 12mn$

Acknowledgment. The second author would like to thank Colin Mallows for introducing him to the problems considered in this paper and for several helpful discussions.

REFERENCES

[BGo] F. Bédard and A. Goupil, *The poset of conjugacy classes and decomposition of products in the symmetric group*, Canad. Math. Bull. (to appear).

[BG] N. Bergeron and A. M. Garsia, *Zonal polynomials and domino tableaux*, preprint.

[B] W. Burnside, *Theory of groups of finite order*, 2nd ed., Dover, New York, 1955.

[G] I. M. Gelfand, *Spherical functions on symmetric Riemann spaces*, Dokl. Akad. Nauk SSSR **70** (1950), 5–8. (Russian)

[Go] R. Godement, *A theory of spherical functions. I*, Trans. Amer. Math. Soc. **73** (1952), 496–556.

[GJ] I. P. Goulden and D. M. Jackson, *The combinatorial relationship between trees, cacti and certain connection coefficients for the symmetric group*, preprint.

[GR] K. I. Gross and D. St. P. Richards, *Special functions of matrix argument I: Algebraic induction, zonal polynomials, and hypergeometric functions*, Trans. Amer. Math. Soc. **301** (1987), 781–811.

[J] A. T. James, *Zonal polynomials of the real positive definite symmetric matrices*, Ann. of Math. **74** (1961), 475–501.

[M1] I. G. Macdonald, *Symmetric functions and Hall polynomials*, Oxford Univ. Press, Oxford, 1979.

[M2] I. G. Macdonald, *Commuting differential operators and zonal spherical functions*, Algebraic Groups, Utrecht 1986 (A. M. Cohen, et. al., eds.), Lecture Notes in Math., vol. 1271, Springer-Verlag, Berlin and New York, 1987, pp. 189–200.

[MW] C. L. Mallows and K. W. Wachter, *Valency enumeration of rooted plane trees*, J. Austral. Math. Soc. **13** (1972), 472–476.

[OU] W. H. Olson and V. R. R. Uppuluri, *Asymptotic distribution of eigenvalues of random matrices*, Sixth Berkeley Symposium on Mathematical Statistics and Probability, vol. III, Univ. of California Press, Berkeley, CA, 1972, pp. 615–644.

[Sl] N. J. A. Sloane, *A handbook of integer sequences*, Academic Press, New York, 1973.

[St] R. P. Stanley, *Some combinatorial properties of Jack symmetric functions*, Adv. in Math. **77** (1989), 76–115.

[Ste] J. R. Stembridge, *On Schur's Q-functions and the primitive idempotents of a commutative Hecke algebra*, preprint.

[T] A. Takemura, *Zonal polynomials*, Institute of Mathematical Statistics, Hayward, CA, 1984.

DEPARTMENT OF MATHEMATICS, UNIVERSITY OF MICHIGAN, ANN ARBOR, MICHIGAN 48109

DEPARTMENT OF MATHEMATICS, MASSACHUSETTS INSTITUTE OF TECHNOLOGY, CAMBRIDGE, MASSACHUSETTS 02139

DEPARTMENT OF MATHEMATICS, UNIVERSITY OF MICHIGAN, ANN ARBOR, MICHIGAN 48109

Contemporary Mathematics
Volume **138**, 1992

Degenerate Principal Series
on Tube Type Domains

KENNETH D. JOHNSON

Dedicated to Sigurdur Helgason in honor of his sixty fifth birthday.

ABSTRACT. Let Ω be an irreducible Hermitian symmetric space of tube type with $G = \mathrm{Aut}_0(\Omega)$. If $\hat{S}$ is the Shilov boundary of Ω, $\hat{S} = G/\tilde{P}$ where $\tilde{P}$ is a maximal parabolic subgroup of G. Suppose $\tilde{P} = \tilde{M}\tilde{A}\tilde{N}$ is a Langlands decomposition and $\chi : \tilde{P} \to \mathbb{C}^*$ is a homomorphism with $\chi(\tilde{M}_0\tilde{N}) = \{1\}$. Let π_χ be the induced principal series representation of G. We determine when π_χ is irreducible, unitarizable, and the nature of the corresponding irreducible subquotient representations when π_χ is reducible.

1. Introduction

Suppose Ω is an irreducible Hermitian symmetric space of tube type. Then $\Omega = G/K$ where G is isomorphic to one of the groups: $SU(n,n)$, $Sp(n,\mathbb{R})$, $SO^*(4n)$, $SO_0(n,2)$, or $E_{7(-25)}$, and K is a maximal compact subgroup of G. If $\hat{S}$ is the Shilov boundary of Ω, then $\hat{S} = G/\tilde{P} = K/K \cap \tilde{P}$ where $\tilde{P}$ is a maximal parabolic subgroup of G. Our main purpose here is the examination of representations of G that are induced by one dimensional representations χ of $\tilde{P}$ where $\chi(K \cap \tilde{P}_0) = 1$.

Fix $G = KAN$ an Iwasawa decomposition of G. We may assume by conjugating if necessary, that $AN \subset \tilde{P}$, and $\tilde{P}$ has Langlands decomposition $\tilde{P} = \tilde{M}\tilde{A}\tilde{N}$ where $\tilde{A} \subset A$, $\tilde{N} \subset N$ and $\tilde{M}/M_0 = \mathbb{Z}_2$. Let $g, k, a, n, \tilde{m}, \tilde{a}$ and $\tilde{n}$ be the respective Lie algebras of $G, K, A, N, \tilde{M}, \tilde{A}$ and $\tilde{N}$.

Let $\chi : \tilde{P} \to \mathbb{C}^*$ be continuous homomorphism with $\chi(\tilde{M}_0\tilde{N}) = \{1\}$, and let H_χ be the space of functions $f : G \to \mathbb{C}$:

1991 *Mathematics Subject Classification*. Primary 33C80, 17B20.

Key words and phrases. Intertwining operators, root systems, Lie algebras, complementary series, invariant polynomials.

This paper is in final form and no version of it will be submitted for publication elsewhere .

1) $f|_K \in \mathcal{L}^2(K)$; and, 2) $f(g\tilde{p}) = \chi(\tilde{p})^{-1}f(g)$ $(\tilde{p} \in \tilde{P})$. The space H_χ is a space of sections of a homogeneous line bundle over $\hat{S}$, and left translation induces a representation π_χ of G on H_χ. That is, for $f \in H_\chi$ and $g, x \in G, \pi_\chi(g)f(x) = f(g^{-1}x)$. In studying the representations π_χ it is essential to parametrize χ in a more convenient fashion.

Since $\chi(\tilde{M}_0) = \{1\}$ either $\chi|_{\tilde{M}} = \sigma$ is trivial or else $\sigma(\tilde{M}) = \{1, -1\}$, and since exp : $\tilde{\mathfrak{a}} \to \tilde{A}$ is an analytic isomorphism with $\dim \tilde{\mathfrak{a}} = 1$ there is an $H_0 \in \tilde{\mathfrak{a}}$ with $B(H_0, H_0) = 1$ (B = Killing form of g) such that $\chi(\tilde{m} \exp sH_0\tilde{n}) = \sigma(\tilde{m})e^{s\nu}(\tilde{m} \in \tilde{M}, s \in \mathbb{R}, \tilde{n} \in \tilde{N})$. We now write $\pi(\sigma, \nu)$ for π_χ and $H(\sigma, \nu)$ for H_χ. Note that $\pi(\sigma, \nu)|_K \cong \pi(\sigma, \nu')|_K$ for any $\nu, \nu' \in \mathbb{C}$. If $L_0 = K \cap \tilde{M}_0$ and $L = K \cap \tilde{M}, \tilde{M}/\tilde{M}_0 \cong L/L_0$. Let $\hat{K}_0 = \{\tau \in \hat{K} : V_\tau^{L_0} \neq (0)\}$; since K/L_0 is a K-symmetric space either $\dim V_\tau^{L_0} = 0$ or $\dim V_\tau^{L_0} = 1$ for any $\tau \in \hat{K}$. As $\sigma(L) = \sigma(\tilde{M})$ and $L/L_0 \cong \tilde{M}/\tilde{M}_0 \cong \mathbb{Z}_2$ we also denote $\sigma|_L$ as σ. If $\hat{K}(\sigma) = \{\tau \in K : \tau(\ell)v = \sigma[\ell]v$ for $\ell \in L, v \in V_\tau^{L_0}\}, \hat{K}_0 = \hat{K}(\sigma_0) \cup \hat{K}(\sigma_1)$ and $\hat{K}(\sigma_0) \cap \hat{K}(\sigma_1) = \emptyset$ where σ_0 is trivial on L and $\sigma_1(L) = \{-1, 1\}$. If $X(\sigma, \nu)$ is the space of K-finite vectors of $H(\sigma, \nu)$ we have, as a K-module,

$$X(\sigma, \nu) = \sum\{V_\tau : \tau \in \hat{K}(\sigma)\}.$$

The representation $\pi(\sigma, \nu)$ induces a representation of g and in fact of $U(g)$, the enveloping algebra of g, on $X(\sigma, \nu)$; we also denote this induced representation by $\pi(\sigma, \nu)$. Working with $X(\sigma, \nu)$ has an advantage over working with $H(\sigma, \nu)$ since we may use purely algebraic techniques to study $X(\sigma, \nu)$. Recall that the irreducibility of $X(\sigma, \nu)$ as a $U(g)$-module is equivalent to the irreducibility of $H(\sigma, \nu)$ as a G-module.

In this paper, we determine which $\pi(\sigma, \nu)$ are irreducible, the existence of complementary series, and the nature of the irreducible subquotients of $\pi(\sigma, \nu)$ when $\pi(\sigma, \nu)$ is reducible. The remainder of the paper contains five sections: section 2 deals with the recollection of basic facts involved in the decomposition of g and some known results about $\hat{K}_0$; section 3 involves a more thorough examination of $\hat{K}_0$ and some essential products of representations; section 4 gives the fundamental equations for the action of special infinitesimal operators; in section 5 we obtain our irreducibility criteria, study intertwining operators and determine complementary series; finally, in section 6, we determine the irreducible subquotients for the non-irreducible representations.

2. Lie algebra decomposition and root systems

If p is the orthogonal complement to k in g with respect to $B, g = k + p$ is a Cartan decomposition of g. The center z of k is 1-dimensional and there is a $Z \in z$ such that $(\text{ad } Z)^2 = -1$ on p. Fixing i a square root of $-1, p_\mathbb{C} = p + ip = p_+ + p_-$ where $\text{ad } Z|_{p_+} = i$ and $\text{ad } Z|_{p_-} = -i$. Then $g_\mathbb{C} = k_\mathbb{C} + p_+ + p_-, [p_\pm, p_\pm] = 0, [p_+, p_-] = k_\mathbb{C}$, and $[k_\mathbb{C}, p_\pm] = p_\pm$. If t is a Cartan subalgebra of k, t is also a Cartan subalgebra of g and $t_\mathbb{C}$ is a Cartan subalgebra of $g_\mathbb{C}$. Set

$h = it$, $h^* = \mathrm{Hom}(h, \mathbb{R})$ and for $\alpha \in h^*$ let

$$g_\alpha = \{x \in g_{\mathbb{C}} : [H, X] = \alpha(H)X \text{ for all } H \in h\}.$$

Call $\alpha \in h^* \sim \{0\}$ a root if $g_\alpha \neq (0)$, and let Δ be the set of all roots. A root α is call compact if $g_\alpha \subset k_{\mathbb{C}}$ and noncompact if $g_\alpha \subset p_{\mathbb{C}}$. If $\Delta_n = \{\alpha \in \Delta : \alpha \text{ is noncompact}\}$ and $\Delta_C = \{\alpha \in \Delta : \alpha \text{ is compact}\}$, then $\Delta = \Delta_C \cup \Delta_n$. There is an ordering on Δ such that

$$\Delta_n^+ = \Delta^+ \cap \Delta_n = \{\alpha \in \Delta_n : g_\alpha \subset p_+\}.$$

Two roots α, β are called strongly orthogonal if neither $\alpha + \beta$ nor $\alpha - \beta$ are roots. Let γ_1 be the lowest positive noncompact root, and for $k > 0$ choose γ_k to be the lowest positive noncompact root strongly orthogonal to γ_j for $j \leq k - 1$. This process defines a maximal set $\{\gamma_1, \ldots, \gamma_r\}$ of strongly orthogonal roots where $r = \mathrm{rank}\ \Omega$ is the dimension of the maximal flat submanifold of Ω.

For $U \in g_{\mathbb{C}}$ let $\overline{U} \in g_{\mathbb{C}}$ denote its conjugate with respect to g. For $\alpha \in \Delta$ let $H_\alpha \in h$ be such that $B(H, H_\alpha) = \alpha(H)$, and furthermore let $\langle,\rangle : h^* \times h^* \to \mathbb{R}$ be dual to $B|_{h \times h}$. For each $\alpha \in \Delta_n^+$ fix $E_\alpha \in g_\alpha$ so that

$$[E_\alpha, \overline{E}_\alpha] = 2H_\alpha/\langle \alpha, \alpha \rangle,$$

and for $1 \leq j \leq r$ set $E_j = E_{\gamma_j}$ and $H_j^0 = H_{\gamma_j}/\langle \gamma_j, \gamma_j \rangle$. For $1 \leq j \leq r$ set $X_j = E_j + \overline{E}_j$, $Y_j = i(E_j - \overline{E}_j)$, $X = \sum_{j=1}^{r} X_j$, and $Y = \sum_{j=1}^{r} Y_j$. Now $\mathfrak{a} = \langle Y_j : 1 \leq j \leq r \rangle$ is a maximal abelian subalgebra of p, $h^- = \langle H_j^0 : 1 \leq j \leq r \rangle$ is a subalgebra of h, and since Ω is of tube type

$$Z_0 = \sqrt{-1} \sum_{j=1}^{r} H_j^0 \text{ and } \mathrm{ad}\ Z_0 = \pm i \text{ on } p_{\pm}.$$

Let $\tilde{U}$ be the centralizer of Y in G, let $\tilde{n}$ be the 2-eigenspace of $\mathrm{ad}\ Y$ in g and $\tilde{N} = \exp \tilde{n}$. Set $\tilde{\mathfrak{a}} = \mathbb{R}Y$ and $\tilde{A} = \exp \tilde{\mathfrak{a}}$. If $[\tilde{U}, \tilde{U}]$ is the commutator subgroup of $\tilde{U}$ and $\tilde{M}$ is the group generated by $[\tilde{U}, \tilde{U}]$ and $K \cap \tilde{U}$ then $\tilde{P} = \tilde{M}\tilde{A}\tilde{N}$ is a maximal parabolic subgroup of G given with its Langlands decomposition. Moreover, $\hat{S} = G/\tilde{P}$.

If $j : h^* \to h^*$ is the restriction map, set $\Delta_0 = j(\Delta) \sim \{0\}$. A result of Moore's states that since Ω is of tube type

$$\Delta_0^+ = j(\Delta^+) \sim \{0\} = \{\gamma_j : 1 \leq j \leq r\} \cup \{\tfrac{1}{2}(\gamma_j \pm \gamma_k) : 1 \leq k < j \leq r\},$$

$$j(\Delta_r^+) = \{\gamma_j : 1 \leq j \leq r\} \cup \{\tfrac{1}{2}(\gamma_j + \gamma_k) : 1 \leq k < j \leq r\},$$

and

$$j(\Delta_C^+) \sim \{0\} = \{\tfrac{1}{2}(\gamma_j - \gamma_k) : 1 \leq k < j \leq r\}.$$

Set $h^+ = \{H \in h : \gamma_j(H) = 0, 1 \leq j \leq r\}$. We now exhibit a parametrization for $\hat{K}_0$.

From [H2] and [J2] we have that $\Lambda \in h^*$ is the highest weight of an irreducible representation with an $L_0 = \tilde{M}_0 \cap K$ fixed vector if and only if

$$\Lambda = -(m_1\gamma_1 + \cdots + m_r\gamma_r)$$

where $m_j - m_{j+1}$ is an integer ≥ 0 for $j < r$ and $2m_r \in \mathbb{Z}$. In keeping with the jargon of Young diagrams we may identify $\hat{K}_0$ with the set

$$\{m \in \mathbb{R}^r : m_j - m_{j+1} \text{ is a nonnegative integer for } j < n \text{ and } 2m_r \in \mathbb{Z}\}.$$

If σ_0 is the trivial representation in L and σ_1 is the representation with $\sigma_1(L) = \{-1, 1\}$ and $\sigma_1(L_0) = \{1\}$, then

$$\hat{K}(\sigma_0) = \{m \in \hat{K}_0 : m_r \in \mathbb{Z}\}, \text{ and}$$

$$\hat{K}(\sigma_1) = \{m \in \hat{K}_0 : m_r \in \frac{1}{2}\mathbb{Z} \sim \mathbb{Z}\}.$$

Recall the result of Hua [Hu] and Schmid [S] (see also [J1]).

THEOREM. *As a $K_{\mathbb{C}}$-module*

$$P(p_+) \cong \sum \{V(m) : m \in \hat{K}(\sigma_0) \text{ and } m_r \geq 0\}.$$

Moreover, the representation with highest weight $\Lambda = -(m_1\gamma_1 + \cdots + m_r\gamma_r)$ occurs with multiplicity one and only in the homogeneous polynomials of degree $m_1 + \cdots + m_r$. If $m \in \hat{K}_0$ and $\tau(m)$ is the irreducible holomorphic representation of $K_{\mathbb{C}}$ with highest weight $-(m_1\gamma_1 + \cdots + m_r\gamma_r)$, then the contragredient representation of $\tau(m)$, $\tau(m)^* = \tau(m^*)$ where $m^* = (-m_r, -m_{r-1}, \ldots, -m_1) \in \hat{K}_0$.

In the following sections it will be necessary to distinguish between $V(m)$ for $m \in \hat{K}_0$, and its realization as a space of functions on K/L_0. Given $v_0 \neq 0$ in $V(m^*)^{L_0}$ and $v \in V(m)$ let f_v be the function on K defined by setting

$$f_v(k) = \langle v, \tau(m^*)(k)v_0 \rangle \quad (k \in K)$$

If $\hat{V}(m) = \langle f_v : v \in V(m) \rangle$, $\hat{V}(m)$ does not depend on our choice of v_0 and

$$L_k f_v(k_0) = f_{\tau(m)(k)v}(k_0) \quad (k, k_0 \in K).$$

The Fourier-Peter-Weyl theorem yields

$$\mathcal{L}^2(K/L_0) \sim \sum \{\hat{V}(m) : m \in \hat{K}_0\}.$$

3. Products of representations

Suppose first only that K is a compact group and L_0 is a closed subgroup. Let $\tau_j : K \to GL(V_j)(j = 1, 2)$ be two complex irreducible representations of K and suppose $\hat{V}_1, \hat{V}_2$ are K-irreducible subspaces of $C(K/L_0)$ with $V_j \cong \hat{V}_j(j = 1, 2)$ as K-modules. Let $\hat{V}_1\hat{V}_2$ be the subspace of $C(K/L)$ spanned by the products of functions in $\hat{V}_1$ and functions in $\hat{V}_2$, and suppose $W \cong \hat{W} \subset \hat{V}_1\hat{V}_2$ is an irreducible K-module. From the Peter-Weyl theorem we have the following result.

PROPOSITION 3.1. *There is a $T \in Hom_K(V_1^* \otimes V_2^*, W^*)$ for which $T(V_1^{*L_0} \otimes V_2^{*L_0}) \neq (0)$.*

Assume now that K and L_0 are as in sections 1 and 2, and suppose $\{e_1, \cdots, e_r\}$ is the standard basis of $\mathbb{R}^r$.

PROPOSITION 3.2. *Suppose $m, m + e_j \in \hat{K}_0$. Then $\hat{V}(m + e_j) \subset \hat{V}(e_1)\hat{V}(m)$ if and only if $\hat{V}(m^* + e_j^*) \subseteq \hat{V}(m^*) \cdot \hat{V}(-e_r)$.*

PROOF. This is immediate since $\hat{V}(-e_1) \cdot \hat{V}(m)$ is dual to $\hat{V}(-e_r) \cdot \hat{V}(m^*)$.

As a corollary to these propositions we state the following result.

COROLLARY. *Suppose $m, m + e_j \in \hat{K}_0$ and $\hat{V}(m + e_j) \subset \hat{V}(e_1) \cdot \hat{V}(m)$. If $0 \neq P_j \in Hom_K(V(e_1) \otimes V(m), V(m+e_j)), 0 \neq Q_j \in Hom_K(V(-e_r) \otimes V(m^*), V(m+e_j)^*)), 0 \neq u \in V(e_1)^{L_0} \otimes V(m)^{L_0}, 0 \neq u^* \in V(-e_r)^{L_0} \otimes V(m^*)^{L_0}$ and $(,) : V(m+e_j) \otimes V(m+e_j)^* \to \mathbb{C}$ is the nondegenerate $\mathbb{C}$-bilinear K-invariant pairing, then $(P_j u, Q_j u^*) \neq 0$. The same consequence holds if we initially assume $\hat{V}(m+e_j)^* \subset \hat{V}(-e_r)\hat{V}(m^*)$.*

Let us set $V(m) = (0) = \hat{V}(m)$ if $m \notin \hat{K}_0$.

THEOREM 3.3. *Suppose $m \in \hat{K}_0$. Then*

(i) $\hat{V}(e_1) \cdot \hat{V}(m) = \sum_{j=1}^{r} \hat{V}(m + e_j)$, *and*

(ii) $\hat{V}(-e_r) \cdot \hat{V}(m) = \sum_{j=1}^{r} \hat{V}(m - e_j)$.

PROOF. Note that (ii) follows from (i) by duality. For $1 \leq j \leq r$ let $p_j \in P(p_+)$ be the highest weight vector of the representation of $K_{\mathbb{C}}$ on $\hat{V}(e_1 + \cdots + e_j)$. Since $\hat{V}(le_1 + \cdots + le_r) = \mathbb{C}p_r^l$; it suffices to prove (i) for the case $m_r \geq 0$ or equivalently when $\hat{V}(m) \subset P(p_+)$. Let $P(p_-)$ be the space of holomorphic polynomials on p_- and suppose $\langle , \rangle$ is a nondegenerate $K_{\mathbb{C}}$-invariant bilinear pairing on $P(p_-) \times P(p_+)$. For $\hat{V}(m) \subset P(p_+)$ set $p(m) = p_1^{m_1 - m_2} \cdots p_{r-1}^{m_{r-1} - m_r} p_r^{m_r}$ and suppose $q(m)$ is a lowest weight vector of $\hat{V}(m^*)$. Our result now follows from the following lemma which will be proved in a later paper.

LEMMA. *If $m \in \hat{K}_0$, $\hat{V}(m) \subset P_{(p+)}$ and $m_{j-1} > m_j$, then*

$$\langle q(m + e_j), B(\overline{E}_j, Z)p(m) \rangle \neq 0.$$

REMARK 1. The referee has pointed out that Theorem 3.3 is an immediate consequence of Theorem 6 of Dieb [D] or Theorem 2 of Lassalle [L]. If $0 \neq \varphi(m) \in \hat{V}(m)^L$, Dieb and Lassalle give a formula for $\varphi(e_1) \cdot \varphi(m)$ as a linear combination of $\varphi(m + e_j)$. Theorem 3.3 follows since the coefficient of $\varphi(m + e_j)$ is nonzero if and only if $\hat{V}(m + e_j) \subset \hat{V}(e_1) \cdot \hat{V}(m)$.

REMARK 2. This theorem is trivial for $S0_0(n, 2)$, and was verified in [J4] for $SU(n, n)$.

4. Fundamental equations and intertwining operators

Recall that the noncompact roots $\gamma_1, \ldots, \gamma_r$ are all of the same length. Hence, if

$$H = \sum_{j=1}^{r} H_j^0, (\text{ad } H)^2 = 1 \text{ on } p_{\mathbb{C}} \text{ and } 0 \text{ on } k_{\mathbb{C}}.$$

Thus $B(H, H) = \dim_{\mathbb{C}} p_{\mathbb{C}} = \frac{r}{||\gamma_1||^2}$, and hence $||\gamma_1||^2 = \frac{r}{\dim p}$. Suppose $H_0 \in \tilde{a}$ is such that $B(H_0, H_0) = 1$. If $c : g_{\mathbb{C}} \to g_{\mathbb{C}}$ is the Cayley transform we have from [K-W] that $c(H_0) = k_0 H$ where $k_0^2 \dim p = 1$. So we fix $k_0 = \frac{1}{\sqrt{\dim p}}$. Now for $j \neq k \leq r$ the $\#\{\alpha \in \Delta_n : j(\alpha) = \gamma_j + \gamma_k\}$ is independent of j and k; set n_0 equal to this number. Then $\dim p = r(2 + n_0(r-1))$.

Now $\overline{E} = \sum_{j=1}^{r} \overline{E}_j \in p_-^{L_0}$ and $E = \sum_{j=1}^{r} E_j \in p_+^{L_0}$. If $m \in \hat{K}_0$, let $p(m) \in \hat{V}(m)$ be the L_0-spherical function with $p(m)(e) = 1$ (here e is the identity of K). Now

(1)
$$B(\overline{E}, \text{Ad } kH_0)p(m)(k) = \sum_{j=1}^{r} a_j(m)p(m + e_j)(k),$$

and

(2)
$$B(E, \text{Ad } kH_0)p(m)(k) = \sum_{j=1}^{r} b_j(m)p(m - e_j)(k)$$

where $a_j(m) \neq 0$ if and only if $p(m + e_j) \neq 0$ and $b_j(m) \neq 0$ if and only if $p(m - e_j) \neq 0$.

Let ℓ be the Lie algebra of L and let $\{u_1, \ldots, u_M\}$ be a basis of ℓ such that $B(u_i, u_j) = -\delta_{ij}$. Suppose $\theta : g \to g$ is the Cartan involution with $\theta = 1$ on k and -1 on p. Let $\{v_j : 1 \leq j \leq N\}$ be a basis of $\tilde{n}$ where $B(v_i, \theta v_j) = -\delta_{ij}$. Then $\{\frac{v_j + \theta v_j}{\sqrt{2}} = w_j : 1 \leq j \leq N\}$ is a basis of $\ell^{\perp} \cap k$ and $B(w_i, w_j) = -\delta_{ij}$. Set

$$\Delta = -\sum_{j=1}^{M} u_j^2 - \sum_{j=1}^{N} w_j^2;$$

and suppose $\Delta = \Lambda(m)$ on $V(m)$. Then

$$\Delta B(\overline{E}, \text{Ad } kH_0)p(m)(k) = \sum_{j=1}^{r} a_j(m)\Lambda(m + e_j)p(m + e_j)(k)$$

$$= (\Lambda(e_1) + \Lambda(m))B(\overline{E}, \text{Ad } kH_0)p(m)(k) - 2\sum_{j=1}^{N} B(\overline{E}, \text{Ad } k[w_j, H_0])p(m)(k : w_j).$$

If f is a function on K, $f(k : w_j) = \frac{d}{d+} f(k \exp + w_j)|_{f=0}$.

Evaluating we obtain

(3)
$$\frac{1}{\sqrt{\dim p}} \sum_{j=1}^{N} B(\overline{E}, \text{Ad } k(v_j - \theta v_j)) p(m)(k : v_j + \theta v_j)$$

$$= \sum_{j=1}^{r} a_j(m)(\Lambda(m + e_j) - \Lambda(e_1) - \Lambda(m)) p(m + e_j)(k).$$

Similarly, we obtain

(4)
$$\frac{1}{\sqrt{\dim p}} \sum_{j=1}^{N} B(E, \text{Ad } k(v_j - \theta v_j)) p(m)(k : v_j + \theta v_j)$$

$$= \sum_{j=1}^{r} b_j(m)(\Lambda(m - e_j) - \Lambda(-e_r) - \Lambda(m)) p(m - e_j)(k).$$

Equations (3) and (4) are the first forms of our fundamental equations.

If $\Lambda = -(m_1 \gamma_1 + \cdots + m_r \gamma_r)$ and $2\rho_k = \sum_{\alpha \in \Delta_c^+} \alpha$, we see that $\Lambda(m) = \langle \Lambda, \Lambda + 2\rho_k \rangle$. Hence, if $2\rho_- = 2\rho_k|_{h^-}$ we have that $\Lambda(m) = \langle \Lambda, \Lambda + 2\rho_- \rangle$. Since $2\rho_- = -n_0 \sum_{j < \ell} \left(\frac{\gamma_j - \gamma_\ell}{2} \right)$, we have that

$$\Lambda(m) = \left[\sum_{j=1}^{r} m_j^2 + n_0 \left(\frac{r + 1 - 2j}{2} \right) m_j \right] \frac{r}{\dim p}.$$

Hence

$$\Lambda(m + e_j) - \Lambda(m) - \Lambda(e_1) = [2m_j + n_0(1 - j)] \frac{r}{\dim p}$$

and

$$\Lambda(m - e_j) - \Lambda(m) - \Lambda(-e_r) = [-2m_j - n_0(r - j)] \frac{r}{\dim p}.$$

Finally, we obtain

(5)
$$\sum_{j=1}^{N} B(\overline{E}, \text{Ad } k(v_j - \theta v_j)) p(m)(k : v_j + \theta v_j)$$

$$= \frac{r}{\sqrt{\dim p}} \sum_{j=1}^{r} a_j(m)(2m_j + n_0(1 - j)) p(m + e_j)(k),$$

and

(6)
$$\sum_{j=1}^{N} B(E, \text{Ad } k(v_j - \theta v_j)) p(m)(k : v_j + \theta v_j)$$

$$= \frac{r}{\sqrt{\dim p}} \sum_{j=1}^{r} b_j(m)(-2m_j + n_0(j - r)) p(m - e_j)(k).$$

Finally, calculating $\pi(\sigma,\nu)(\overline{E})$ and $\pi(\sigma,\nu)(E)$ we have for $m \in \hat{K}(\sigma)$

$$\pi_\nu(\overline{E})p(m) = -p(m)(k : \operatorname{Ad} k^{-1}\overline{E})$$

$$= -B(\operatorname{Ad} k^{-1}\overline{E}, H_0)p(m)(k : H_0) + \sum_{j=1}^{N} B(\operatorname{Ad} k^{-1}\overline{E}, v_j)p(m)(k : \theta v_j)$$

$$= \nu \sum_{j=1}^{r} a_j(m)p(m+e_j)(k) + \frac{1}{2}\sum_{j=1}^{N} B(\operatorname{Ad} k^{-1}\overline{E}, v_j - \theta v_j)p(m)(k : v_j + \theta v_j).$$

Finally setting $\nu = \nu_0 \frac{r}{2\sqrt{\dim p}}$ we have

$$(7) \quad \pi(\sigma,\nu)(\overline{E})p(m) = \frac{r}{2\sqrt{\dim p}} \sum_{j=1}^{r} a_j(m)(\nu_0 + 2m_j + n_0(1-j))p(m+e_j),$$

and

$$(8) \quad \pi(\sigma,\nu)(E)p(m) = \frac{r}{2\sqrt{\dim p}} \sum_{j=1}^{r} b_j(m)(\nu_0 - 2m_j + n_0(j-r))p(m-e_j).$$

An immediate consequence of (7) and (8) is the following.

THEOREM 4.1. $\pi(\sigma,\nu)$ is irreducible if and only if whenever $m, m+e_j \in \hat{K}(\sigma)$ $\nu_0 + 2m_j + n_0(1-j) \neq 0$ and $\nu_0 - 2m_j - 2 - n_0(r-j) \neq 0$.

Suppose $A : X(\sigma,\nu) \to X(\sigma,\lambda)$ intertwines the representations $\pi(\sigma,\nu)$ and $\pi(\sigma,\lambda)$. In particular, for any $m \in \hat{K}(\sigma)$,

$$\pi(\sigma,\lambda)(\overline{E}) \circ A(p(m)) = A \circ \pi(\sigma,\nu)(\overline{E})(p(m))$$

and

$$\pi(\sigma,\lambda)(E) \circ A(p(m)) = A \circ \pi(\sigma,\nu)(E)(p(m)).$$

Now since A intertwines the representation of K we have for each $m \in \hat{K}(\sigma)$ there is an $\alpha(m) \in \mathbb{C}$ such that $A|_{\hat{V}(m)} = \alpha(m)$. Putting this information into equations (7) and (8) we see that if $m, m + e_j \in \hat{K}(\sigma)$.

$$[\lambda_0 + 2m_j + n_0(1-j)]\alpha(m) = [\nu_0 + 2m_j + n_0(1-j)]\alpha(m+e_j),$$
$$[\lambda_0 - 2m_j - 2 + n_0(j-r)]\alpha(m+e_j) = [\nu_0 - 2m_j - 2 + n_0(j-r)]\alpha(m).$$

Hence

$$[\lambda_0 - 2m_j - 2 + n_0(j-r)][\lambda_0 + 2m_j + n_0(1-j)]$$
$$= [\nu_v - 2m_j - 2 + n_0(j-r)][\nu_0 + 2m_j + n_0(1-j)].$$

Solving we see that either

$$\lambda_0 = \nu_0 \quad \text{and} \quad A = cI$$

or else $\lambda_0 = 2\rho_0 - \nu_0$ where $2\rho(H_0) = \operatorname{tr} \operatorname{ad} H_0|_{\bar{n}} = 2\rho_0 \frac{r}{2\sqrt{\dim p}}$ ($2\rho_0 = 2 + n_0(r-1)$).

5. Irreducibility and complementary series

Before elaborating further on which $\pi(\sigma, \nu)$ are irreducible or unitarizable, it will be useful to have the following chart where we assume $n > 1$ in all cases

G	$Sp(n, \mathbb{R})$	$SU(n, n)$	$SO^*(4n)$	$SO_0(n, 2)$	$E_{7(-25)}$
r	n	n	n	2	3
n_0	1	2	4	$n - 2$	8

If $r = 1, n_0 = 0$. We can now apply theorem 4.1 to give precise irreducibility criteria for each of our groups.

THEOREM 5.1. *Suppose σ is trivial.*

(1) *For n_0 even, $\pi(\sigma, \nu)$ is irreducible if and only if ν_0 is not an even integer.*

(2) *For n_0 odd, $\pi(\sigma, \nu)$ is irreducible if and only if ν_0 is not an integer.*

Suppose σ is not trivial.

(1') *For n_0 even, $\pi(\sigma, \nu)$ is irreducible if and only if ν_0 is not an odd integer.*

(2') *For n_0 odd, $\pi(\sigma, \nu)$ is irreducible if and only if ν_0 is not an integer.*

REMARK. The irreducibility criterion given here for $SU(n, n)$ differs from that given in [**J4**] because of a change in the normalization factor.

Recall the following result of Stern.

THEOREM [ST]. *There is a nondegenerate G-invariant sesquilinear pairing*

$$\langle \, , \, \rangle : H(\sigma, \nu) \times H(\sigma, 2\rho - \overline{\nu}) \to \mathbb{C}.$$

Moreover, for $f_1 \in H(\sigma, \nu)$ and $f_2 \in H(\sigma, 2\rho - \overline{\nu})$,

$$\langle f_1, f_2 \rangle = \int_K f_1(k) \overline{f_2(k)} dk.$$

Recall also the following result of Harish-Chandra Theorem [**H-C**]. *The representation $\pi(\sigma, \nu)$ is unitarizable if and only if there is a g-intertwining operator $A : X(\sigma, \nu) \to X(\sigma, 2\rho - \overline{\nu})$ such that $\langle f, Af \rangle > 0$ for all $0 \neq f \in X(\sigma, \nu)$.* An obvious consequence of Stern's theorem is the following.

COROLLARY. *The representation $\pi(\sigma, \nu)$ is unitary if and only if $\nu - \rho$ is an imaginary number.*

A consequence of our calculations and Harish-Chandra's result is the following.

COROLLARY. *The representation $\pi(\sigma, \nu)$ is unitarizable if and only if ν is real and there is an intertwining operator*

$$A : X(\sigma, \nu) \to X(\sigma, 2\rho - \nu) \quad \text{such that} \quad \langle f, Af \rangle > 0$$

for any $0 \neq f \in X(\sigma, \nu)$.

Suppose A is a g-intertwining operator of $\pi(\sigma, \nu)$ and $\pi(\sigma, 2\rho - \nu)$ with $Ap(m) = \alpha(m)p(m)$ for $m \in \hat{K}(\sigma)$. If $m + e_j, m \in \hat{K}(\sigma)$ we have

$$\alpha(m + e_j) = \left[-\frac{\nu_0 + 2 + 2m_j + n_0(r - j)}{\nu_0 + 2m_j + n_0(1 - j)} \right] \alpha(m).$$

We see that $\pi(\sigma, \nu)$ is unitarizable if we always have $\alpha(m + e_j)/\alpha(m) > 0$ whenever $m + e_j, m \in \hat{K}(\sigma)$, or equivalently if

$$[\nu_0 + 2m_j + n_0(1 - j)][-\nu_0 + 2 + 2m_j + n_0(r - j)] > 0$$

whenever $m, m + e_j \in \hat{K}(\sigma)$. Writing $\nu_0 = \rho_0 + \tau_0$ and $\varphi_j(m) = 2m_j + 1 + \frac{n_0}{2}(r - 2j + 1)$ we have

$$-\tau_0^2 + \varphi_j(m)^2 = [\nu_0 + 2m_j + n_0(1 - j)][-\nu_0 + 2 + 2m_j + n_0(r - j)].$$

Let $\mu(\sigma) = \inf\{|\varphi_j(m)| : m, m + e_j \in \hat{K}(\sigma)\}$. The following proposition is now obvious.

PROPOSITION 5.1. *The representation $\pi(\sigma, \nu)$ is unitarizable if and only if ν is real and $|\nu_0 - \rho_0| < \mu(\sigma)$. In particular, if $\mu(\sigma) > 0$, there is a complementary series.*

The following tables give values of $\mu(\sigma)$ for prescribed values of r and n_0

Table 1 (σ trivial)

n_o	$2k + 1$	1	$4k$	$4k + 2$	$4k + 2$
r	even	odd	arbitrary	even	odd
$\mu(\sigma)$	$1/2$	0	1	0	1

Note that if n_0 is odd and r is odd, $n_0 = 1$.

Table 2 (σ nontrivial)

n_o	$2k + 1$	1	$4k$	$4k + 2$	$4k + 2$
r	even	odd	arbitrary	even	odd
$\mu(\sigma)$	$1/2$	0	0	1	0

We now have the following resulting table.

Table 3

G	σ	$\mu(\sigma)$
$Sp(2n,\mathbb{R})$	Trivial or Nontrivial	$1/2$
$(n \geq 1)\ Sp(2n,\mathbb{R})$	Trivial or Nontrivial	0
$SU(2n,2n)$	Trivial	0
$SU(2n,2n)$	Nontrivial	1
$SU(2n+1,2n+1)$	Trivial	1
$SU(2n+1,2n+1)$	Nontrivial	0
$SO^*(4n)$ or $E_{7(-25)}$	Trivial	1
$SO^*(4n)$ or $E_{7(-25)}$	Nontrivial	0
$SO_0(2n+1,2)$	Trivial or Nontrivial	$1/2$
$SO_0(4n,2)$	Trivial	0
$SO_0(4n,2)$	Nontrivial	1
$SO_0(4n+2,2)$	Trivial	1
$SO_0(4n+2,2)$	Nontrivial	0

As a consequence of table 3 and theorem 5.1 we state

COROLLARY. $\mu(\sigma) > 0$ if and only if $\pi(\sigma,\rho)$ is irreducible. In other words, complementary series occur if and only if $\pi(\sigma,\rho)$ is irreducible.

6. Irreducible composition factors

Suppose $\pi(\sigma,\nu)$ is reducible. Set

$$\ell_j(\sigma,\nu) = \{m \in \hat{K}(\sigma) : 2m_j \leq -\nu_0 - n_0(1-j)\}$$
$$r_j(\sigma,\nu) = \{m \in \hat{K}(\sigma) : 2m_j \geq \nu_0 + n_0(j-r)\}$$
$$L_j(\sigma,\nu) = \langle \hat{V}(m) : m \in \ell_j(\sigma,\nu)\rangle$$
$$R_j(\sigma,\nu) = \langle \hat{V}(m) : m \in r_j(\sigma,\nu)\rangle$$

PROPOSITION 6.1. (1) Suppose σ is trivial. Then, if $\nu_0 + n_0(1-j) \in 2\mathbb{Z}$, $L_j(\sigma,\nu)$ is $\mathfrak{g}$-invariant, and if $\nu_0 + n_0(j-r) \in 2\mathbb{Z}$, $R_j(\sigma,\nu)$ is $\mathfrak{g}$-invariant.

(2) Suppose σ is nontrivial. Then, if $\nu_0 + n_0(1-j) \in \mathbb{Z} \sim 2\mathbb{Z}$, $L_j(\sigma,\nu)$ is $\mathfrak{g}$-invariant, and if $\nu_0 + n_0(j-r) \in \mathbb{Z} \sim 2\mathbb{Z}$, $R_j(\sigma,\nu)$ is $\mathfrak{g}$-invariant.

This follows immediately from equations 5.7 and 5.8.

Since $m_j \geq m_{j+1}$ for any $m \in \hat{K}_0$ we have

$$L_1(\sigma, \nu) \subset \cdots \subset L_r(\sigma, \nu)$$
$$R_1(\sigma, \nu) \supset \cdots \supset R_r(\sigma, \nu).$$

Set $L_0(\sigma, \nu) = R_{r+1}(\sigma, \nu) = 0$, and $L_{r+1}(\sigma, \nu) = R_0(\sigma, \nu) = X(\sigma, \nu)$. For $V \subset X(\sigma, \nu)$ set $V^\perp = \{g \in X(\sigma, 2\rho - \nu) : \langle f, g \rangle = 0 \text{ for all } f \in V\}$. Observe that $L_j(\sigma, \nu)^\perp = R_j(\sigma, 2\rho - \nu)$ and $R_j(\sigma, \nu)^\perp = L_j(\sigma, 2\rho - \nu)$. We now consider two separate cases for n_0.

Case 1 (n_0 is even): Then all the spaces $L_j(\sigma, \nu)$ and $R_j(\sigma, \nu)$ are $\mathfrak{g}$-invariant. Set

$$U_{jk}(\sigma, \nu) = (L_j \cap R_k)/(L_{j-1} \cap R_k + L_j \cap R_{k+1})$$
$$u_{jk}(\sigma, \nu) = \ell_j \cap r_k \sim (\ell_{j-1} \cap r_k \cup \ell_j \cap r_{k+1}).$$

Using equations 5.7 and 5.8 we see that $p(m)$ is a cyclic vector for any $m \in u_{jk}(\sigma, \nu)$. Hence we have the following result.

THEOREM 6.1. *If n_0 is even and $\pi(\sigma, \nu)$ is reducible, the $\mathfrak{g}$-module $U_{jk}(\sigma, \nu)$ is either (0) or irreducible.*

PROPOSITION 6.2. *Suppose n_0 is even and $\pi(\sigma, \nu)$ reducible. Then (i) for $j \leq k$, $U_{jk} \neq (0)$ if and only if $2\nu_0 \leq n_0(n - 1 + j - k)$, (ii) for $j - k > 1$, $U_{jk} \neq (0)$ if and only if $2\nu_0 \geq (n-1+j-k)$, and (iii) for $k = j-1$, $U_{jj-1} \neq (0)$. (Provided $1 \leq j \leq r + 1$ and $0 \leq k \leq r$.)*

PROOF. If $j \geq k, u_{jk}(\sigma, \nu) \neq$ if there is an $m \in \hat{K}(\sigma)$ such that $-\nu_0 - n_0(1 - j) \geq 2m_j \geq 2m_k \geq \nu_0 + n_0(k - r)$ while $2m_{j-1} > -\nu_0 - n_0(2 - j)$ and $2m_{k+1} < \nu_0 + n_0(k + 1 - r)$ but this can always be done whenever

$$2\nu_0 \leq n_0(r - 1 + j - k).$$

As $U_{jk}(\sigma, \nu)^* \cap X(\sigma, 2\varphi - \nu) = U_{k+1,j-1}(\sigma, 2\rho - \nu)$, (ii) follows from (i). To prove (iii) we must show there is an $m \in \hat{K}(\sigma)$ such that $2m_{j-1} \geq \nu_0 + n_0(j - 1 - r), 2m_j \leq -\nu_0 - n_0(1 - j),$

$$2m_{j-1} > -\nu_0 - n_0(2 - j), 2m_j < \nu_0 + n_0(j - r).$$

If m_{j-1} is sufficiently large and m_j is sufficiently negative we see $U_{jj-1}(\sigma, \nu) \neq (0)$.

REMARK. In [J4], the result for $U_{jj-1}(\sigma, \nu) \neq (0)$ was incorrectly stated.

Case 2: Suppose n_0 is odd. If $L_j(\sigma, \nu)$ is g-invariant L_{j-1} is not $\mathfrak{g}$-invariant when $j > 1, L_{j+1}$ is not g-invariant when $j < r$, but $L_{j\pm 2}$ is $\mathfrak{g}$-invariant. A similar result holds for the spaces R_j.

We conclude with the following result.

THEOREM 6.3. *Suppose n_0 is odd and L_j and R_k are $\mathfrak{g}$-invariant. If $W_{jk} = L_j \cap R_k/(L_{j-2} \cap R_k + L_j \cap R_{k+2}) \neq (0)$, then W_{jk} is an irreducible $\mathfrak{g}$-module. Assume $L_{-1} = 0 = R_{r+2}$.*

It is an easy matter to determine when $W_{jk} \neq (0)$, but we will not do so here.

REFERENCES

[D]   H.Dieb, *Thesis*, Université Louis Pasteur (1988).

[FK]   J. Faraut and A. Koranyi, *Function spaces and reproducing kernels on bounded symmetric domains*, J. of Functional Analysis **88** (1990), 64–89.

[H-C] Harish-Chandra, *Representations of semisimple Lie groups, IV*, Proc. Natl. Acad. Sci. **37** (1951), 691–694.

[H1]   S. Helgason, *Differential geometry and symmetric spaces*, Academic Press, New York.

[H2]   _____, *A duality for symmetric spaces with applications to group representations*, Adv. in Math. **5** (1970), 1–154.

[Hua] L. K. Hua, *Harmonic Analysis of Functions of Several Complex Variables in Classical Domains*, Vol. 6 Transl. of Math. Monographs (1963), Amer. Math. Soc., Providence, R.I..

[J1]   K. D. Johnson, *On a ring of invariant polynomials on a Hermitian symmetric space*, J. of Algebra **67** (1980), 72–81.

[J2]   _____, *A strong generalization of Helgason's theorem*, Trans. A.M.S. **304** (1987), 171–192.

[J3]   _____, *A constructive approach to tensor product decompositions*, J. Reine Angew. Math. **388** (1988), 129–148.

[J4]   _____, *Degenerate principal series and compact groups*, Math. Ann. **287** (1990), 703–718.

[K-W] A. Koranyi and J. A. Wolf, *Realization of Hermitian symmetric spaces as generalized half planes*, Ann. of Math. **81** (1965), 265–288.

[L]   M. Lassalle, *Coefficients du binome généralisés*, C. R. Acad. Sci. Paris 310 série I (1990), 257–260.

[S]   W. Schmid, *Die Randwerte holomopher Functionen auf hermitsch symmetrichen Raumen*, Invent. Math. **9** (1969–1970), 61–80.

[St]   A. I. Stern, *Completely irreducible class 1 representations of real semi-simple Lie groups*, Sov. Math. **10** (1969), 1254–1557.

DEPARTMENT OF MATHEMATICS, UNIVERSITY OF GEORGIA, ATHENS, GA 30602

Contemporary Mathematics
Volume **138**, 1992

Askey-Wilson Polynomials
for Root Systems of Type BC

TOM H. KOORNWINDER

ABSTRACT. This paper introduces a family of Askey-Wilson type orthogonal polynomials in n variables associated with a root system of type BC_n. The family depends, apart from q, on 5 parameters. For $n = 1$ it specializes to the four-parameter family of one-variable Askey-Wilson polynomials. For any n it contains Macdonald's two three-parameter families of orthogonal polynomials associated with a root system of type BC_n as special cases.

1. Introduction

In recent years, some families of orthogonal polynomials associated with root systems were introduced. The families studied by Heckman & Opdam [6], [4], [5] become Jacobi polynomials for root system BC_1. The families studied by Macdonald (see [11] for root system A_n and [12] for general root systems) become continuous q-ultraspherical polynomials for root system A_1 and continuous q-Jacobi polynomials for root system BC_1 (see Askey & Wilson [1, §4]). For all root systems Macdonald's polynomials tend to the Heckman-Opdam polynomials as q tends to 1.

This paper introduces a family of Askey-Wilson type polynomials for root system BC_n which depends, apart from q, on 5 parameters. For $n = 1$ it specializes to the four-parameter family of Askey-Wilson polynomials. For any n it contains Macdonald's two three-parameter families as special cases: for the pair (BC_n, B_n) directly and for the pair (BC_n, C_n) when q is replaced by q^2. Moreover, the weight function integrated over the orthogonality domain was explicitly evaluated by Gustafson [3] as a generalization of Selberg's beta integral.

1991 *Mathematics Subject Classification*. Primary 33D45, 33D80, 17B20.

Key words and phrases. Askey-Wilson polynomials, Macdonald's orthogonal polynomials associated with root systems, root system of type BC.

This paper is in final form and no version of it will be submitted for publication elsewhere .

The proofs in this paper are very much inspired by Macdonald's proofs in [12], in particular by his proofs in case of root systems E_8, F_4, G_2, where there is no minuscule fundamental weight available.

The contents of this paper are as follows. Section 2 summarizes Macdonald's results. The special case BC_n of these results is discussed in §3 and the further specialization to BC_1 in §4. The long section 5 introduces Askey-Wilson polynomials for root system BC_n, shows that these polynomials are eigenfunctions of a certain difference operator and establishes the full orthogonality of the polynomials. Finally, in §6, special cases and open problems are discussed

ACKNOWLEDGEMENTS. The research presented here was essentially done several years ago, in December 1987, while the author was visiting the IRMA at the University of Abidjan, Côte d'Ivoire. I thank prof. S. Touré for his hospitality. I thank prof. I. G. Macdonald for sending me his informal preprints at such an early stage. The observation that the Askey-Wilson polynomials for root system BC_n also include Macdonald's polynomials for the pair (BC_n, C_n), was recently made by J. F. van Diejen. I thank G. Heckman and the referee for useful comments to an earlier version of this paper.

2. Summary of Macdonald's results

In this section we summarize Macdonald's [12] results on orthogonal polynomials associated with root systems. See Humphreys [7] and Bourbaki [2, Chap.6] for preliminaries on root systems. Let V be a finite dimensional real vector space with inner product $\langle . , . \rangle$. Write $|v| := \langle v, v \rangle^{1/2}$ for the norm of $v \in V$. Write

$$v^\vee := 2v/|v|^2, \quad 0 \neq v \in V.$$

Let R be a not necessarily reduced root system spanning V. Let S be a reduced root system in V such that the set of lines $\{\mathbb{R}\alpha \mid \alpha \in R\}$ equals $\{\mathbb{R}\alpha \mid \alpha \in S\}$. Then the pair (R, S) is called admissible and R and S have the same Weyl group W. Now, for each $\alpha \in R$, there is a (unique) $u_\alpha > 0$ such that $\alpha_* := u_\alpha^{-1}\alpha \in S$. Assume that R is irreducible. It can be arranged, after possible dilation of R and S, that u_α takes values in $\{1, 2\}$ or in $\{1, 3\}$.

Let $0 < q < 1$. Put $q_\alpha := q^{u_\alpha}$. Let $\alpha \mapsto t_\alpha$ be a W-invariant function on R, taking values in $(0, 1)$ (for convenience). Then t_α only depends on $|\alpha|$. Put $t_\alpha := 1$ if $\alpha \in V \backslash R$. Let $k_\alpha \geq 0$ be such that $q_\alpha^{k_\alpha} = t_\alpha$.

Let R^+ be a choice for the set of positive roots in R. Let

$$Q := \mathbb{Z}\text{-Span}(R), \quad Q^+ := \mathbb{Z}_+\text{-Span}(R^+).$$

Here, and throughout the paper, $\mathbb{Z}_+ := \{0, 1, 2, \dots\}$. Let

$$P := \{\lambda \in V \mid \langle \lambda, \alpha^\vee \rangle \in \mathbb{Z} \ \forall \alpha \in R\}, \quad P^+ := \{\lambda \in V \mid \langle \lambda, \alpha^\vee \rangle \in \mathbb{Z}_+ \ \forall \alpha \in R^+\}$$

be respectively the weight lattice of R and the cone of dominant weights. Define a partial order on P by $\lambda \geq \mu$ iff $\lambda - \mu \in Q^+$.

For $\lambda \in P$ let e^λ be the function on V defined by

$$e^\lambda(x) := e^{i\langle \lambda, x \rangle}, \quad x \in V.$$

Extend this holomorphically to $V + iV$. If f is a function on V then put $(wf)(x) := f(w^{-1}x)$ for $w \in W$, $x \in V$. Hence $we^\lambda = e^{w\lambda}$. Let A be the complex linear span of the e^λ ($\lambda \in P$). Let A^W denote the space of W-invariants of A. Put

$$m_\lambda := |W_\lambda|^{-1} \sum_{w \in W} e^{w\lambda} = \sum_{\mu \in W\lambda} e^\mu, \quad \lambda \in P^+.$$

Here W_λ denotes the stabilizer of λ in W. The m_λ ($\lambda \in P^+$) form a basis of A^W. Note that $\overline{m_\lambda(x)} = m_\lambda(-x)$ ($\lambda \in P^+$, $x \in V$). If $-\mathrm{id} \in W$ then $f(x) = f(-x)$ for $f \in A^W$. In particular, we will then have that m_λ is real-valued on V.

Let $R^\vee := \{\alpha^\vee \mid \alpha \in R\}$ be the root system dual to R. Let $Q^\vee := \mathbb{Z}$-Span$(R^\vee)$. Then $T := V/(2\pi Q^\vee)$ is a torus. Let $\dot{x}$ be the image in T of $x \in V$. Let $d\dot{x}$ be the normalized Haar measure on T. For $\lambda \in P$ the function $\dot{x} \mapsto e^\lambda(x)$ is well-defined on T. For $a, a_1, \dots, a_k \in \mathbb{C}$ put

$$(a;q)_\infty := \prod_{j=0}^\infty (1 - aq^j), \quad (a_1, \dots, a_k; q)_\infty := \prod_{i=1}^k (a_i; q)_\infty.$$

Define

$$\Delta := \prod_{\alpha \in R} \frac{(t_{2\alpha}^{1/2} e^\alpha; q_\alpha)_\infty}{(t_\alpha t_{2\alpha}^{1/2} e^\alpha; q_\alpha)_\infty}, \quad \Delta^+ := \prod_{\alpha \in R^+} \frac{(t_{2\alpha}^{1/2} e^\alpha; q_\alpha)_\infty}{(t_\alpha t_{2\alpha}^{1/2} e^\alpha; q_\alpha)_\infty}.$$

Then $\Delta = \Delta^+ \overline{\Delta^+}$. Define a hermitian inner product on A^W by

$$\langle f, g \rangle := |W|^{-1} \int_T f(\dot{x}) \overline{g(\dot{x})} \, \Delta(x) \, d\dot{x}.$$

DEFINITION 2.1. For $\lambda \in P^+$ let $P_\lambda \in A^W$ be characterized by the two conditions

(i) $P_\lambda = m_\lambda + \sum_{\mu < \lambda} u_{\lambda,\mu} \, m_\mu$ for certain complex coefficients $u_{\lambda,\mu}$;

(ii) $\langle P_\lambda, m_\mu \rangle = 0$ if $\mu < \lambda$.

THEOREM 2.2.

$$\langle P_\lambda, P_\mu \rangle = 0 \quad \text{if } \lambda \neq \mu.$$

Define

$$(T_v f)(x) := f(x - i(\log q)v), \quad x, v \in V,$$

for functions f being analytic on a suitable subset of $V + iV$ containing V. Hence

$$T_v e^\lambda = q^{\langle v, \lambda \rangle} e^\lambda, \quad \lambda \in P.$$

Let $\sigma \in V$ be such that $\langle \sigma, \alpha_* \rangle$ takes just two values 0 and 1 as α runs through R^+ (σ is a so-called minuscule fundamental weight for $S^\vee$). Such σ exists for all

S not being of type E_8, F_4 or G_2. In these last three cases we can choose σ such that $\langle\sigma,\alpha_*\rangle$ takes values 0, 1 and 2 as α runs through R^+. Now put

$$\Phi_\sigma := \frac{T_\sigma\Delta^+}{\Delta^+},$$

$$E_\sigma f := |W_\sigma|^{-1}\sum_{w\in W} w(\Phi_\sigma\, T_\sigma f),$$

$$D_\sigma f := |W_\sigma|^{-1}\sum_{w\in W} w(\Phi_\sigma(T_\sigma f - f)),$$

$$\widetilde{m}_\sigma(\lambda) := |W_\sigma|^{-1}\sum_{w\in W} q^{\langle w\sigma,\lambda\rangle},$$

$$\rho_k := \tfrac{1}{2}\sum_{\alpha\in R^+} k_\alpha\,\alpha.$$

THEOREM 2.3. *D_σ maps A^W into itself. The P_λ are eigenfunctions of D_σ with eigenvalue*

$$q^{\langle\sigma,\rho_k\rangle}\left(\widetilde{m}_\sigma(\lambda+\rho_k) - \widetilde{m}_\sigma(\rho_k)\right).$$

If S is not of type E_8, F_4 or G_2 then E_σ maps A^W into itself, the P_λ are also eigenfunctions of E_σ with eigenvalue

$$q^{\langle\sigma,\rho_k\rangle}\,\widetilde{m}_\sigma(\lambda+\rho_k).$$

and

$$D_\sigma = E_\sigma - E_\sigma(1)$$

with $E_\sigma(1)$ scalar.

3. The case $R = BC_n$

Identify V with $\mathbb{R}^n$ and let $\varepsilon_1,\dots,\varepsilon_n$ be its standard basis. Consider in V the root systems

$R := \{\pm\varepsilon_j\}\cup\{\pm2\varepsilon_j\}\cup\{\pm\varepsilon_i\pm\varepsilon_j\}_{i<j}$ of type BC_n,

$S_B := \{\pm\varepsilon_j\}\cup\{\pm\varepsilon_i\pm\varepsilon_j\}_{i<j}$ of type B_n.

$S_C := \{\pm\varepsilon_j\}\cup\{\tfrac{1}{2}(\pm\varepsilon_i\pm\varepsilon_j)\}_{i<j}$ of type C_n.

Then S_B and S_C are reduced, R, S_B and S_C have the same Weyl groups and in the mappings $\alpha\mapsto u_\alpha^{-1}\alpha$ of R onto S_B and onto S_C, u_α take the values 1 and 2.

Note that $R^\vee = R$ and that the weight lattice P and the root lattice Q of R are both given by

$$P = Q = \{m_1\varepsilon_1 + \cdots + m_n\varepsilon_n \mid m_1,\dots,m_n\in\mathbb{Z}\}.$$

Take

$$R^+ := \{\varepsilon_j\}\cup\{2\varepsilon_j\}\cup\{\varepsilon_i\pm\varepsilon_j\}_{i<j}.$$

Then

$$P^+ = \{m_1\varepsilon_1 + \cdots + m_n\varepsilon_n \mid m_1 \geq m_2 \geq \ldots \geq m_n \geq 0, \quad m_1,\ldots,m_n \in \mathbb{Z}\},$$
$$Q^+ = \{m_1(\varepsilon_1 - \varepsilon_2) + \cdots + m_{n-1}(\varepsilon_{n-1} - \varepsilon_n) + m_n\varepsilon_n \mid m_1,\ldots,m_n \in \mathbb{Z}_+\}.$$

The torus $T := V/(2\pi Q^\vee)$ becomes $\mathbb{R}^n/(2\pi\mathbb{Z}^n)$. Recall that we have a partial ordering on P such that $\lambda \geq \mu$ iff $\lambda - \mu \in Q^+$.

For the pair (R, S_B) we have

$$q_{\pm\varepsilon_j} = q, \quad q_{\pm 2\varepsilon_j} = q^2, \quad q_{\pm\varepsilon_i \pm \varepsilon_j} = q,$$

and there are three different parameters t_α, which we write as

$$a := t_{\pm\varepsilon_j}, \quad b := t_{\pm 2\varepsilon_j}, \quad t := t_{\pm\varepsilon_i \pm \varepsilon_j}.$$

(Recall that $t_\alpha = 1$ if $\alpha \notin R$.) Thus

$$(3.1) \qquad\qquad \Delta^+ = \Delta_1^+ \Delta_2^+,$$

where

$$(3.2) \qquad \Delta_1^+ := \prod_{j=1}^{n} \frac{(b^{1/2}e^{\varepsilon_j};q)_\infty}{(a\,b^{1/2}e^{\varepsilon_j};q)_\infty} \frac{(e^{2\varepsilon_j};q^2)_\infty}{(b\,e^{2\varepsilon_j};q^2)_\infty}$$
$$= \prod_{j=1}^{n} \frac{(e^{2\varepsilon_j};q)_\infty}{(q^{1/2}e^{\varepsilon_j}, -q^{1/2}e^{\varepsilon_j}, a\,b^{1/2}e^{\varepsilon_j}, -b^{1/2}e^{\varepsilon_j};q)_\infty}$$

and

$$(3.3) \qquad \Delta_2^+ := \prod_{\alpha = \varepsilon_i \pm \varepsilon_j;\, i<j} \frac{(e^\alpha;q)_\infty}{(t\,e^\alpha;q)_\infty}.$$

For the pair (R, S_C) we have

$$q_{\pm\varepsilon_j} = q, \quad q_{\pm 2\varepsilon_j} = q^2, \quad q_{\pm\varepsilon_i \pm \varepsilon_j} = q^2,$$

and there are three different parameters t_α, which we write as

$$a := t_{\pm\varepsilon_j}, \quad b := t_{\pm 2\varepsilon_j}, \quad t := t_{\pm\varepsilon_i \pm \varepsilon_j}.$$

Thus (3.1) holds with

$$(3.4) \qquad \Delta_1^+ := \prod_{j=1}^{n} \frac{(b^{1/2}e^{\varepsilon_j};q)_\infty}{(a\,b^{1/2}e^{\varepsilon_j};q)_\infty} \frac{(e^{2\varepsilon_j};q^2)_\infty}{(b\,e^{2\varepsilon_j};q^2)_\infty}$$
$$= \prod_{j=1}^{n} \frac{(e^{2\varepsilon_j};q^2)_\infty}{(a\,b^{1/2}e^{\varepsilon_j}, q\,a\,b^{1/2}e^{\varepsilon_j}, -b^{1/2}e^{\varepsilon_j}, -q\,b^{1/2}e^{\varepsilon_j};q^2)_\infty}$$

and

$$(3.5) \qquad \Delta_2^+ := \prod_{\alpha = \varepsilon_i \pm \varepsilon_j;\, i<j} \frac{(e^\alpha;q^2)_\infty}{(t\,e^\alpha;q^2)_\infty}.$$

Since $-\mathrm{id} \in W$ in case of root system BC_n, m_λ will be real-valued and we can read for condition (ii) of Definition 2.1 that

$$\int_T P_\lambda(x)\, m_\mu(x)\, \Delta(x)\, d\dot{x} = 0 \quad \text{if } \mu < \lambda.$$

For the element σ of §2 we can take ε_1 in the case S_B and $\varepsilon_1 + \varepsilon_2 + \cdots + \varepsilon_n$ in the case S_C. In both cases σ is minuscule. So Theorem 2.3 is valid with these choices of σ. In particular, in the case S_B the polynomial P_λ is eigenfunction of D_{ε_1} with eigenvalue

$$(3.6) \qquad \sum_{j=1}^n \left(ab\, t^{2n-j-1}(q^{\lambda_j}-1) + t^{j-1}(q^{-\lambda_j}-1) \right).$$

The choice $\sigma := 2\varepsilon_1$ in the case S_C would give values 0, 1 and 2 for $\langle \sigma, \alpha_* \rangle$ as α runs through R^+. It will turn out in §6.1 that, in case S_C, the P_λ are not only eigenfunctions of $E_{\varepsilon_1 + \cdots + \varepsilon_n}$ but also of $D_{2\varepsilon_1}$.

4. The case $R = BC_1$

For $n = 1$ the two root systems S_B and S_C coincide and the results of §3 specialize as follows. We have $T = \mathbb{R}/(2\pi\mathbb{Z})$, $P = Q = \mathbb{Z}$, $P^+ = \mathbb{Z}_+$, the partial order on P is the ordinary total order on $\mathbb{Z}$,

$$m_l(x) = \begin{cases} e^{ilx} + e^{-ilx}, & l = 1, 2, \ldots, \\ 1, & l = 0. \end{cases}$$

and

$$\Delta^+(x) = \frac{(e^{2ix}; q)_\infty}{(q^{1/2}e^{ix}, -q^{1/2}e^{ix}, a\,b^{1/2}e^{ix}, -b^{1/2}e^{ix}; q)_\infty}$$

$$= \frac{(e^{2ix}; q^2)_\infty}{(a\,b^{1/2}e^{ix}, q\,a\,b^{1/2}e^{ix}, -b^{1/2}e^{ix}, -q\,b^{1/2}e^{ix}; q^2)_\infty}.$$

The inner product for W-invariant functions f, g becomes an integral over the period of 2π-periodic even functions, so it can be written as

$$\langle f, g \rangle = \frac{1}{2\pi} \int_0^\pi f(x)\, \overline{g(x)}\, \Delta(x)\, dx.$$

Askey-Wilson polynomials $p_n(y; a, b, c, d \mid q)$ ($n \in \mathbb{Z}_+$) are defined, up to a constant factor, as polynomials of degree n in y which satisfy the orthogonality relations
(4.1)

$$\int_0^\pi (p_n p_m)(\cos x; a, b, c, d \mid q) \left| \frac{(e^{2ix}; q)_\infty}{(ae^{ix}, be^{ix}, ce^{ix}, de^{ix}; q)_\infty} \right|^2 dx = 0, \quad n \neq m.$$

See Askey & Wilson [1]. Here a, b, c, d are real, or if complex, appear in complex conjugate pairs, and $|a|, |b|, |c|, |d| \leq 1$, but the pairwise products of a, b, c, d are not ≥ 1. When the condition $|a|, |b|, |c|, |d| \leq 1$ on the parameters is dropped, finitely many discrete terms have to be added to the orthogonality relation (4.1).

When we compare the expression for $\Delta^+(x)$ with the Askey-Wilson weight function we see that Macdonald's polynomials for root system BC_1 coincide, up to a constant factor, with Askey-Wilson polynomials

$$p_l(\cos x; q^{1/2}, -q^{1/2}, ab^{1/2}, -b^{1/2} \,|\, q).$$

By Askey & Wilson [1, (4.16), (4.17), (4.20)] the continuous q-Jacobi polynomials in M. Rahman's notation can be expressed in terms of Askey-Wilson polynomials by

$$
\begin{aligned}
P_l^{(\alpha,\beta)}(\cos x; q) &= \text{const. } p_l(\cos x; q^{1/2}, -q^{1/2}, q^{\alpha+1/2}, -q^{\beta+1/2} \,|\, q) \\
&= \text{const. } p_l(\cos x; q^{\alpha+1/2}, q^{\alpha+3/2}, -q^{\beta+1/2}, -q^{\beta+3/2} \,|\, q^2).
\end{aligned}
$$

Thus, if we put $a := q^\alpha$, $b := q^{2\beta}$, then Macdonald's polynomials for root system BC_1 coincide, up to a constant factor, with continuous q-Jacobi polynomials $P_l^{(\alpha+\beta-1/2, \beta-1/2)}(\cos x; q)$. This observation was already made by Macdonald [12, §9].

If $n = 1$ then, with $\sigma := 1$,

$$\Phi_\sigma(x) = \frac{(1 - a b^{1/2} e^{ix})(1 + b^{1/2} e^{ix})}{1 - e^{2ix}}.$$

Thus, if we write

$$R_l(e^{ix}) := P_l(x)$$

then Theorem 2.3 yields

$$\Phi_\sigma(-x) R_l(q^{-1} e^{ix}) + \Phi_\sigma(x) R_l(q e^{ix}) = (abq^l + q^{-l}) R_l(e^{ix}).$$

Compare this with Askey & Wilson [1, (5.7), (5.8), (5.9)]:

$$
\begin{aligned}
A(-x)\left(R_l(q^{-1} e^{ix}) - R_l(e^{ix})\right) + A(x)\left(R_l(q e^{ix}) - R_l(e^{ix})\right) \\
= -(1 - q^{-l})(1 - q^{l-1} abcd) R_l(e^{ix}),
\end{aligned}
$$

where

$$R_l(e^{ix}) := \text{const. } p_l(\cos x; a, b, c, d \,|\, q)$$

and

$$A(x) := \frac{(1 - ae^{ix})(1 - be^{ix})(1 - ce^{ix})(1 - de^{ix})}{(1 - e^{2ix})(1 - qe^{2ix})}.$$

If $c = q^{1/2}$, $d = -q^{1/2}$ (the continuous q-Jacobi case) then

$$A(x) = \frac{(1 - ae^{ix})(1 - be^{ix})}{1 - e^{2ix}}$$

and

$$A(x) + A(-x) = 1 - ab,$$

so

$$A(-x) R_l(q^{-1} e^{ix}) + A(x) R_l(q e^{ix}) = (q^{-l} - q^l ab) R_l(e^{ix}).$$

Thus Macdonald's difference equation for P_l in case $R = BC_1$ coincides with the continuous q-Jacobi case of the difference equation for Askey-Wilson polynomials.

5. Askey-Wilson polynomials for root system BC_n

We use the notation of §2 and §3. Let

$$R_1^+ := \{2\varepsilon_j\}_{j=1,\ldots,n}, \quad R_2^+ := \{\varepsilon_i \pm \varepsilon_j\}_{1 \le i < j \le n},$$

$$R_1 := R_1^+ \cup (-R_1^+), \quad R_2 := R_2^+ \cup (-R_2^+),$$

$$R_\ell^+ := R_1^+ \cup R_2^+, \quad R_\ell := R_1 \cup R_2 = R_\ell^+ \cup (-R_\ell^+).$$

Let R be the root system of type BC_n of §3. Then $R_\ell = \{\alpha \in R \mid 2\alpha \notin R\}$, a root system of type C_n in V with subsystems R_1 of type nA_1 and R_2 of type D_n. (The subscript ℓ stands for 'long'.) Let W be the Weyl group of R_ℓ. It is a semidirect product of the group of permutations of the coordinates and the group of sign changes of the coordinates. Let ρ, ρ_1, ρ_2 denote half the sum of the positive roots of R_ℓ, R_1, R_2, respectively. Then $\rho = \rho_1 + \rho_2$ and

$$\rho_1 = \varepsilon_1 + \varepsilon_2 + \cdots + \varepsilon_n, \quad \rho_2 = (n-1)\varepsilon_1 + (n-2)\varepsilon_2 + \cdots + \varepsilon_{n-1}.$$

Let P, P^+ and the partial order $\le$ be as in §3. Write $\varepsilon(w) := \det(w)$ $(w \in W)$. Let $A^{W,\varepsilon}$ consist of all $f \in A$ such that $wf = \varepsilon(w)f$ $(w \in W)$. Write

$$J_\lambda := \sum_{w \in W} \varepsilon(w) e^{w\lambda}, \quad \lambda \in P.$$

The $J_{\lambda+\rho}$ $(\lambda \in P^+)$ form a basis of $A^{W,\varepsilon}$. In particular, put

$$\delta := J_\rho = \prod_{\alpha \in R_\ell^+} (e^{\frac{1}{2}\alpha} - e^{-\frac{1}{2}\alpha}) = e^\rho \prod_{\alpha \in R_\ell^+} (1 - e^{-\alpha})$$

and

$$\chi_\lambda := \delta^{-1} J_{\lambda+\rho}, \quad \lambda \in P.$$

The χ_λ $(\lambda \in P^+)$ are in A^W and form a basis of A^W. We have

$$\chi_\lambda = m_\lambda + \sum_{\mu \in P^+; \, \mu < \lambda} a_{\lambda,\mu} \, m_\mu, \quad \lambda \in P^+,$$

for certain complex $a_{\lambda,\mu}$.

Fix $q \in (0,1)$ and $a, b, c, d, t \in \mathbb{C}$. Let

$$(5.1) \qquad \Delta^+ := \prod_{\alpha \in R_1^+} \frac{(e^\alpha; q)_\infty}{(ae^{\frac{1}{2}\alpha}, be^{\frac{1}{2}\alpha}, ce^{\frac{1}{2}\alpha}, de^{\frac{1}{2}\alpha}; q)_\infty} \prod_{\alpha \in R_2^+} \frac{(e^\alpha; q)_\infty}{(te^\alpha; q)_\infty}$$

and

$$(5.2) \qquad \Delta(x) := \Delta^+(x)\,\Delta^+(-x).$$

We are now ready to introduce Askey-Wilson polynomials for root system BC_n.

DEFINITION 5.1. Assume a, b, c, d are real, or if complex, appear in conjugate pairs, and that $|a|, |b|, |c|, |d| \leq 1$, but the pairwise products of a, b, c, d are not ≥ 1. Assume $-1 < t < 1$. Let $T := [-\pi, \pi]^n \subset V$. For $\lambda \in P^+$ define $P_\lambda \in A^W$ by the two conditions

   (i) $P_\lambda = m_\lambda + \sum_{\mu \in P+; \, \mu < \lambda} u_{\lambda, \mu} \, m_\mu$ for certain coefficients $u_{\lambda, \mu}$;
   (ii) $\int_T P_\lambda(x) \, m_\mu(x) \, \Delta(x) \, dx = 0$ for $\mu \in P^+$, $\mu < \lambda$.

We will generalize the case $R = BC_n$ of Theorem 2.2 by showing that the P_λ are orthogonal on T with respect to the weight function Δ. The proof will be based on two lemmas, the first one giving the action of a suitable difference operator on the m_λ, and the second one showing self-adjointness of this operator with respect to Δ on T, when acting on A^W.

Let $\sigma := \varepsilon_1$, similarly as in §3 for the pair (BC_n, B_n). Define Φ_σ and D_σ as in §2, with Δ^+ being given by (5.1). Thus

$$(5.3) \qquad \Phi_\sigma := \frac{T_\sigma \Delta^+}{\Delta^+},$$

and

$$(5.4) \qquad D_\sigma f := |W_\sigma|^{-1} \sum_{w \in W} w(\Phi_\sigma \, (T_\sigma f - f))$$

$$= |W_\sigma|^{-1} \sum_{w \in W} (w \Phi_\sigma) \, (T_{w\sigma} f - f), \quad f \in A^W.$$

LEMMA 5.2.

$$D_\sigma m_\lambda = \sum_{\mu \in P+; \, \mu \leq \lambda} a_{\lambda, \mu} \, m_\mu$$

with

$$(5.5) \qquad a_{\lambda, \lambda} = \sum_{j=1}^{n} \left(q^{-1} \, abcd \, t^{2n-j-1} \, (q^{\lambda_j} - 1) + t^{j-1} \, (q^{-\lambda_j} - 1) \right).$$

Here a, b, c, d, t may be arbitrarily complex.

PROOF. It will be convenient to replace a, b, c, d in the expression (5.1) for Δ^+ by a, $-b$, $q^{\frac{1}{2}}c$, $-q^{\frac{1}{2}}d$, respectively. Thus

$$(5.6) \qquad \Delta^+ := \prod_{\alpha \in R_1^+} \frac{(e^\alpha; q)_\infty}{(ae^{\frac{1}{2}\alpha}, -be^{\frac{1}{2}\alpha}, q^{\frac{1}{2}}ce^{\frac{1}{2}\alpha}, -q^{\frac{1}{2}}de^{\frac{1}{2}\alpha}; q)_\infty} \prod_{\alpha \in R_2^+} \frac{(e^\alpha; q)_\infty}{(te^\alpha; q)_\infty}.$$

By substitution of (5.6) in (5.3) we obtain

$$
\begin{aligned}
\Phi_\sigma &= \frac{(1-ae^{\varepsilon_1})(1+be^{\varepsilon_1})(1-q^{\frac{1}{2}}ce^{\varepsilon_1})(1+q^{\frac{1}{2}}de^{\varepsilon_1})}{(1-e^{2\varepsilon_1})(1-qe^{2\varepsilon_1})} \prod_{\alpha=\varepsilon_1\pm\varepsilon_l;\, l=2,\dots,n} \frac{1-te^{\alpha}}{1-e^{\alpha}} \\
&= abcd\, t^{2(n-1)} \frac{(1-a^{-1}e^{-\varepsilon_1})(1+b^{-1}e^{-\varepsilon_1})}{(1-e^{-2\varepsilon_1})} \\
&\quad \times \frac{(1-q^{-\frac{1}{2}}c^{-1}e^{-\varepsilon_1})(1+q^{-\frac{1}{2}}d^{-1}e^{-\varepsilon_1})}{(1-q^{-1}e^{-2\varepsilon_1})} \prod_{\alpha=\varepsilon_1\pm\varepsilon_l;\, l=2,\dots,n} \frac{1-t^{-1}e^{-\alpha}}{1-e^{-\alpha}} \\
&= abcd\, t^{2(n-1)} \prod_{\alpha\in R_2^+} \frac{1-t^{-\langle\sigma,\alpha\rangle}e^{-\alpha}}{1-e^{-\alpha}} \prod_{j=1}^{n}\Bigg[\frac{(1-a^{-\langle\sigma,\varepsilon_j\rangle}e^{-\varepsilon_j})(1+b^{-\langle\sigma,\varepsilon_j\rangle}e^{-\varepsilon_j})}{(1-e^{-2\varepsilon_j})} \\
&\quad \times \frac{(1-q^{-\frac{1}{2}}c^{-\langle\sigma,\varepsilon_j\rangle}e^{-\varepsilon_j})(1+q^{-\frac{1}{2}}d^{-\langle\sigma,\varepsilon_j\rangle}e^{-\varepsilon_j})}{(1-q^{-1}e^{-2\varepsilon_j})}\Bigg].
\end{aligned}
$$

Hence

$$\Phi_\sigma = \delta^{-1}\delta_q^{-1}\,\Psi_\sigma,$$

where

$$
\begin{aligned}
(5.7)\quad \Psi_\sigma &:= (abcd)^{\langle\sigma,\rho_1\rangle}\, t^{\langle\sigma,2\rho_2\rangle}\, e^{\rho+2\rho_1} \prod_{\alpha\in R_2^+} (1-t^{-\langle\sigma,\alpha\rangle}e^{-\alpha}) \\
&\quad \times \prod_{j=1}^{n}\big[(1-qe^{-2\varepsilon_j})(1-a^{-\langle\sigma,\varepsilon_j\rangle}e^{-\varepsilon_j})(1+b^{-\langle\sigma,\varepsilon_j\rangle}e^{-\varepsilon_j}) \\
&\qquad \times (1-q^{-\frac{1}{2}}c^{-\langle\sigma,\varepsilon_j\rangle}e^{-\varepsilon_j})(1+q^{-\frac{1}{2}}d^{-\langle\sigma,\varepsilon_j\rangle}e^{-\varepsilon_j})\big]
\end{aligned}
$$

and δ_q is the following element of A^W:

$$
\begin{aligned}
(5.8)\qquad \delta_q &:= \prod_{j=1}^{n}(q^{-\frac{1}{2}}e^{\varepsilon_j}-q^{\frac{1}{2}}e^{-\varepsilon_j})(q^{\frac{1}{2}}e^{\varepsilon_j}-q^{-\frac{1}{2}}e^{-\varepsilon_j}) \\
&= e^{2\rho_1}\prod_{j=1}^{n}(1-qe^{-2\varepsilon_j})(1-q^{-1}e^{-2\varepsilon_j}).
\end{aligned}
$$

If $E\subset(\tfrac{1}{2}R_1)\cup R_2$ then write

$$\|E\| := \sum_{\alpha\in E}\alpha.$$

Expansion of (5.7) yields

$$(5.9)\quad \Psi_\sigma = \sum_{E_0,\dots,E_4\subset\frac{1}{2}R_1^+}\ \sum_{F\subset R_2^+} c_{E_0,\dots,E_4,F}\, e^{\rho+2\rho_1-2\|E_0\|-\|E_1\|-\dots-\|E_4\|-\|F\|},$$

where

$$
\begin{aligned}
(5.10)\quad c_{E_0,\dots,E_4,F} &:= (-1)^{|E_0|+|E_1|+|E_3|+|F|}\, q^{|E_0|-\frac{1}{2}|E_3|-\frac{1}{2}|E_4|} \\
&\quad \times a^{\langle\sigma,\rho_1-\|E_1\|\rangle}\, b^{\langle\sigma,\rho_1-\|E_2\|\rangle}\, c^{\langle\sigma,\rho_1-\|E_3\|\rangle}\, d^{\langle\sigma,\rho_1-\|E_4\|\rangle}\, t^{\langle\sigma,2\rho_2-\|F\|\rangle}.
\end{aligned}
$$

Now we can rewrite (5.4) as

$$D_\sigma f = \delta^{-1} \delta_q^{-1} \, \widetilde{D}_\sigma f$$

with

(5.11) $$\widetilde{D}_\sigma f := |W_\sigma|^{-1} \sum_{w \in W} \varepsilon(w) \, (w\Psi_\sigma) \, (T_{w\sigma} f - f).$$

Consider (5.11) with $f := m_\lambda$ $(\lambda \in P^+)$ and substitute (5.9). Then

$$\widetilde{D}_\sigma m_\lambda = |W_\sigma|^{-1} |W_\lambda|^{-1} \sum_{w_1, w_2 \in W} \; \sum_{E_0, \ldots, E_4 \subset \frac{1}{2} R_1^+} \; \sum_{F \subset R_2^+} c_{E_0, \ldots, E_4, F} \, \varepsilon(w_1)$$
$$\times \left(q^{\langle w_1 \sigma, w_2 \lambda \rangle} - 1 \right) e^{w_1 (\rho + 2\rho_1 - 2\|E_0\| - \|E_1\| - \cdots - \|E_4\| - \|F\|) + w_2 \lambda}.$$

Put $w_2 = w_1 w$. Then

(5.12)
$$\widetilde{D}_\sigma m_\lambda = |W_\sigma|^{-1} |W_\lambda|^{-1} \sum_{w \in W} \; \sum_{E_0, \ldots, E_4 \subset \frac{1}{2} R_1^+} \; \sum_{F \subset R_2^+} c_{E_0, \ldots, E_4, F} \left(q^{\langle \sigma, w\lambda \rangle} - 1 \right)$$
$$\times J_{w\lambda + \rho + 2\rho_1 - 2\|E_0\| - \|E_1\| - \cdots - \|E_4\| - \|F\|}.$$

Hence $\widetilde{D}_\sigma m_\lambda \in A^{W,\varepsilon}$. Now the J-function in (5.12) is either 0 or $\varepsilon(w') \delta \chi_\nu$, where $w' \in W$, $\nu \in P^+$ and

$$w'(\nu + \rho) = w\lambda + \rho + 2\rho_1 - 2\|E_0\| - \|E_1\| - \cdots - \|E_4\| - \|F\|,$$

so that

(5.13) $$\nu + \rho = (w')^{-1} w\lambda$$
$$+ (w')^{-1} (3\rho_1 - 2\|E_0\| - \|E_1\| - \cdots - \|E_4\|) + (w')^{-1} (\rho_2 - \|F\|).$$

Now
(5.14)
$$(w')^{-1} (3\rho_1 - 2\|E_0\| - \|E_1\| - \cdots - \|E_4\|) = (w')^{-1} \sum_{j=1}^n k_j \varepsilon_j = \sum_{j=1}^n k_j' \varepsilon_j \le 3\rho_1$$

with $k_j, k_j' \in \{-3, -2, -1, 0, 1, 2, 3\}$,

(5.15) $$(w')^{-1} (\rho_2 - \|F\|) = (w')^{-1} \sum_{\alpha \in R_2^+} k_\alpha \alpha = \sum_{\alpha \in R_2^+} k_\alpha' \alpha \le \rho_2$$

with $k_\alpha, k_\alpha' = \pm \frac{1}{2}$, and

(5.16) $$(w')^{-1} w\lambda \le \lambda.$$

Substitution of (5.14), (5.15), (5.16) in (5.13) yields

(5.17) $$\nu + \rho \le \lambda + 3\rho_1 + \rho_2 = \lambda + \rho + 2\rho_1.$$

Hence

$$(5.18) \qquad \tilde{D}_\sigma m_\lambda = \sum_{\nu \in P+;\, \nu \leq \lambda + 2\rho_1} b_\nu J_{\nu+\rho}$$

for certain coefficients b_ν.

In order to compute $b_{\lambda+2\rho_1}$ observe that equality in (5.17) holds iff equality holds in (5.14), (5.15), (5.16), i.e., iff $(w')^{-1}w \in W_\lambda$ and

$$(5.19) \qquad ||E_0|| = ||E_1|| = \cdots = ||E_4|| = \tfrac{1}{2}(\rho_1 - w'\rho_1),$$

$$(5.20) \qquad ||F|| = \rho_2 - w'\rho_2.$$

Hence, by (5.12),
(5.21)
$$b_{\lambda+2\rho_1} = |W_\sigma|^{-1} |W_\lambda|^{-1} \sum_{w,w' \in W;\, (w')^{-1}w \in W_\lambda} \varepsilon(w')\, c_{E_0,\ldots,E_4,F}\left(q^{\langle \sigma, w\lambda \rangle} - 1\right),$$

where $E_0,\ldots,E_4,F$ are determined by (5.19), (5.20). It follows from (5.19), (5.20) that

$$2||E_0|| + ||F|| = \rho - w'\rho, \quad \text{hence} \quad (-1)^{|E_0|+|F|} = \varepsilon(w').$$

Substitution of (5.19), (5.20) into (5.10) now yields:

$$c_{E_0,\ldots,E_4,F} = \varepsilon(w')\,(abcd)^{\frac{1}{2}(1+((w')^{-1}\sigma,\rho_1))}\, t^{n-1+((w')^{-1}\sigma,\rho_2)}.$$

When we substitute this last expression into (5.21) then we obtain

$$b_{\lambda+2\rho_1} = |W_\sigma|^{-1} |W_\lambda|^{-1}$$
$$\times \sum_{w,w' \in W;\, (w')^{-1}w \in W_\lambda} (abcd)^{\frac{1}{2}(1+((w')^{-1}\sigma,\rho_1))}\, t^{n-1+((w')^{-1}\sigma,\rho_2)}\left(q^{((w')^{-1}\sigma,\lambda)} - 1\right)$$
$$= |W_\sigma|^{-1} \sum_{w \in W} (abcd)^{\frac{1}{2}(1+(w\sigma,\rho_1))}\, t^{n-1+(w\sigma,\rho_2)}\left(q^{(w\sigma,\lambda)} - 1\right).$$

Hence

$$(5.22) \qquad b_{\lambda+2\rho_1} = \sum_{j=1}^{n} \sum_{\varepsilon=\pm1} (abcd)^{\frac{1}{2}(1+\varepsilon)}\, t^{n-1+\varepsilon(n-j)}\left(q^{\varepsilon\lambda_j} - 1\right),$$

which is (5.5), when we take in account the replacement made for a, b, c, d.

Next we show that, for $f \in A^W$, $\tilde{D}_\sigma f$ given by (5.11) is divisible by δ_q. In view of (5.8) this will follow if we can show that, for each $w \in W$, $(w\Psi_\sigma)(T_{w\sigma}f - f)$ is divisible by the $4n$ prime factors $1 \pm q^{\pm\frac{1}{2}}e^{-\varepsilon_j}$. By (5.7), all but the two factors $1 \pm q^{-\frac{1}{2}}e^{-w\sigma}$ are divisors of $w\Psi_\sigma$. We will show that these two factors are divisors of $T_{w\sigma}f - f$. Write f as a Laurent polynomial $F(e^{\varepsilon_1},\ldots,e^{\varepsilon_n})$, invariant under the transformations $e^{\varepsilon_j} \mapsto e^{-\varepsilon_j}$. If $w\sigma = \varepsilon_j$ then

$$T_{w\sigma}f - f = F(e^{\varepsilon_1},\ldots,qe^{\varepsilon_j},\ldots,e^{\varepsilon_n}) - F(e^{\varepsilon_1},\ldots,e^{\varepsilon_j},\ldots,e^{\varepsilon_n})$$

becomes 0 for $e^{\varepsilon j} = \pm q^{-\frac{1}{2}}$, hence it is divisible by $1 \pm q^{-\frac{1}{2}} e^{-\varepsilon j}$. A similar argument is valid for $w\sigma = -\varepsilon_j$.

By (5.18),

$$(5.23) \qquad \delta^{-1} \tilde{D}_\sigma m_\lambda = \sum_{\nu \in P^+; \, \nu \le \lambda + 2\rho_1} c_\nu \, m_\nu$$

for certain coefficients c_ν, with $c_{\lambda+2\rho_1} = b_{\lambda+2\rho_1}$ given by (5.22). Also, $\delta^{-1} \tilde{D}_\sigma m_\lambda$ will still be divisible by δ_q. By (5.8), $\delta_q \in A^W$ with highest term $m_{2\rho_1}$. Hence $D_\sigma m_\lambda = \delta^{-1} \delta_q^{-1} \tilde{D}_\sigma m_\lambda$ will be in A^W with highest term $b_{\lambda+2\rho_1} m_\lambda$. $\square$

We have

$$\Delta = \prod_{\alpha \in R_1} \frac{(e^\alpha; q)_\infty}{(ae^{\frac{1}{2}\alpha}, be^{\frac{1}{2}\alpha}, ce^{\frac{1}{2}\alpha}, de^{\frac{1}{2}\alpha}; q)_\infty} \prod_{\alpha \in R_2} \frac{(e^\alpha; q)_\infty}{(te^\alpha; q)_\infty}.$$

Hence

$$\Delta(x) = (w\Delta)(x) = (w\Delta^+)(x)\,(w\Delta^+)(-x), \quad w \in W.$$

LEMMA 5.3. *With the assumptions of Definition 5.1 we have*

$$(5.24) \qquad \int_T (D_\sigma f)(x)\, g(x)\, \Delta(x)\, dx = \int_T f(x)\,(D_\sigma g)(x)\, \Delta(x)\, dx, \quad f, g \in A^W.$$

PROOF. Since $-\mathrm{id} \in W$, $f(x) = f(-x)$ and $g(x) = g(-x)$. By (5.4) and (5.3), formula (5.24) can be equivalently written as

$$(5.25) \quad \sum_{w \in W} \int_T (T_{w\sigma}(w\Delta^+))(x)\, ((T_{w\sigma} f)(x) - f(x))\,(w\Delta^+)(-x)\, g(-x)\, dx$$

$$= \sum_{w \in W} \int_T (w\Delta^+)(x)\, f(x)\,(T_{w\sigma}(w\Delta^+))(-x)\,((T_{w\sigma} g)(-x) - g(-x))\, dx.$$

Since T, f and g are W-invariant, formula (5.25) will be implied by the two identities

$$\int_T (\Delta^+ f)(x - i(\log q)\sigma)\,(\Delta^+ g)(-x)\, dx = \int_T (\Delta^+ f)(x)\,(\Delta^+ g)(-x - i(\log q)\sigma)\, dx$$

and

$$\sum_{w \in W} (T_\sigma \Delta^+)(w^{-1}x)\, \Delta^+(-w^{-1}x) = \sum_{w \in W} \Delta^+(w^{-1}x)\,(T_\sigma \Delta^+)(-w^{-1}x).$$

The second identity is obvious, since $-\mathrm{id} \in W$. For the first identity observe that the integral

$$\int_C (\Delta^+ f)(z - i(\log q), x_2, \ldots, x_n)\,(\Delta^+ g)(-z, -x_2, \ldots, -x_n)\, dz$$

over the contour

$$C = [-\pi, \pi] \cup [\pi, \pi + i\log q] \cup [\pi + i\log q, -\pi + i\log q] \cup [-\pi + i\log q, -\pi]$$

vanishes by Cauchy's theorem. (By the assumptions on a, b, c, d, t there are no singularities inside the contour.) Now the result follows, since $\Delta^+ f$ and $\Delta^+ g$ are invariant under translations by $2\pi\sigma$. $\square$

It follows now immediately from Lemmas 5.2 and 5.3 that:

THEOREM 5.4. $D_\sigma P_\lambda = a_{\lambda,\lambda} P_\lambda$ with $a_{\lambda,\lambda}$ given by (5.5).

Now we are ready for the main theorem.

THEOREM 5.5. If $\lambda, \mu \in P^+$, $\lambda \neq \mu$, then

$$\int_T P_\lambda(x)\, P_\mu(x)\, \Delta(x)\, dx = 0.$$

PROOF. All integrals

$$\int_T m_\lambda(x)\, m_\mu(x)\, \Delta(x)\, dx$$

are continuous in a, b, c, d, t. Hence the coefficients $u_{\lambda,\mu}$ in Definition 5.1 are continuous in a, b, c, d, t. This implies that

$$\int_T P_\lambda(x)\, P_\mu(x)\, \Delta(x)\, dx$$

is continuous in a, b, c, d, t. By Theorem 5.4 and Lemma 5.3,

$$\int_T P_\lambda(x)\, P_\mu(x)\, \Delta(x)\, dx = 0$$

if $a_{\lambda,\lambda} \neq a_{\mu,\mu}$. Fix distinct λ and μ it follows from (5.5) that, for fixed nonzero a, b, c, d, the eigenvalues $a_{\lambda,\lambda}$ and $a_{\mu,\mu}$ are distinct as polynomials in t. This implies the orthogonality of P_λ and P_μ for a, b, c, d, t in a dense subset of the parameter domain under consideration. Hence, by continuity, the theorem follows. $\square$

The method of proof in this last theorem is different from the method used in similar situations by Macdonald [12]. While Macdonald leaves the parameters fixed and shows that equality of eigenvalues for all q implies (in most cases) equality of weights, the above proof leaves q fixed and shows that equality of eigenvalues for all parameter values implies equality of weights.

6. Discussion of results

6.1. Special cases. When we compare (5.1) with (3.1), (3.2) and (3.3), then it is clear that Askey-Wilson polynomials for root system BC_n with a, b, c, d, t replaced by $q^{\frac{1}{2}}, -q^{\frac{1}{2}}, ab^{\frac{1}{2}}, -b^{\frac{1}{2}}, t$ become Macdonald's polynomials for the pair (BC_n, B_n). The operator D_σ given by (5.4) then specializes to the operator for which Theorem 2.3 is valid in case (BC_n, B_n), and the eigenvalue (5.5) specializes to the eigenvalue in Theorem 2.3, cf. (3.6). We can also work then with E_σ instead of D_σ.

Next, when we compare (5.1) with (3.1), (3.4) and (3.5) then it is clear that our polynomials with a, b, c, d, t, q replaced by $ab^{\frac{1}{2}}, qab^{\frac{1}{2}}, -b^{\frac{1}{2}}, -qb^{\frac{1}{2}}, t, q^2$ become Macdonald's polynomials for the pair (BC_n, C_n). The operator D_σ given by (5.4) then becomes the operator $D_{2\varepsilon_1}$ for the pair (BC_n, C_n). Theorem 2.3 does not say anything about eigenfunctions of this operator, but Theorem 5.4 implies that Macdonald's polynomials for the pair (BC_n, C_n) are eigenfunctions of $D_{2\varepsilon_1}$. This corresponds nicely with the cases E_8, F_4, G_2 of Theorem 2.3, where $\langle \sigma, \alpha_* \rangle$ takes values $0, 1, 2$ as α runs through R^+ and we have to work with D_σ instead of E_σ. It would be interesting to consider if P_λ might also be eigenfunction of D_σ for other "quasi-minuscule" σ.

Comparison of (5.1) and (4.1) makes it evident that the BC_n Askey-Wilson polynomials reduce to the one-variable Askey-Wilson polynomials for $n = 1$.

6.2. A Selberg-type integral and a conjectured quadratic norm. Let Δ be given by (5.2) and (5.1) and let the parameters satisfy the inequalities of Definition 5.1. Gustafson [3, (2)] evaluated the Selberg type integral

$$\frac{1}{(2\pi)^n} \int_0^{2\pi} \cdots \int_0^{2\pi} \Delta(x)\, dx = 2^n n! \prod_{j=1}^n \frac{(t, t^{n+j-2}abcd; q)_\infty}{(t^j, q, abt^{j-1}, act^{j-1}, \ldots, cdt^{j-1}; q)_\infty}.$$

On the other hand, Macdonald [12, (12.6)] conjectured an explicit expression for the quadratic norm $\langle P_\lambda, P_\lambda \rangle$ for polynomials P_λ associated with any admissible pair (R, S). It can be shown that Macdonald's conjecture in case $\lambda = 0$ and $(R, S) = (BC_n, B_n)$ coincides with Gustafson's formula for $(a, b, c, d) = (q^{\frac{1}{2}}, -q^{\frac{1}{2}}, ab^{\frac{1}{2}}, -b^{\frac{1}{2}})$. In October 1991, when prof. Macdonald was visiting The Netherlands, first the author has given a conjectured expression for $\langle P_\lambda, P_\lambda \rangle / \langle 1, 1 \rangle$, where P_λ is an Askey-Wilson polynomial for root system BC_n, and next Macdonald [13] has rewritten this as a conjectured expression for $\langle P_\lambda, P_\lambda \rangle$. On the same occasion, Macdonald [13] has also extended his other conjectures in [12, §12] to the BC_n Askey-Wilson case.

6.3. The Askey-Wilson hierarchy for BC_n. It is very probable that all specializations and limit cases of one-variable Askey-Wilson polynomials have their analogues in the case of BC_n. Someone should certainly write down the orthogonality relations and difference operators with explicit eigenvalues for all these specializations. In some cases these explicit formulas may be rigorously proved by straightforward limit transition from the general Askey-Wilson case. In other cases, the limit transition may only give a formal proof and, for a rigorous derivation, the proofs of the present paper will have to be imitated.

q-Racah-type polynomials for root system BC_n should also be obtained. Here analytic continuation from the BC_n Askey-Wilson polynomials will be needed and residues, possibly higher dimensional, will have to be taken. Similar problems will arise when the condition $|a|, |b|, |c|, |d| \le 1$ is dropped in Definition 5.1. In the corresponding one-variable case discrete terms are then added to the orthogonality relations.

6.4. Quantum group interpretations. It is known from work by Koornwinder [9], [10], Koelink [8] and Noumi & Mimachi [15], [16] that one-variable Askey-Wilson polynomials have an interpretation on the quantum group $SU_q(2)$. Noumi [14] announces an interpretation of Macdonald's polynomials for root system A_{n-1} as zonal spherical functions on the quantum analogues of the homogeneous spaces $GL(n)/SO(n)$ and $GL(2n)/Sp(2n)$. According to Noumi, this was already done for the quantum analogue of $SL(3)/SO(3)$ by Ueno & Takebayashi. It would be interesting to find quantum group interpretations of Macdonald's polynomials in case of all root systems, and also of the BC_n Askey-Wilson polynomials considered in the present paper.

REFERENCES

1. R. Askey and J. Wilson, *Some basic hypergeometric orthogonal polynomials that generalize Jacobi polynomials*, Mem. Amer. Math. Soc. no. 319 (1985).
2. N. Bourbaki, *Groupes et algèbres de Lie*, Chapitres 4, 5 et 6, Hermann, 1968.
3. R. A. Gustafson, *A generalization of Selberg's beta integral*, Bull. Amer. Math. Soc. (N. S.) **22** (1990), 97–105.
4. G. J. Heckman, *Root systems and hypergeometric functions. II*, Compositio Math. **64** (1987), 353–373.
5. _____, *An elementary approach to the hypergeometric shift operators of Opdam*, Invent. Math. **103** (1991), 341–350.
6. G. J. Heckman and E. M. Opdam, *Root systems and hypergeometric functions. I*, Compositio Math. **64** (1987), 329–352.
7. J. E. Humphreys, *Introduction to Lie algebras and representation theory*, Springer, 1972.
8. H. T. Koelink, *The addition formula for continuous q-Legendre polynomials and associated spherical elements on the SU(2) quantum group related to Askey-Wilson polynomials*, Univ. of Leiden, Dept. of Math., preprint W-90-11, 1990.
9. T. H. Koornwinder, *Orthogonal polynomials in connection with quantum groups*, Orthogonal polynomials: theory and practice (P. Nevai, ed.), NATO ASI Series C 294, Kluwer, 1990, pp. 257–292.
10. _____, *Askey-Wilson polynomials as zonal spherical functions on the SU(2) quantum group*, CWI Rep. AM-R9013, preprint, 1990.
11. I. G. Macdonald, *A new class of symmetric functions*, Séminaire Lotharingien de Combinatoire, 20e Session (L. Cerlienco and D. Foata, eds.), Publication de l'Institut de Recherche Mathématique Avancée 372/S-20, Strasbourg, 1988, pp. 131–171.
12. _____, *Orthogonal polynomials associated with root systems*, preprint, 1988.
13. _____, *Some conjectures for Koornwinder's orthogonal polynomials*, informal manuscript, October 1991.
14. M. Noumi, *Macdonald's symmetric polynomials as zonal spherical functions on some quantum homogeneous spaces*, informal summary, March 1992.
15. M. Noumi and K. Mimachi, *Askey-Wilson polynomials and the quantum group $SU_q(2)$*, Proc. Japan Acad. Ser. A Math. Sci. 66 (1990), 146–149.
16. _____, *Rogers' q-ultraspherical polynomials on a quantum 2-sphere*, Duke Math. J. **63** (1991), 65–80.

CWI, P.O. BOX 4079, 1009 AB AMSTERDAM, THE NETHERLANDS

Current address: UNIVERSITY OF AMSTERDAM, FACULTY OF MATHEMATICS AND COMPUTER SCIENCE, PLANTAGE MUIDERGRACHT 24, 1018 TV AMSTERDAM, THE NETHERLANDS
E-mail: thk@fwi.uva.nl

Contemporary Mathematics
Volume **138**, 1992

Some Special Functions in the Fock Space

RAY A. KUNZE

Introduction

There is a surprising amount of fairly recent literature [1], [2], [3] concerning analytic questions centering around the old and well understood topic of how the Fock space associated with a homogeneous bounded symmetric domain decomposes under the action of a maximal compact subgroup of the automorphism group of the domain [4], [5], [6]. This problem has been studied from a variety of viewpoints. Here we attempt to set the stage for a systematic attack on this and similar problems from the point of view of special functions. In particular, we study an integral similar to the integral for generalized Bessel functions [7], [8], [9] that is relevant for this class of problems. Our initial results apply to the more general context considered in the first part of [10].

1. The General Setting

The operator-valued Bessel functions considered in [7] and [8] were studied by analytic techniques from real variables. They are essentially determined by the following data: a compact subgroup of the orthogonal group on Euclidean space and an irreducible unitary representation of the subgroup; the Bessel function is then given by an integral over the subgroup that involves the representation and yields an integral kernel that implements the Fourier transform in the subspace of all square integrable functions that transform according to the given representation. However, a rather different and perhaps richer theory is available if one replaces the Euclidean space by its complexification and considers the natural holomorphic extensions of the Bessel function integrals [9]; the extended Bessel functions then relate in a natural way to a space of entire functions, the Fock space, on the complexification. Here we shall consider a more general class of special functions.

1991 *Mathematics Subject Classification*. Primary 22E30, 33C80, Secondary 22E45.

This paper is in final form and no version of it will be submitted for publication elsewhere .

1.1 The basic assumptions. Let G be a reductive complex affine algebraic group, U a compact real form of G, and $a \to a^*$, $a \in G$ the conjugate regular involutory anti-automorphism of G such that $u^* = u^{-1}$ for $u \in U$. Suppose that $(a, z) \to a\, z$ is an algebraic linear action of G on a complex finite dimensional inner product space S such that

$$(1) \qquad\qquad (a\, z \,|\, w) = (z \,|\, a^* w)$$

for all $a \in G$ and all z, w in S. In particular, (1) implies that U acts unitarily on S.

An example, involving a special case of the general problem mentioned above, is given by $GL(n, \mathbb{C})$ acting on the space S of $n \times n$ complex symmetric matrices, the action being

$$(2) \qquad\qquad a \cdot z = aza^t.$$

Here we take

$$(3) \qquad\qquad (z|w) = tr(zw^*)$$

where * denotes conjugate transpose. In this case, U is just $U(n)$.

Another example with the same underlying space and action is given by the pair $O(n, \mathbb{C})$, $O(n)$.

1.2 The function I_λ. Assume that λ is an irreducible finite dimensional algebraic representation of G on a complex Hilbert space $\mathcal{H}(\lambda)$ such that

$$(1) \qquad\qquad \lambda(a^*) = \lambda(a)^*, \quad a \in G$$

The λ-*type function for the transformation group* (G, U) is the operator-valued function I_λ on $S \times S$ defined by

$$(2) \qquad\qquad I_\lambda(z, w) = \int_U e^{(uz|w)} \lambda(u) du,$$

du denoting normalized Haar measure on U.

The integral in (2) is similar to the integral defining the generalized Bessel function. Specifically, suppose there is a U-stable real form V of S on which the inner product is real-valued; then

$$(z, w) \to (z|\overline{w})$$

$\overline{w}$ denoting the conjugation defined by V, is a nondegenerate G-invariant symmetric bilinear form on S. In this case, I_λ may be expressed in terms of the extended Bessel function

$$(3) \qquad\qquad J_\lambda(z, w) = \int_U e^{i(z|(\overline{uw}))} \lambda(u) du$$

As one easily sees, it is then true that

$$(4) \qquad\qquad I_\lambda(z, w) = J_\lambda(\overline{-iw}|z)$$

and hence that J_λ is given in terms of I_λ by

$$(5) \qquad J_\lambda(z,w) = I_\lambda(w, \overline{iz})$$

In the first of the above examples, there is no $U(n)$-invariant real form of S; however, in the second, the subspace of real symmetric matrices is $O(n)$-invariant, and the inner product is real-valued there.

The λ-type function has a number of special properties, the simplest of which will be stated next.

1.3 Proposition. *The function $I_\lambda(z,w)$ is holomorphic as a function of z and anti-holomorphic (or conjugate holomorphic) as a function of w. It is hermitian in the sense that*

$$(1) \qquad I_\lambda(z,w)^* = I_\lambda(w,z)$$

for all z, w in S, and transforms under G according to

$$(2) \qquad I_\lambda(az, bw) = \lambda(b^{-1})^* I_\lambda(z,w)\lambda(a)^{-1}.$$

PROOF. We omit the proof of the first statement. For any z, w in S,

$$I_\lambda(z,w)^* = \int_U e^{(w|uz)}\lambda(u^{-1})du$$

and making the transformation $u \to u^{-1}$, we obtain (1). For a in U

$$I_\lambda(az, w) = \int_U e^{(uaz|w)}\lambda(u)du$$

and making the transformation $u \to ua^{-1}$, we see that

$$(3) \qquad I_\lambda(az, w) = I_\lambda(z,w)\lambda(a^{-1})$$

Now for fixed z, w both sides of (3) are holomorphic as functions of a on G. Therefore, (3) holds for all a in G. This and (1) now imply the remainder of (2).

These functions have a special mean value property.

1.4 Proposition. *If ϵ denotes the trivial one-dimensional representation of G then*

$$(1) \qquad \int_U I_\lambda(z + uv, w)du = I_\epsilon(v, w)\,I_\lambda(z,w)$$

for all v, w, z in S.

PROOF. We have

$$\int \left(\int e^{(sz + suv|w)}\lambda(s)ds \right) du = \int \left(\int e^{(sz|w)}e^{(suv|w)}\lambda(s)du \right) ds$$

$$= \int e^{(uv|w)} \left(\int e^{(sz|w)}\lambda(s)ds \right) du.$$

1.5. For $v \in S$, we set

$$D_v F(z) = \frac{d}{dt} F(z + tv)$$

the derivative with respect to the complex variable t being evaluated at 0. Then there is a unique conjugate-linear isomorphism $f \to f(D)$ of the algebra of polynomial functions on S with the algebra generated by the derivations D_v such that for any linear functional $z \to (z|v)$

$$(D|v) = D_v .$$

1.6 Proposition. *If f is a G-invariant polynomial function on S then for fixed w*

$$f(D)I_\lambda(\cdot, w) = \overline{f(w)}I_\lambda(\cdot, w).$$

PROOF. Differentiating with respect to z, we see that

$$D_v I_\lambda(z, w) = \int_U D_v e^{(uz|w)} \lambda(u) du = \int_U \overline{(u^{-1}w|v)} \, e^{(uz|w)} \, \lambda(u) du$$

which implies

$$(1) \qquad\qquad f(D)I_\lambda(z, w) = \int_U \overline{f(u^{-1}w)} \, e^{(uz|w)} \, \lambda(u) du$$

for any polynomial function f.

In particular, if $f(w) = 0$ for all G-invariant polynomials f with 0 constant term, then $I_\lambda(\cdot, w)$ is G-*harmonic* in the sense that

$$(2) \qquad\qquad f(D)I_\lambda(\cdot, w) = 0$$

for all such f.

2. The Fock Space

Here we are concerned with some particular Hilbert spaces of holomorphic functions, their reproducing kernels, and some general features of the way U and G act in these spaces.

2.1. Let $\mathcal{K}$ be a complex inner product space of finite dimension, $\mathcal{P}(S, \mathcal{K})$ the space of polynomial functions on S with values in $\mathcal{K}$, and $\mathcal{H}(S, \mathcal{K})$ the associated *Fock* space of all holomorphic $f : S \to \mathcal{K}$ such that

$$\int_S |f(z)|^2 \, e^{-(z|z)} dz < \infty$$

where dz denotes Lebesgue measure on S normalized so that

$$\int_S e^{-(z|z)} \, dz = 1.$$

When $\mathcal{K} = \mathbb{C}$, we write $\mathcal{P}(S)$ for $\mathcal{H}(S, \mathcal{K})$ and $\mathcal{H}(S)$ for $\mathcal{H}(S, \mathcal{K})$. We shall assume the many well-known properties of these spaces that are summarized by the following statement.

2.2 Theorem. *If the Fock space $\mathcal{H}(S, \mathcal{K})$ is equipped with the inner product*

$$(1) \qquad (f|g) = \int_U (f(z) \,|\, g(z))\, e^{-(z|z)} dz$$

then the point evaluations

$$(2) \qquad E_z : f \to f(z), \quad z \in S$$

are continuous linear maps, $\mathcal{H}(S, \mathcal{K})$ is a Hilbert space isomorphic to $\mathcal{H}(S) \otimes \mathcal{K}$, and

$$(3) \qquad E_z\, E_w^* = e^{(z|w)} I_{\mathcal{K}}$$

for all z, w in S. If g is any function on S with values in K that is square-integrable with respect to the Gaussian measure in 2.1, then the equation

$$(4) \qquad f(z) = \int_U e^{(z|w)}\, g(w)\, e^{-(w|w)} dw, \quad z \in S$$

defines the orthogonal projection of g on the subspace $\mathcal{H}(S, \mathcal{K})$. In particular, the reproducing kernel for $\mathcal{H}(S, \mathcal{K})$ is given by the function

$$(5) \qquad Q(z, w) = e^{(z|w)}$$

In addition, $\mathcal{P}(S, \mathcal{K})$ is a dense linear subspace, and there is a canonical conjugate linear map

$$f \to f(D), \quad f \in \mathcal{P}(S, \mathcal{K})$$

that assigns to each f a differential operator

$$f(D) : \mathcal{P}(S, \mathcal{K}) \to \mathcal{P}(S)$$

such that for any f, g in $\mathcal{P}(S, \mathcal{K})$, $h \in \mathcal{P}(S)$, and $z \in S$,

$$(6) \quad (g(D)h(D)f)(z) = ((hg)(D)f)(z) = \int_S e^{(z|w)}(f(w)|g(w))\overline{h}(w)e^{-(w|w)} dw,$$

$h(D)$ being the operator defined in 1.5; so the Fock space inner product is given algebraically on the polynomial subspace $\mathcal{P}(S, \mathcal{K})$ by

$$(7) \qquad (f|g) = g(D)f(0).$$

On this subspace, for $h \in \mathcal{P}(S)$ the adjoint of the differential operator $h(D)$ is multiplication by h. Moreover, the equation

$$(8) \qquad (L(u)f)(z) = f(u^{-1}z), \quad u \in U, z \in S$$

defines a continuous unitary representation of U on $\mathcal{H}(S, \mathcal{K})$.

In (8), if one replaces u by an arbitrary element a in G, the resulting function $L(a)f$ is again holomorphic but generally not an element of $\mathcal{H}(S, \mathcal{K})$. However, G does act on various dense subspaces, e.g., on $\mathcal{P}(S, \mathcal{K})$.

Since point evaluations are continuous, every continuous linear operator on $\mathcal{H}(S, \mathcal{K})$ is given by a uniquely determined integral kernel.

2.3 Proposition. *For each continuous linear operator B on $\mathcal{H}(S, \mathcal{K})$, the equation*

(1) $$Q_B(z, w) = E_z B E_w^*$$

defines an operator-valued integral kernel Q_B such that

$$w \rightarrow Q_B(z, w)^* \phi$$

lies in $\mathcal{H}(S, \mathcal{K})$ for all $z \in S$, $\phi \in \mathcal{K}$, and

(2) $$Bf(z) = \int_S Q_B(z, w) f(w) e^{-(w|w)} dw, \quad z \in S,$$

for all $f \in \mathcal{H}(S, \mathcal{K})$. It is unique in the sense that if M is a similar integral kernel satisfying (2), then $M = Q_B$.

PROOF. The first statement follows from (1). To prove (2) let $f \in \mathcal{H}(S, \mathcal{K})$ and $\phi \in \mathcal{K}$. Then for arbitrary $z \in S$,

$$
\begin{aligned}
(Bf(z) \mid \phi) &= (f \mid B^* E_z^* \phi) \\
&= \int_S \left(f(w) \mid E_w B^* E_z^* \phi \right) e^{-(w|w)} dw \\
&= \int_S \left(Q_B(z, w) f(w) \mid \phi \right) e^{-(w|w)} dw \\
&= \left(\int_S Q_B(z, w) f(w) e^{-(w|w)} dw \mid \phi \right).
\end{aligned}
$$

Since this holds for every ϕ in $\mathcal{K}$, we see that (2) is true. If M has the cited properties, then

$$\int_S \left(f(w) \mid Q_B(z, w)^* \phi \right) e^{-(w|w)} dw = \int_S \left(f(w) \mid M(z, w)^* \phi \right) e^{-(w|w)} dw$$

for all $\phi \in S$ and $f \in \mathcal{H}(S, \mathcal{K})$. It follows that $M = Q_B$.

2.4. Recall that if R is a continuous unitary representation of U on a Hilbert space $\mathcal{H}$, then the operator

(1) $$T_\lambda = deg\,(\lambda) \int_U \overline{tr(\lambda(u))}\, R(u) du$$

is the orthogonal projection that maps $\mathcal{H}$ onto the closed subspace of all U-finite vectors of type λ . Because point evaluations are continuous linear maps on $\mathcal{H}(S, \mathcal{K})$ and integration commutes with such maps, it is easy to see that the orthogonal projection on the subspace of functions of type λ is given by the following result.

2.5 Proposition. *For the representation L of U on $\mathcal{H}(S, \mathcal{K})$, the projection onto the space of functions of type λ is given pointwise for arbitrary $f \in \mathcal{H}(S, \mathcal{K})$ by*

$$T_\lambda f(z) = deg\,(\lambda) \int_U tr(\lambda(u)) f(uz) du, \quad z \in S.$$

Next we show that the integral kernel corresponding to T_λ is a scalar (multiple of the identity) specified by the special function I_λ.

2.6 Corollary. *The integral kernel Q_λ for the projection T_λ is defined by the formula*

(1) $$Q_\lambda(z, w) = deg(\lambda)\, tr(I_\lambda(z, w))$$

and has the property that

(2) $$Q_\lambda(az, w) = Q_\lambda(z, a^* w)$$

for all $a \in G$ and z, w in S. It is the reproducing kernel for the subspace of functions of type λ, and the functions

(3) $$f(z) = \sum_i Q_\lambda(z, w_i) \phi_i$$

where $w_1, w_2, \ldots$ is any finite sequence of points in S and $\phi_1, \phi_2, \ldots$ a corresponding sequence of vectors in $\mathcal{K}$ form a dense linear subspace on which G acts by

(4) $$L(a) f(z) = \sum_i Q_\lambda(z, (a^*)^{-1} w_i) \phi_i, \quad a \in G.$$

PROOF. By **2.5** and **2.2**

$$T_\lambda f(z) = deg\,(\lambda) \int_U tr\,\lambda(u) f(uz) du$$

$$= deg\,(\lambda) \int_U tr\,\lambda(u) \Big(\int_S e^{(uz|w)} f(w)\, e^{-(w|w)} dw \Big)\, du$$

$$= deg\,(\lambda) \int_S f(w)\, \Big(\int_U tr\,\lambda(u) e^{(uz|w)} du \Big) e^{-(w|w)} dw$$

$$= deg\,(\lambda) \int_S \big(tr(I_\lambda(z, w)) \big)\, f(w) e^{-(w|w)} dw.$$

To complete the proof of (1), it now suffices, by **2.3**, to show that the function

$$w \to tr\big(I_\lambda(z, w)^* \big) \phi$$

lies in $\mathcal{H}(S, \mathcal{K})$ for each $z \in S$ and $\phi \in \mathcal{K}$. For this it is enough to show that

$$w \to \int_U e^{(uw|z)} tr(\lambda(u)) du$$

lies in $\mathcal{H}(S)$ for every $z \in S$. But this follows at once from **2.5**. The second statement (2) follows from **1.2**(1), **1.3**(2), and properties of the trace function; (3) is a standard property of reproducing kernels and (4) follows from (2).

2.7 The F_λ transform. For $f \in \mathcal{H}(S)$ the formula for $T_\lambda f$ in **2.5** can be written in the form

$$(1) \qquad T_\lambda f(z) = deg(\lambda) \, tr \int_U f(uz)\lambda(u)du$$

This suggests that for this case one should consider the holomorphic operator-valued functions $F_\lambda f$ defined for $f \in \mathcal{H}(S)$ by

$$(2) \qquad F_\lambda f(z) = \int_U f(uz)\lambda(u)du, \quad z \in S$$

Note that $I_\lambda(\cdot, w)$ is the image of $z \to e^{(z|w)}$ under the map F_λ defined by (2). By the argument in **1.3** the functions $F_\lambda f$ transform under an element $a \in G$ by

$$(3) \qquad F_\lambda f(az) = F_\lambda f(z)\,\lambda(a^{-1})$$

Moreover, it is easy to check that for $u \in U$

$$(4) \qquad F_\lambda L(u)f(z) = \lambda(u)F_\lambda f(z)$$

i.e., that F_λ converts translations into multiplications. All of this suggests that F_λ may be viewed as an alternate form of the projection T_λ . For this purpose, let $\mathcal{O}(S, \lambda)$ be the space of holomorphic functions $F : S \to Hom_{\mathbb{C}}(\mathcal{H}(\lambda), \mathcal{H}(\lambda))$ which transform under G by (3) such that

$$(5) \qquad \|F\|_\lambda^2 = deg(\lambda) \int_S tr(F(z)^* F(z))e^{-(z|z)}dz < \infty.$$

If $Hom_{\mathbb{C}}(\mathcal{H}(\lambda), \mathcal{H}(\lambda))$ is equipped with the Hilbert-Schmidt inner product normalized so that

$$(6) \qquad (A|B) = deg\,(\lambda)\,tr(B^* A)$$

it then follows that $\mathcal{O}(S, \lambda)$ is a subspace of the Fock space $\mathcal{H}(S, Hom_{\mathbb{C}}(\mathcal{H}(\lambda), \mathcal{H}(\lambda))$. Now we have the following result.

2.8 Proposition. *The F_λ transform is a U-equivariant partial isometry of $\mathcal{H}(S)$ onto $\mathcal{O}(S, \lambda)$ such that $F_\lambda^* F_\lambda = T_\lambda$. Its adjoint is the unitary map F_λ^* of $\mathcal{O}(S, \lambda)$ onto $T_\lambda \mathcal{H}(S)$ that is given by*

$$(1.) \qquad (F_\lambda^* F)(z) = deg(\lambda)\,tr(F(z), \quad z \in S$$

PROOF. Let $f \in \mathcal{H}(S)$ and define $F = F_\lambda f$ by **7.2**(2). Then, denoting the Hilbert-Schmidt norm by the subscript 2, we have

$$\|F(z)\|_2 \le \int_U |f(uz)| \, \|\lambda(u)\|_2 \, du.$$

From this and the Schwarz inequality, we see that

$$\|F(z)\|_2^2 \leq (deg\,(\lambda))^2 \int_U |f(uz)|^2 du.$$

It follows that

(2) $$\int_S \|F(z)\|_2^2 \, e^{-(z|z)} dz \leq (deg(\lambda))^2 \int_U \Big(\int_S |f(uz)|^2 e^{-(z|z)} dz \Big) du.$$

In (2), the measure preserving transformation $z \to u^{-1}z$ eliminates the variable u and shows that the F_λ is a continuous map of $\mathcal{H}(S)$ into $\mathcal{O}(S, \lambda)$.

Now suppose $F \in \mathcal{O}(S, \lambda)$ and define f by the right side of (1). Then for $u \in U$

(3) $$f(uz) = deg(\lambda)\,tr(\lambda(u)^* F(z)).$$

It then follows by the Plancherel theorem for U, or by direct computation from the orthogonality relations, that

(4) $$\int_U |f(uz)|^2 du = deg\,(\lambda)\,tr(F(z)^* F(z)).$$

Now using (3) and the Fubini theorem, we see that

(5) $$\|F\|_\lambda^2 = \int_U \Big(\int_S |f(uz)|^2 e^{-(z|z)} dz \Big) du = \|f\|^2.$$

Next, we argue that $F_\lambda f = F$. For this, let ϕ, ψ be elements of $\mathcal{H}(\lambda)$. Then by (2)

(6) $$(F_\lambda f(z)\phi|\psi) = deg(\lambda) \int_U tr(\lambda(u)^* F(z))\,(\lambda(u)\phi|\psi)du.$$

Now let $\epsilon_1, ..., \epsilon_n$ be an orthonormal basis for $\mathcal{H}(S)$. Then using this base to compute the trace in (5) and applying the orthogonality relations one sees that

$$(F_\lambda f(z)\phi|\psi) = \sum_i deg(\lambda) \int_U (\lambda(u)\phi|\psi)\overline{(\lambda(u)\epsilon_i|F(z)\epsilon_i)}\,du$$

$$= \sum_i (\phi|\epsilon_i)\,(F(z)\epsilon_i|\psi) = (F(z)\phi|\psi).$$

Therefore, $F_\lambda f = F$. It is now easy to complete the proof.

Since F_λ is norm decreasing, it follows in particular that

(7) $$\|I_\lambda(\cdot, w)\|_\lambda^2 \leq \int_S |e^{(z|w)}|^2 \, e^{-(z|z)} dz = e^{(w|w)}.$$

2.9 Proposition. *The F_λ transform is also given by the integral*

(1)
$$F_\lambda f(z) = \int_S f(w) I_\lambda(z, w) e^{-(w|w)} . dw$$

PROOF. Since $I_\lambda(\cdot, w)$ lies in $\mathcal{O}(S, \lambda)$, it follows from the Schwarz inequality that the integral in (1) is absolutely convergent. Now from the definition, **2.2**(4), and the Fubini theorem, we see that

$$
\begin{aligned}
F_\lambda f(z) &= \int_U f(uz)\lambda(u)du \\
&= \int_U \left(\int_S e^{(uz|w)} f(w) e^{-(w|w)} dw \right) \lambda(u) du \\
&= \int_S f(w) \left(\int_U e^{(uz|w)} \lambda(u) du \right) e^{-(w|w)} dw.
\end{aligned}
$$

Since the functions in $\mathcal{O}(S, \lambda)$ have values in $Hom_{\mathbb{C}}(\mathcal{H}(\lambda), \mathcal{H}(\lambda))$, the product of any two is well-defined, holomorphic, and, by the Schwarz inequality, absolutely integrable with respect to the Gaussian measure. It follows that I_λ determines the reproducing kernel for the space $\mathcal{O}(S, \lambda)$ in the following way.

2.10 Proposition. *If $F \in \mathcal{O}(S, \lambda)$ then*

$$F(w) = \int_S F(z) I_\lambda(z, w)^* e^{-(z|z)} dz$$

for every $w \in S$.

PROOF. For $F \in \mathcal{O}(S, \lambda)$ and $u \in U$, we have

$$F(w) = \int_S e^{(w|z)} F(z) e^{-(z|z)} dz = \int_S e^{(uw|z)} F(z) \lambda(u) e^{-(z|z)} dz$$

and the desired formula results from integration over U.

2.11 Corollary.

(1)
$$I_\lambda(w, v) = \int_S I_\lambda(z, v) \, I_\lambda(z, w)^* e^{-(z|z)} dz,$$

(2)
$$deg(\lambda) \, tr(I_\lambda(w, v)) = \int_S (I_\lambda(z, v) \,|\, (I_\lambda(z, w)) e^{-(z|z)} dz,$$

and

(3)
$$\int_S |(I_\lambda(z, w)\phi|\psi)|^2 e^{-(z|z)} dz \le deg(\lambda) \, tr(I_\lambda(w, w)) \, |\phi|^2 \, |\psi|^2$$

for all ϕ and ψ in $\mathcal{H}(\lambda)$.

As we shall see, the function I_λ plays yet another role in that it also determines the multiplicity of λ in the space $T_\lambda \mathcal{H}(S)$. Before considering this question, some general comments are in order.

2.12. If $\mathcal{H}$ is a complex inner product space, we let $\overline{\mathcal{H}}$ denote the complex vector space that one obtains from $\mathcal{H}$ by multiplying the complex structure by -1, i.e., by redefining scalar multiplication so that for $c \in \mathbb{C}$ and $f \in \mathcal{H}$

$$(1) \qquad\qquad c \cdot f = \overline{c}f$$

Then the complex conjugate of the inner product on $\mathcal{H}$ is an inner product on $\overline{\mathcal{H}}$, and with this inner product, $\overline{\mathcal{H}}$ is again an inner product space; if $\mathcal{H}$ is a Hilbert space, then $\overline{\mathcal{H}}$ is a Hilbert space canonically isomorphic to the dual of $\mathcal{H}$. We shall call $\overline{\mathcal{H}}$ the *conjugate of* $\mathcal{H}$.

Thus, we may consider the space $\overline{S}$ and the associated Fock spaces $\mathcal{H}(\overline{S}, \mathcal{K})$. It is easy to see that a function on $\overline{S}$ is holomorphic iff its complex conjugate is holomorphic on S. In fact, the map

$$f \to \overline{f}, \quad f \in \mathcal{H}(S)$$

which takes a function to its complex conjugate is a unitary map of $\mathcal{H}(S)$ onto the conjugate of $\mathcal{H}(\overline{S})$. It follows that the Hilbert space $\mathcal{H}(\overline{S}, \mathcal{K})$ is the set of all maps $f : S \to \mathcal{K}$ such that

$$z \to (\phi|f(\dot{z}))$$

lies in $\mathcal{H}(S)$ for every $\phi \in \mathcal{K}$. The reproducing kernel $\overline{Q}$ for this space is evidently given by the formula

$$(2) \qquad\qquad \overline{Q}(z, w) = e^{(w|z)}$$

cf., **2.2**(5).

2.13. To further study $T_\lambda \mathcal{H}(S)$ it is convenient to have a description of the space $Hom_U(\lambda, L)$ of operators that intertwine the restriction of λ to U with the representation L of U on $\mathcal{H}(S)$. It turns out that each such U intertwining operator is automatically a G intertwining operator. The precise statement requires some terminology. Let λ' denote the representation that agrees with λ on U and is *dual* to λ in the sense that

$$(1) \qquad\qquad \lambda'(a) = \lambda(a^*)^{-1}, \quad a \in G$$

Equivalently, if we set

$$(2) \qquad\qquad a' = (a^*)^{-1}, \quad a \in G$$

then the map $a \to a'$ is a conjugate holomorphic involutory automorphism of G and $\lambda'(a) = \lambda(a')$. Let $\mathcal{H}(\overline{S}, \lambda')$ denote the subspace of all f in $\mathcal{H}(\overline{S}, \mathcal{H}(\lambda))$ such that

$$(3) \qquad\qquad f(az) = \lambda'(a)f(z)$$

for all $a \in G$ and $z \in S$. A map satisfying (3), will be called a λ'-*covariant on* $\overline{S}$.

2.14 Theorem. *For every λ satisfying the conditions in* **1.2**

(1) $$Hom_U(\lambda, L) = Hom_G(\lambda, L).$$

When equipped with the unnormalized Hilbert-Schmidt inner product

(2) $$(A|B) = tr(B^*A)$$

the intertwining space $Hom_G(\lambda, L)$ is a Hilbert space, and there is a canonical unitary map $A \to A(\cdot)$ of $Hom_G(\lambda, L)$ onto the conjugate of $\mathcal{H}(\overline{S}, \lambda')$ such that

(3) $$A\phi(z) = (\phi|A(z)), \quad z \in S.$$

PROOF. Let $A \in Hom_U(\lambda, L)$. Then for fixed z

$$\phi \to (A\phi)(z), \quad \phi \in \mathcal{H}(\lambda)$$

is a continuous linear functional on $\mathcal{H}(\lambda)$; it is determined in the usual way by a unique vector $A(z)$ in $\mathcal{H}(\lambda)$. Thus

$$(A\phi)(z) = (\phi|A(z))$$

for every $\phi \in \mathcal{H}(\lambda)$, and the function $A(\cdot) : S \to \mathcal{H}(\lambda)$ lies in $\mathcal{H}(\overline{S}, \mathcal{H}(\lambda))$. In addition, for $u \in U$

(4) $$A\phi(uz) = (L(u^{-1})A\phi)(z) = (AL(u^{-1})\phi)(z) = (\lambda(u^{-1})\phi|A(z))$$

for arbitrary $\phi \in \mathcal{H}(\lambda)$. For fixed z, the functions $a \to A\phi(az)$ and $a \to (\lambda(a^{-1})\phi|A(z))$ are holomorphic on G. By (4), they agree on the compact real form U; hence, they agree on G. It follows that

$$L(a)A\phi = A\lambda(a^{-1})\phi$$

for every $a \in G$ and $\phi \in \mathcal{H}(\lambda)$. This implies (1). Now for $a \in G$, we have

$$(\phi|A(az)) = (L(a^{-1})A\phi)(z) = (\lambda(a^{-1})\phi|A(z)) = (\phi|\lambda'(a)A(z))$$

for every $\phi \in \mathcal{H}(\lambda)$. Thus, $A(\cdot)$ is a λ'-*covariant*. If f is any λ'-covariant, the argument may be reversed to show that there is an element $A \in Hom_G(\lambda, L)$ such that $f = A(\cdot)$; to compute the norm of f, let $\epsilon_1, ..., \epsilon_n$ be an orthonormal base for $\mathcal{H}(\lambda)$. Then

$$\int_S |f(z)|^2 e^{-(z|z)} dz = \int_S \left(\sum_i |(f(z)|\epsilon_i)|^2 \right) e^{-(z|z)} dz$$

and since

$$A\epsilon_i(z) = (\epsilon_i \,|\, f(z))$$

it follows that

$$\int_S |f(z)|^2 e^{-(z|z)} dz = \sum_i (A\epsilon_i \,|\, A\epsilon_i) = tr(A^*A).$$

Hence, $A \to A(\cdot)$ is a unitary map of $Hom_G(\lambda, L)$ onto the conjugate of $\mathcal{H}(\overline{S}, \lambda')$. Completeness follows from standard Hilbert-Schmidt theory: a Cauchy sequence of Hilbert-Schmidt operators converges in the uniform norm, and this implies convergence in the Hilbert-Schmidt norm as well.

Since the multiplicity of λ in the subspace $T_\lambda \mathcal{H}(S)$ is precisely the dimension of $Hom_G(\lambda, L)$, we see that this multiplicity is the dimension of $\overline{\mathcal{H}}(\overline{S}, \lambda')$, the conjugate of $\mathcal{H}(\overline{S}, \lambda')$. In fact, the functions of type λ are specified by the λ'-covariants in the following way.

2.15 Theorem. *The canonical linear map of $\mathcal{H}(\lambda) \otimes \overline{\mathcal{H}}(\overline{S}, \lambda')$ to $T_\lambda \mathcal{H}(S)$ such that*

$$\phi \otimes A(\cdot) \to \sqrt{deg(\lambda)}\, A\,\phi$$

is unitary and intertwines $\lambda \otimes I$ with the restriction of L to $T_\lambda \mathcal{H}(S)$.

PROOF. Let $f = A(\cdot)$. Because $\mathcal{H}(\lambda)$ is finite-dimensional, the bilinear map

$$(\phi, f) \to A\,\phi$$

factors through the tensor product in the usual way. We also have

$$deg(\lambda) \int_S |A\,\phi(z)|^2 e^{-(z|z)} dz = \int_S deg(\lambda)\,|(f(z)\,|\,\phi)|^2 e^{-(z|z)} dz$$

$$= \int_S \left(deg(\lambda) \int_U |(\lambda(u)f(z)\,|\,\phi)|^2 du \right) e^{-(z|z)} dz$$

$$= |\phi|^2 \int_S |f(z)|^2 e^{-(z|z)} dz$$

by the Schur orthogonality relations. To show surjectivity, suppose now that f $\in T_\lambda \mathcal{H}(S)$. As in **2.8**, set $F = F_\lambda f$. Then

(1) $$F(z) = \int_U f(uz)\lambda(u)du, \quad z \in S,$$

(2) $$f(z) = deg(\lambda)\, tr(F(z)), \quad z \in S,$$

and by **2.8**(4) and **2.8**(5),

(3) $$\int_S |f(z)|^2 e^{-(z|z)} dz = deg(\lambda) \int_S tr(F(z)^* F(z)) e^{-(z|z)} dz.$$

It follows that the matrix entries

$$z \to (F(z)\phi\,|\,\psi)$$

of F are functions in $\mathcal{H}(S)$. Let $\epsilon_1, ..., \epsilon_n$ be an orthonormal base for $\mathcal{H}(\lambda)$ and set $f_i(z) = \sqrt{deg(\lambda)}\, F^*(z)\epsilon_i$. Then by (3) and the observation just made concerning matrix entries, one sees that $f_i \in \mathcal{H}(\overline{S}, \lambda')$ and, by (2), that

$$\sum_i \epsilon_i \otimes f_i \to f.$$

The final statement is a matter of straightforward verification. enddemo

From this result, we obtain the following *orthogonality relations*.

2.16 Corollary. *For any vectors* $\phi_1, \psi_1, \phi_2, \psi_2$ *in* $\mathcal{H}(\lambda)$ *and points* w_1, w_2 *in* S,

$$deg(\lambda) \int_S \big(I_\lambda(z, w_1)\phi_1|\psi_1\big)\overline{\big(I_\lambda(z, w_2)\phi_2|\psi_2\big)}\, e^{-(z|z)} dz$$
$$= (I_\lambda(w_2, w_1)\psi_2|\psi_1)(\phi_1|\phi_2).$$

PROOF. If $w \in S$ and $\psi \in \mathcal{H}(\lambda)$, it is easy to see that the equation

$$(1) \qquad\qquad A\phi(z) = \big(I_\lambda(z, w)\phi|\psi\big), \quad \phi \in \mathcal{H}(\lambda)$$

defines an operator $A \in Hom_G(\lambda, L)$ and that the associated λ'-covariant $A(\cdot)$ is given by

$$(2) \qquad\qquad A(z) = I_\lambda(w, z)\psi, \quad z \in S.$$

Let A_1 be the operator defined by the pair w_1, ψ_1 and A_2 the operator defined by w_2, ψ_2. Then by **2.15**

$$(3) \qquad deg(\lambda)(A_1\phi_1|A_2\phi_2) = (\phi_1 \otimes A_1(\cdot)|\phi_2 \otimes A_2(\cdot)) = (\phi_1|\phi_2)(A_2(\cdot)|A_1(\cdot)).$$

Now from (2)

$$(4) \qquad (A_2(\cdot)|A_1(\cdot)) = \int_S \big(I_\lambda(w_2, z)\psi_2|I_\lambda(w_1, z)\psi_1\big)e^{-(z|z)} dz$$
$$= \int_S \big(I_\lambda(z, w_1)I_\lambda(w_2, z)\psi_2|\psi_1\big)e^{-(z|z)} dz$$
$$= \big(I_\lambda(w_2, w_1)\psi_2|\psi_1\big)$$

the last equality following from **2.11**(1). Substituting this into (3), we obtain the desired relation.

2.17 Corollary. *Let* $w \in S$, $\phi \in \mathcal{H}(\lambda)$, *and* $A_{w,\psi}$ *the intertwining operator defined, as above in* **2.16***(1), by the pair* w, ψ. *Then*

$$(1) \qquad\qquad \|A_{w,\psi}\|^2 = (I_\lambda(w, w)\psi|\psi)$$

and $A_{w,\psi}$ *is either 0 or injective.*

PROOF. The last statement follows from Schur's lemma and (1) from **2.16**(4).

2.18 Corollary. *The operator* $I_\lambda(w, w)$ *is always non-negative.*

2.19 Corollary. *Let* $w \in S$. *Then the following statements are equivalent.*

(1)   *For every non zero* $\phi \in \mathcal{H}(\lambda)$, *there exists* $z \in S$ *such that* $I_\lambda(z, w)\phi \neq 0$.
(2) $I_\lambda(w, w) \neq 0$.
(3)   *There is a vector* $\psi \in \mathcal{H}(\lambda)$ *such that* $A_{w,\psi}$ *is injective.*

PROOF. (1) $\implies$ (2): By **2.11**(2), $deg(\lambda)\, tr I_\lambda(w,w) = \|I_\lambda(\cdot, w)\|^2$ and this is $\neq 0$ by (1). Hence, $I_\lambda(w,w) \neq 0$.

(2) $\implies$ (3): Because $I_\lambda(w,w)$ is nonnegative and $\neq 0$, there exists $\psi \in \mathcal{H}(\lambda)$ such that $(I_\lambda(w,w)\psi|\psi) > 0$. Hence, by **2.16**, $A_{w,\psi}$ is injective.

(3) $\implies$ (1): This is obvious.

Next, we show that the adjoint of the function I_λ is the reproducing kernel for the space $\mathcal{H}(\overline{S}, \lambda')$ of λ'-covariants.

2.20 Theorem. *The integral kernel for the orthogonal projection of $\mathcal{H}(\overline{S}, \mathcal{H}(\lambda))$ onto $\mathcal{H}(\overline{S}, \lambda')$ is I_λ^* . In fact, if g is any $\mathcal{H}(\lambda)$-valued function on S that is square-integrable with respect to the Gaussian measure, then the equation*

(1) $$f(w) = \int_S I_\lambda(z,w)g(z)e^{-(z|z)}dz$$

defines the orthogonal projection of g on $\mathcal{H}(\overline{S}, \lambda')$.

PROOF. To clarify the statement, we recall, cf. **4.2**, that

$$I_\lambda(z,w) = I_\lambda(w,z)^*.$$

Now suppose that g is square-integrable and set

(2) $$h(w) = \int_S e^{(z|w)}g(z)e^{-(z|z)}dz.$$

Then, by **2.2**, h is the orthogonal projection of g on $\mathcal{H}(\overline{S}, \mathcal{H}(\lambda))$. Let

(3) $$k(w) = \int_U \lambda(u)h(u^{-1}w)du.$$

Then k is λ-covariant and holomorphic on $\overline{S}$; moreover

$$|k(w)| \leq \int_U |\lambda(u)h(u^{-1}w)|du \leq \left(\int_U |h(u^{-1}w)|^2 du \right)^{\frac{1}{2}}.$$

Therefore

$$\int_S |k(w)|^2 e^{-(w|w)}dw \leq \int_U \int_S |h(u^{-1}w)|^2 e^{-(w|w)}dw \leq \int_S |h(w)|^2 e^{-(w|w)}dw;$$

so $k \in \mathcal{H}(\overline{S}, \lambda)$. We also have

$$\begin{aligned}
k(w) &= \int_U \lambda(u)\left(\int_S e^{(uz|w)}g(z)e^{-(z|z)}dz \right)du \\
&= \int_S \left(\int_U e^{(uz|w)}\lambda(u)du \right)g(z)e^{-(z|z)}dz \\
&= \int_S I_\lambda(z,w)g(z)e^{-(z|z)}dz.
\end{aligned}$$

Here the interchange in the order of integration is justified by a simple application of Fubini's theorem which also shows that the integral in (1) is absolutely convergent. If we start with $f \in \mathcal{H}(\overline{S}, \lambda')$ then in the above,

$$k(w) = \int_U \lambda(u)f(u^{-1}w)du = \int_U f(w)du = f(w)$$

which shows that the map defined by (1) is a projection with range $\mathcal{H}(\overline{S}, \lambda')$. If f is defined by (1) and h is square-integrable, it follows that

$$\int_S (f(w)|h(w))e^{-(w|w)}dw = \int_S \Big(\int_S (I_\lambda(z,w)g(z)|h(w))e^{-(z|z)}dz \Big)e^{-(w|w)}dw$$

$$= \int_S \Big(\int_S (g(z)|I_\lambda(w,z)h(w))e^{-(w|w)}dw \Big)e^{-(z|z)}dz$$

$$= \int_S \Big(g(z)\, |\, \int_S I_\lambda(w,z)h(w)e^{-(w|w)}dw \Big)e^{-(z|z)}dz.$$

Therefore, the projection defined by (1) is self adjoint.

From the theorem and standard kernel function theory one obtains the following result.

2.21 Corollary. *The λ' covariants of the form*

(1) $$f(z) = \sum_i I_\lambda(z_i, z)\psi_i$$

where $z_1, z_2, \ldots$ is any finite sequence of points in S and $\psi_1, \psi_2, \ldots$ a corresponding sequence of vectors in $\mathcal{H}(\lambda)$ form a dense linear subspace $\mathcal{H}_0(\overline{S}, \lambda')$ of $\mathcal{H}(\overline{S}, \lambda')$.

2.22 Proposition. *Fix a point w in S. Then evaluation at w,*

$$E_w : f \to f(w), \quad f \in \mathcal{H}(\overline{S}, \lambda')$$

maps the space $\mathcal{H}(\overline{S}, \lambda')$ of λ' covariants onto a subspace $\mathcal{H}(w, \lambda)$ of $\mathcal{H}(\lambda)$ consisting of vectors that are fixed by the restriction of λ' to the stabilizer of w. Moreover

(1) $$\mathcal{H}(w, \lambda) = range\,(I_\lambda(w, w)),$$

(2) $$\mathcal{H}(w, \lambda)^\perp = ker(\,I_\lambda(w, w)),$$

*and the intertwining operator $A_{w,\psi}$ in **2.16**(1) is injective for non-zero $\psi \in \mathcal{H}(w, \lambda)$ and 0 for $\psi \in \mathcal{H}(w, \lambda)^\perp$.*

PROOF. If H is the stabilizer of w in G and $f \in \mathcal{H}(\overline{S}, \lambda')$, then

(3) $$\lambda'(a)f(w) = f(aw) = f(w)$$

for all $a \in H$; so the vectors in $\mathcal{H}(w, \lambda)$ are fixed by the $\lambda'(a)$ with $a \in H$. The first observation in proving (1) is that **2.21** and the continuity of point evaluations implies $E_w(\mathcal{H}_0(\overline{S}, \lambda')) = \mathcal{H}(w, \lambda)$. It follows that $\psi \in \mathcal{H}(w, \lambda)^\perp$ iff $A_{w,\psi} = 0$. By **2.17**

$$\|A_{w,\psi}\|^2 = (I_\lambda(w, w)\psi|\psi).$$

Because $I_\lambda(w, w)$ is a non-negative operator, $(I_\lambda(w, w)w|w) = 0$ iff $I_\lambda(w, w)\psi = 0$. This proves (2), and (1) follows since the range of a self adjoint operator on a finite dimensional space is the orthogonal complement of its kernel. The final statement is now a consequence of the fact that $A_{w,\psi}$ is either 0 or injective.

From this we get the following general result concerning the multiplicity of λ in L.

2.22 Corollary.

$$dim\,(Hom_G(\lambda, L)) \geq rank\,(I_\lambda(w, w)).$$

There are many cases in which equality holds in **2.22**. In particular, this is the case for the Fock space associated to homogeneous bounded symmetric domain, as the ambient vector space is then one for which the following applies [11].

2.23 Theorem. *Suppose there is a point $w \in S$ with a dense G orbit. Then the space $\mathcal{H}(\overline{S}, \lambda')$ of λ' covariants is canonically isomorphic, via evaluation at w, to the range $\mathcal{H}(w, \lambda)$ of $I_\lambda(w, w)$. The map*

$$\psi \rightarrow A_{w,\psi}\,, \quad \psi \in \mathcal{H}(w, \lambda)$$

is a conjugate linear isomorphism of $\mathcal{H}(w, \lambda)$ onto $Hom_G(\lambda, L)$, and I_λ is a polynomial function on $S \times \overline{S}$.

PROOF. Since Gw is dense in S, each function in $\mathcal{H}(\overline{S}, \lambda')$ is determined by its value at w; so evaluation at w is injective. For each $\eta \in \mathcal{H}(w, \lambda)$, there is a unique $\psi \in \mathcal{H}(w, \lambda)$ such that η is the value of the λ' covariant

$$z \rightarrow I_\lambda(w, z)\psi\,, \quad z \in S$$

at w. Since this is the covariant $A_{w,\psi}(\cdot)$ it follows that the linear map

$$\psi \rightarrow A_{w,\psi}(\cdot)\,, \quad \psi \in \mathcal{H}(w, \lambda)$$

is an isomorphism of $\mathcal{H}(w, \lambda)$ onto $\mathcal{H}(\overline{S}, \lambda')$; so $\psi \rightarrow A_{w,\psi}$ is a conjugate linear isomorphism of $\mathcal{H}(w, \lambda)$ onto $Hom_G(\lambda, L)$. Now we indicate how one obtains the final statement. For an integer $n \geq 0$, let

(1) $$I_\lambda^n(z, w) = \frac{1}{n!} \int_U (uz|w)^n \lambda(u)du$$

Then

(2) $$I_\lambda(z, w) = \sum_n I_\lambda^n(z, w).$$

the non-zero I_λ^n are linearly independent, polynomial, and have properties similar to I_λ . Therefore, since the multiplicity of λ in L is finite, there are at most a finite number of n for which $I_\lambda^n \neq 0$.

REFERENCES

1. J. Faraut and A. Koranyi, *Function spaces and reproducing kernels on bounded symmetric domains*, J. Functional Analysis **88** (1990), 64-87.

2. M. Lassalle, *Noyau de Szegö, K-types et algebres de Jordan*, C. R. Acad. Sci. Paris **310** (1990), 253-256.

3. J. Faraut, *Prolongement analytique des series de Taylor spheriques*, Contemporary Math. (1992), in this volume.

4. W. Schmid, *Die randwerte holomorpher funktionen auf Hermitesch symmetrischen raumen*, Invent. Math. **9 year 1969**, 61-80.

5. M. Takeuchi, *Polynomial representations associated with symmetric bounded domains*, Osaka J. Math **10** (1973 pages 441-475).

6. K. D. Johnson, *On a ring of invariant polynomials on a Hermitian symmetric space*, Journal of Algebra **67** (1980), 72-81.

7. K. I. Gross and R. A. Kunze, *Bessel functions and representation theory I*, J. Functional Analysis **22** (1976), 73-105.

8. _____, *Bessel functions and representation theory II : Holomorphic discrete series and metaplectic representations*, J. Functional Analysis **25** (1977), 1-49.

9. R. A. Kunze, *Generalized Bessel functions in the Fock space*, Suppl. Rendiconti, Circ. Mat. Palermo **1** (1981), 163-169.

10. B. Konstant, *Lie group representations on polynomial rings*, American J. Math. **85** (1963), 327-404.

11. M. Sato and T. Kimura, *A classification of irreducible prehomogeneous vector spaces and their relative invariants*, Nagoya Math. J. **65** (1977), 1-55.

DEPARTMENT OF MATHEMATICS, UNIVERSITY OF GEORGIA, ATHENS, GEORGIA 30602

Matrice de Hua et Polynômes de Jack

MICHEL LASSALLE

1. Introduction

Dans l'espace vectoriel E des matrices symétriques réelles, soit Ω le cône des matrices définies positives. Le groupe $G = GL(n, \mathbb{R})$ opère sur Ω par la loi naturelle $g \cdot x = gxg^t$. On sait que l'analyse harmonique des fonctions G-invariantes sur Ω se réduit à l'étude de certains polynômes sur E, appelés "polynômes zonaux" et étudiés notamment en analyse statistique multivariée [8].

Ces polynômes ne sont pas connus explicitement. Dans son livre [1] Hua proposait de les exprimer comme combinaison linéaire de caractères du groupe unitaire $U(n)$. Il montrait ([1], p. 131-136) que le calcul des coefficients de cette décomposition se réduit à celui d'intégrales sur $U(n)$. Cependant Hua concluait que sa méthode ne permettait pas d'obtenir des formules générales.

Le but de cet article est de revenir sur ce problème, que nous allons aborder dans le cadre plus général des polynômes de Jack. Soit α un nombre réel positif. Les polynômes de Jack $J_\lambda^{(\alpha)}(x_1, \ldots, x_n)$ sont une famille de polynômes symétriques de n variables, associés à toute partition λ de longueur inférieure ou égale à n. On dispose ainsi d'une base de l'anneau Λ_n des polynômes symétriques de n variables.

Considérons alors deux nombres réels positifs α et β. Les familles de polynômes de Jack $\{J_\lambda^{(\alpha)}\}$ et $\{J_\lambda^{(\beta)}\}$ respectivement associées à α et β forment deux bases de Λ_n. Nous appelons "matrice de Hua" $H(\alpha, \beta)$ la matrice de changement de base ainsi définie. Le cas particulier $\alpha = 2, \beta = 1$ correspond à la situation initialement envisagée par Hua.

Nous posons ici le problème d'étudier (et si possible d'expliciter) les éléments matriciels de $H(\alpha, \beta)$. Nous présentons un début de réponse à ce problème en déterminant les plus élémentaires de ces éléments.

1991 *Mathematics Subject Classification*. Primary 33C50, 33C80, Secondary 22E30.

Key words and phrases. Polynômes zonaux, nombres de Kostka, l'anneau des polynômes symétriques, fonction de Schur, la formule de Pieri, la formule du binôme généralisée.

The final version of this paper will be submitted for publication elsewhere .

Il est remarquable que les expressions ainsi obtenues s'expriment en termes combinatoires. Ce qui explique les difficultés recontrées par la méthode purement analytique de Hua.

2. Polynômes de Jack

Nous rappelons d'abord la terminologie et les notations de la théorie des fonctions symétriques et des partitions, pour laquelle nous renvoyons le lecteur à Macdonald ([5], [7]).

Une partition λ est une suite décroissante finie d'entiers positifs. On dit que le nombre n d'entiers non nuls est la longueur de λ. On note $\lambda = (\lambda_1, \dots, \lambda_n)$ et $n = \ell(\lambda)$. On dit que $|\lambda| = \sum_i \lambda_i$ est le poids de λ, et pour tout entier i que $m_i(\lambda) = \operatorname{card}(j : \lambda_j = i)$ est la multiplicité de i dans λ. On pose

$$z_\lambda = \prod_{i \geq 1} i^{m_i} m_i! \ .$$

On note λ' la partition conjugée de λ définie par $\lambda_i' = \operatorname{card}(j : \lambda_j \geq i)$. On identifie λ avec son diagramme $\{(i,j) : 1 \leq i \leq \ell(\lambda), 1 \leq j \leq \lambda_i\}$ et λ' avec le diagramme $\{(i,j) : (j,i) \in \lambda\}$.

On écrit $\lambda \geq \mu$ si $|\lambda| = |\mu|$ et si on a $\lambda_1 + \lambda_2 + \cdots + \lambda_k \geq \mu_1 + \mu_2 + \cdots + \mu_k$ pour tout $k \geq 1$. On définit ainsi un ordre partiel sur les partitions de même poids.

Considérons n indéterminées indépendantes $x_1, \dots, x_n$. On note Λ_n l'anneau des polynômes symétriques en $x_1, \dots, x_n$ à coefficients *réels*. Soit λ une partition de longueur $\leq n$. Le polynôme symétrique monomial m_λ est défini par

$$m_\lambda(x_1, \dots, x_n) = \sum_\sigma x_1^{\sigma_1} \cdots x_n^{\sigma_n},$$

où la somme s'effectue sur toutes les permutations distinctes de $\lambda_1, \dots, \lambda_n$. Les polynômes $\{m_\lambda, \ell(\lambda) \leq n\}$ forment une base de Λ_n.

Pour tout entier r, on note 1^r la partition $(1, \dots, 1)$ de longueur r et (r) sa conjugée de longueur un. On pose $e_r = m_{1^r}$ et $p_r = m_{(r)}$. A chaque partition λ, on associe les polynômes symétriques

$$e_\lambda = \prod_{i=1}^{\ell(\lambda)} e_{\lambda_i} = \prod_{i \geq 1} e_i^{m_i(\lambda)} \ ,$$

$$p_\lambda = \prod_{i=1}^{\ell(\lambda)} p_{\lambda_i} = \prod_{i \geq 1} p_i^{m_i(\lambda)} \ .$$

Chacune des familles $\{e_\lambda, \ell(\lambda) \leq n\}$ et $\{p_\lambda, \ell(\lambda) \leq n\}$ forme une base de Λ_n.

Soit maintenant $x = (x_1, x_2, \dots)$ un ensemble infini d'indéterminées. On appelle fonction symétrique $f(x)$ la donnée pour tout entier n d'un polynôme symétrique $f_n(x_1, \dots, x_n) \in \Lambda_n$, telle que

$$f_{n+1}(x_1, \dots, x_n, 0) = f_n(x_1, \dots, x_n).$$

On définit ainsi l'anneau Λ des fonctions symétriques à coefficients *réels*. Chacune des familles m_λ, e_λ, p_λ précédemment introduites définit une base de Λ.

Soit α un nombre réel positif. On munit Λ d'un produit scalaire $< \cdot, \cdot >_\alpha$ défini par

$$< p_\lambda, p_\mu >_\alpha = \delta_{\lambda\mu} z_\lambda \alpha^{\ell(\lambda)},$$

avec $\delta_{\lambda\mu} = 1$ si $\lambda = \mu$ et sinon $\delta_{\lambda\mu} = 0$. Le résultat fondamental suivant est dû à Macdonald [7].

THÉORÈME. *Pour toute partition λ il existe une fonction symétrique unique $J_\lambda^{(\alpha)} \in \Lambda$, telle que l'on ait:*

(P1) *Si $\lambda \neq \mu$ on a*

$$< J_\lambda^{(\alpha)}, J_\mu^{(\alpha)} >_\alpha = 0.$$

(P2) *Si l'on pose*

$$J_\lambda^{(\alpha)} = \sum_\mu v_\mu^\lambda(\alpha) m_\mu,$$

on a $v_\mu^\lambda(\alpha) \neq 0$ seulement si $\lambda \geq \mu$.

(P3) *Si $|\lambda| = k$ on a $v_{1^k}^\lambda = k!$.*

La famille $\{J_\lambda^{(\alpha)}\}$ forme une base de Λ.

Soit λ une partition telle que $\ell(\lambda) \leq n$. La restriction $J_\lambda^{(\alpha)}(x_1, \ldots, x_n) \in \Lambda_n$ est un polynôme symétrique homogène de degré $|\lambda|$ qu'on appelle polynôme de Jack. La famille $\{J_\lambda^{(\alpha)}, \ell(\lambda) \leq n\}$ forme une base de Λ_n.

Les quantités $v_\lambda^\lambda(\alpha)$ et $j_\lambda^{(\alpha)} = < J_\lambda^{(\alpha)}, J_\lambda^{(\alpha)} >_\alpha$ ont été évaluées par Stanley ([9], p. 97). On a

$$v_\lambda^\lambda(\alpha) = \prod_{(i,j) \in \lambda} (\lambda_j' - i + 1 + \alpha(\lambda_i - j))$$

$$j_\lambda^{(\alpha)} = \prod_{(i,j) \in \lambda} (\lambda_j' - i + 1 + \alpha(\lambda_i - j))(\lambda_j' - i + \alpha(\lambda_i - j + 1)).$$

En ce qui concerne les polynômes de Jack, nous renvoyons le lecteur à Macdonald ([6], [7]) et à Stanley [9].

3. Matrice de Hua

On fixe maintenant, et pour toute la suite de cet article, deux nombres réels positifs α et β. La matrice de Hua $H(\alpha, \beta)$ est la matrice de changement de base formée des éléments $h_{\mu;\beta}^{\lambda;\alpha}$ définis par

$$J_\lambda^{(\alpha)} = \sum_\mu h_{\mu;\beta}^{\lambda;\alpha} \frac{J_\mu^{(\beta)}}{j_\mu^{(\beta)}}.$$

En vertu de la propriété (P1) on a évidemment

$$h_{\mu;\beta}^{\lambda;\alpha} = < J_\lambda^{(\alpha)}, J_\mu^{(\beta)} >_\beta .$$

La propriété (P2) implique immédiatement les deux résultats suivants.

PROPOSITION 1. *La matrice $H(\alpha, \beta)$ est triangulaire supérieure, c'est à dire qu'on a $h_{\mu;\beta}^{\lambda;\alpha} \neq 0$ seulement si $\mu \leq \lambda$.*

PROPOSITION 2. *On a*

$$h_{\lambda;\beta}^{\lambda;\alpha} = \frac{v_\lambda^\lambda(\alpha)}{v_\lambda^\lambda(\beta)} j_\lambda^{(\beta)} = \prod_{(i,j)\in\lambda} (\lambda_j' - i + 1 + \alpha(\lambda_i - j))(\lambda_j' - i + \beta(\lambda_i - j + 1)).$$

Notre premier résultat est une propriété de dualité fondamentale.

THÉORÈME 1. *Pour toutes les partitions λ, μ on a*

$$h_{\mu;\beta}^{\lambda;\alpha} = (\alpha\beta)^{|\lambda|} h_{\lambda';1/\alpha}^{\mu';1/\beta} .$$

On observera ([5], p. 6) que les conditions $\lambda \geq \mu$ et $\mu' \geq \lambda'$ sont équivalentes. Nous formulons la conjecture suivante.

CONJECTURE. *Les éléments matriciels $h_{\mu;\beta}^{\lambda;\alpha}$ sont des polynômes en α et β, à coefficients dans $\mathbb{Z}$.*

La suite de cet article est consacrée à présenter des résultats partiels qui confirment cette conjecture.

4. Premiers résultats

Nous traitons d'abord le cas particulier où λ est de longueur un, c'est à dire $\lambda = (k)$ avec k entier positif.

THÉORÈME 2. *Pour tout entier $k \geq 0$ et toute partition μ de poids k, on a*

$$h_{\mu;\beta}^{(k);\alpha} = k! \prod_{(i,j)\in\mu} (\beta - \alpha(i-1) + \alpha\beta(j-1)).$$

Par la propriété de dualité, ceci équivaut à

THÉORÈME 3. *Pour toute partition λ avec $|\lambda| = k$, on a*

$$h_{1^k;\beta}^{\lambda;\alpha} = k! \prod_{(i,j)\in\lambda} (\beta + i - 1 - \alpha(j-1)).$$

Dans le cas particulier $\beta = 1$, le théorème 2 avait déjà été obtenu par Stanley ([9], p. 80). Nous traitons ensuite le cas où λ et μ sont de longueur deux.

THÉORÈME 4. *Soient $\lambda = (k, \ell)$ et $\mu = (r, s)$ avec $\mu \leq \lambda$. On a*

$$h_{(r,s);\beta}^{(k,\ell);\alpha} = \beta^r \frac{(k-\ell)!s!}{(s-\ell)!} \prod_{i=1}^{r-\ell}(1 + \alpha(i-1)) \prod_{i=1}^{s-\ell}(\beta - \alpha + \alpha\beta(i-1))$$

$$\cdot \prod_{i=1}^{\ell}[(1 + \alpha(i-1))(1 + \beta(r-i+1))(2 + \alpha(k-i))].$$

Par dualité on en déduit le

THÉORÈME 5. *Soit* $\lambda = (2^r, 1^s)$. *Toute partition* $\mu \leq \lambda$ *s'écrit* $\mu = (2^\ell, 1^m)$
avec $\ell \leq r$ *et* $2\ell + m = 2r + s$. *On a*

$$h_{(2^\ell,1^m);\beta}^{(2^r,1^s);\alpha} = \frac{m!r!}{(r-\ell)!} \prod_{i=1}^{r+s-\ell} (\beta + i - 1) \prod_{i=1}^{r-\ell} (\beta - \alpha + i - 1)$$

$$\cdot \prod_{i=1}^{\ell} [(\beta + i - 1)(\alpha + r + s - i + 1)(2\beta + m + i - 1)].$$

Dans le cas particulier $\beta = 1$, le théorème 5 avait été conjecturé par Stanley
([9], p. 105).

5. Le cas des équerres

Nous considérons maintenant le cas particulier où λ et μ sont des équerres.

THÉORÈME 6. *Soient* $\lambda = (k, 1^\ell)$ *et* $\mu = (r, 1^s)$ *avec* $\mu \leq \lambda$, *c'est à dire* $r \leq k$
et $r + s = k + \ell$. *On a*

$$h_{(r,1^s);\beta}^{(k,1^\ell);\alpha} = \frac{(k-1)!s!}{(s-\ell)!} \beta^{r-1} \prod_{i=1}^{\ell} (\beta + i - 1) \prod_{i=1}^{r-1} (1 + \alpha(i-1)) \prod_{i=2}^{s-\ell+1} (\beta - \alpha(i-1))$$

$$\cdot ((\ell + \beta k)(1 + \alpha(r-1)) + \ell(s + \beta r)).$$

Dans tous les résultats que nous avons présentés jusqu'ici, $h_{\mu;\beta}^{\lambda;\alpha}$ possède la
propriété remarquable d'être décomposable en facteurs du premier degré en α
(ou β). Cette propriété *n'est pas générale.*

Même lorsque λ est une équerre, des facteurs irréductibles de degré > 1 en α
ou β apparaissent dès que $|\lambda| \geq 6$. On a par exemple,

$$h_{(3,2,1);\beta}^{(5,1);\alpha} = 24\beta^3(\alpha + 1)(\beta - \alpha)$$

$$\cdot (10\alpha^2\beta^2 + 40\alpha\beta^2 - 41\alpha^2\beta + 22\beta^2 - 15\alpha\beta - 14\alpha^2 - 10\alpha + 8\beta).$$

Cependant on va voir que $h_{\mu;\beta}^{\lambda;\alpha}$ demeure décomposable en facteurs du premier
degré dans les quatre cas où les deux conditions suivantes sont satisfaites:

- λ ou μ est une équerre,
- α ou β est égal à un.

Les expressions obtenues possèdent d'élégantes formulations combinatoires.

THÉORÈME 7. *Soient* λ *une partition arbitraire et* $\mu = (r, 1^s)$ *avec* $\mu \leq \lambda$.
Pour $\beta = 1$ *on a*

$$h_{\mu;1}^{\lambda;\alpha} = |\lambda| s! \frac{(\lambda_1 - 1)!}{(\lambda_1 - r)!} \prod_{(i,j) \in \lambda} (\lambda_j' - i + 1 + \alpha(\mu_i - j)).$$

Comme $\mu \leq \lambda$ implique $r \leq \lambda_1$ et $\ell(\mu) \geq \ell(\lambda)$, μ_i est bien défini pour $(i, j) \in \lambda$,
de sorte que l'expression a un sens. On comparera ce résultat au théorème 3.
Par dualité on en déduit le

THÉORÈME 8. *Soient* $\lambda = (r, 1^s)$ *et* μ *une partition telle que* $\mu \leq \lambda$. *Pour* $\alpha = 1$ *on a*

$$h_{\mu;\beta}^{\lambda;1} = |\mu|(r-1)! \frac{(\ell(\mu)-1)!}{(\ell(\mu)-s-1)!} \prod_{(i,j)\in\mu} (\lambda_j' - i + \beta(\mu_i - j + 1)).$$

Pour toute partition $\lambda = (\lambda_1, \ldots, \lambda_n)$ on note $\lambda^{\#} = (\lambda_2, \ldots, \lambda_n)$ et $\check{\lambda} = (\lambda_1 - 1, \lambda_2 - 1, \ldots, \lambda_n - 1)$. On a $(\lambda^{\#})' = (\lambda')\check{}$.

THÉORÈME 9. *Soient* λ *une partition arbitraire et* $\mu = (r, 1^s)$ *avec* $\mu \leq \lambda$. *Pour* $r \neq 1$ *et* $\alpha = 1$ *on a*

$$h_{\mu;\beta}^{\lambda;1} = \frac{(r-1)!s!}{(\lambda_1-r)!} \beta^{r-1} (\beta r + r + s - \lambda_1) \prod_{j=1}^{\lambda_1} (\lambda_1 + \lambda_j' - j) \prod_{i=1}^{\lambda_1 - r} (\beta - i) \prod_{(i,j)\in\lambda^{\#}} (\beta + i - j).$$

Par dualité ce résultat est équivalent au

THÉORÈME 10. *Soient* $\lambda = (r, 1^s)$ *et* μ *une partition telle que* $\mu \leq \lambda$. *Pour* $s \neq 0$ *et* $\beta = 1$ *on a*

$$h_{\mu;1}^{\lambda;\alpha} = \frac{(r-1)!s!}{(\ell(\mu)-s-1)!} (s + 1 + \alpha(r + s - \ell(\mu)))$$

$$\cdot \prod_{i=1}^{\ell(\mu)} (\ell(\mu) + \mu_i - i) \prod_{j=1}^{\ell(\mu)-s-1} (1 - \alpha j) \prod_{(i,j)\in\check{\mu}} (1 + \alpha(j - i)).$$

Observons que pour $\alpha = 1$ ou $\beta = 1$, la propriété de décomposition de $h_{\mu;\beta}^{\lambda;\alpha}$ en facteurs du premier degré *n'est pas générale*. On a par exemple

$$h_{(4,2,2);\beta}^{(6,2);1} = \frac{4}{5} 8! \beta^4 (\beta - 1)(18\beta^3 - 5\beta^2 + 2).$$

6. Passage à la limite

Par un résultat de Macdonald ([9], p. 109) on sait qu'on a

$$\lim_{\alpha\to 0} J_\lambda^{(\alpha)} = (\prod_{i=1}^{\lambda_1} \lambda_i'!) e_{\lambda'} \quad ,$$

$$\lim_{\alpha\to 0} \alpha^{|\lambda|-\ell(\lambda)} J_\lambda^{(1/\alpha)} = (\prod_{i=1}^{\ell(\lambda)} (\lambda_i - 1)! \prod_{i\geq 1} m_i(\lambda)!) m_\lambda.$$

Nous introduisons les matrices de changement de base définies par

$$J_\lambda^{(\alpha)} = \sum_\mu v_\mu^\lambda(\alpha) m_\mu = \sum_\mu c_\mu^\lambda(\alpha) e_\mu.$$

Le résultat suivant montre que ces matrices sont des limites de la matrice de Hua.

THÉORÈME 11. *On a*

$$\lim_{\beta \to 0} \beta^{|\mu|} h^{\lambda;\alpha}_{\mu;1/\beta} = (\prod_{i=1}^{\ell(\mu)} \mu_i!) v^{\lambda}_{\mu}(\alpha),$$

$$\lim_{\beta \to 0} (1/\beta)^{\ell(\mu)} h^{\lambda;\alpha}_{\mu';\beta} = (\prod_{i \geq 1} m_i(\mu)!) \, (\prod_{i=1}^{\ell(\mu)} (\mu_i - 1)!) c^{\lambda}_{\mu}(\alpha).$$

La conjecture de Macdonald ([6], p. 196) selon laquelle $v^{\lambda}_{\mu}(\alpha)$ est un polynôme en α apparaît ainsi comme une conséquence de la conjecture que nous avons énoncéeau §3.

Par passage à la limite $\beta \to 0$ ou $1/\beta \to 0$, les résultats que nous avons présentés précédemment permettent d'obtenir des formules intéressantes. Notons d'abord les deux corrollaires suivants du théorème 2.

PROPOSITION 3. *Pour tout entier $k \geq 0$ et toute partition μ de poids k, on a*

$$v^{(k)}_{\mu}(\alpha) = \frac{k!}{\prod_{i \geq 1} \mu_i!} \prod_{(i,j) \in \mu} (1 + \alpha(j-1)),$$

$$c^{(k)}_{\mu}(\alpha) = (-\alpha)^{k-\ell(\mu)} \frac{k!}{\prod_{i \geq 1} m_i(\mu)!} \prod_{i=1}^{\ell(\mu)} (1 + \alpha(i-1)).$$

Ces expressions avaient déjà été obtenues par Stanley ([9], p.80). La proposition suivante est un corollaire du théorème 5.

PROPOSITION 4. *Soit $\lambda = (2^r, 1^s)$. Toute partition $\mu \leq \lambda$ s'écrit $\mu = (2^{\ell}, 1^m)$ avec $\ell \leq r$ et $2\ell + m = 2r + s$. On a*

$$v^{\lambda}_{\mu}(\alpha) = \frac{m! r!}{(r-\ell)!} \prod_{i=1}^{\ell} (\alpha + r + s - i + 1),$$

$$c^{\lambda}_{\mu'}(\alpha) = m \frac{r!}{(r-\ell)!} (r + s - \ell - 1)! \prod_{i=1}^{r-\ell} (-\alpha + i - 1) \prod_{i=1}^{\ell} (\alpha + r + s - i + 1).$$

La première expression avait déjà été donnée par Stanley ([9], p. 105). En vertu du théorème 6 on a la

PROPOSITION 5. *Soient $\lambda = (k, 1^{\ell})$ et $\mu = (r, 1^s)$ avec $\mu \leq \lambda$. On a*

$$v^{\lambda}_{\mu}(\alpha) = \frac{(k-1)! s!}{(s-\ell)! r!} (\ell r + k + k\alpha(r-1)) \prod_{i=1}^{r-1} (1 + \alpha(i-1)),$$

$$c^{\lambda}_{\mu'}(\alpha) = (-\alpha)^{s-\ell} \frac{(k-1)! \ell!}{(r-1)!} (s + 1 + \alpha(r-1)) \prod_{i=1}^{r-1} (1 + \alpha(i-1)).$$

Enfin le théorème 4 possède le corollaire suivant.

PROPOSITION 6. *Soient* $\lambda = (k, \ell)$ *et* $\mu = (r, s)$ *avec* $\mu \le \lambda$. *On a*

$$v_\mu^\lambda(\alpha) = \frac{(k-\ell)!}{(r-\ell)!(s-\ell)!} \prod_{i=1}^{r-\ell}(1 + \alpha(i-1)) \prod_{i=1}^{s-\ell}(1 + \alpha(i-1))$$

$$\cdot \prod_{i=1}^{\ell}[(1 + \alpha(i-1))(2 + \alpha(k-i))],$$

$$c_{\mu'}^\lambda(\alpha) = \frac{(k-\ell)!}{(r-s)!(s-\ell)!}(-\alpha)^{s-\ell}\prod_{i=1}^{r-\ell}(1 + \alpha(i-1))$$

$$\cdot \prod_{i=1}^{\ell}[(1 + \alpha(i-1))(2 + \alpha(k-i))].$$

Observons pour terminer que les théorèmes 8 et 9 conduisent à une expression explicite pour certains nombres de Kostka. Rappelons ([5], p. 56) que ceux-ci sont définis par

$$s_\lambda = \sum_{\mu \le \lambda} K_{\lambda\mu} m_\mu,$$

où s_λ est la fonction de Schur ([5], p. 24).

PROPOSITION 7.

a) *Pour* $\lambda = (r, 1^s)$ *et* $\mu \le \lambda$, *on a*

$$K_{\lambda\mu} = \binom{\ell(\mu) - 1}{s}.$$

b) *Pour* $\lambda = (\lambda_1, \ldots, \lambda_n)$ *arbitraire et* $\mu = (r, 1^s)$ *avec* $r \ne 1$ *et* $\mu \le \lambda$, *on a*

$$K_{\lambda\mu} = \frac{s!}{(\lambda_1 - r)!}\frac{1}{h_{\lambda^\#}},$$

où l'on note $\lambda^\# = (\lambda_2, \ldots, \lambda_n)$ *et* $h_{\lambda^\#}$ *sa longueur-équerre* ([5], p. 9).

7. Conjecture de Macdonald

Macdonald [6] a formulé les conjectures que les éléments matriciels $v_\mu^\lambda(\alpha)$ et $c_\mu^\lambda(\alpha)$ définis précédemment sont des polynômes en α. Les deux assertions sont bien sûr équivalentes, puisque la matrice de changement de base $\{m_\mu\} \to \{e_\mu\}$ est indépendante de α.

Dans le cas particulier où λ est de longueur deux, nous sommes en mesure de démontrer cette conjecture de Macdonald. En fait nous obtenons bien d'avantage puisque nous donnons une expression analytique totalement explicite pour $c_\mu^\lambda(\alpha)$.

DÉFINITION. Soient deux entiers positifs r et s, avec $s \ge r$. On dit qu'un multi-entier $p = \{p_i, i \ge 1\}$ est un choix de type (r, s) si on a $\sum_{i \ge 1} p_i = r$ et $\sum_{i \ge 1} i p_i = s$. On note $\mathcal{C}(r, s)$ l'ensemble des choix de type (r, s).

REMARQUE. Tout choix p définit une partition μ en posant $m_i(\mu) = p_i, i \geq 1$. On obtient ainsi une bijection entre $\mathcal{C}(r,s)$ et les partitions de longueur r et de poids s.

THÉORÈME 12. *Soient* $\lambda = (k, \ell)$ *et* $\mu \leq \lambda$. *On note* $\ell(\mu)$ *et* $\{m_i(\mu), i \geq 1\}$ *la longueur et les multiplicités de* μ. *Alors on a*

$$J_\lambda^{(\alpha)} = \sum_\mu \frac{x_\mu}{\prod_{i \geq 1} m_i(\mu)!} e_\mu,$$

où x_μ *est donné par*

$$x_\mu = (-1)^\ell (-\alpha)^{k - \ell(\mu)} \ell! \sum_{r=0}^{\ell} \sum_{s=r}^{\ell} \sum_{p \in \mathcal{C}(r,s)} (k + \ell - 2s) \frac{(k - s - 1)!}{(\ell - s)!}$$

$$\cdot \prod_{i=1}^{\ell(\mu)-r} (1 + \alpha(i-1)) \prod_{i=1}^{r} (1 + \alpha(i-1)) \prod_{i=0}^{\ell-s-1} (i\alpha - 1)$$

$$\cdot \prod_{i=1}^{s} (1 + \alpha(k - i + 1)) \prod_{i \geq 1} \binom{m_i(\mu)}{p_i}.$$

Nous espérons pouvoir étendre ce résultat au cas d'une partition λ arbitraire. Pour $\ell = 0$ on retrouve le résultat de Stanley énoncé á la proposition 3.

8. Remarque finale

Les démonstrations des résultats présentés ici paraîtront très prochainement [4]. Notre méthode utilise les résultats que nous avons annoncés dans trois précédentes notes [2,3]. En particulier nous avons besoin de la formule de Pieri généralisée [2] et de la formule du binôme généralisée [3] que nous avons obtenues pour les polynômes de Jack.

Nous soulignons que notre méthode fournit un algorithme simple qui permet de calculer rapidement n'importe quel élément matriciel $h_{\mu;\beta}^{\lambda;\alpha}$ pourvu que les partitions λ et μ soient données explicitement. Cependant c'est seulement dans les cas particuliers présentés ci-dessus que nous avons pu jusqu'ici démontrer des formules générales.

A titre d'exemple nous donnons en appendice les sous-matrices de Hua

$$H_n(\alpha, \beta) = \left\{ h_{\mu;\beta}^{\lambda;\alpha} \right\}_{|\lambda|=n}$$

pour $n \leq 5$.

REMERCIEMENTS. Les résultats qui précédent n'auraient pu être obtenus sans lesmoyens informatiques que le Groupe de Combinatoire de l'Université de Bordeaux-I a mis à ma disposition. Je remercie tout particulièrement Maylis Delest, Serge Dulucq et Gérard Viennot de leur aide amicale.

Appendice: Les matrices $H_n(\alpha, \beta)$ pour $n \leq 5$

	1
1	β

	2	1^2
2	$2\beta^2(\alpha+1)$	$2\beta(\beta-\alpha)$
1^2		$2\beta(\beta+1)$

	3	21	1^3
3	$6\beta^3(\alpha+1)$ $(2\alpha+1)$	$6\beta^2(\alpha+1)$ $(\beta-\alpha)$	$6\beta(\beta-\alpha)$ $(\beta-2\alpha)$
21		$\beta^2(2\beta+1)$ $(\alpha+2)$	$6\beta(\beta+1)$ $(\beta-\alpha)$
1^3			$6\beta(\beta+1)$ $(\beta+2)$

	4	31	2^2
4	$24\beta^4(\alpha+1)$ $(2\alpha+1)(3\alpha+1)$	$24\beta^3(\beta-\alpha)$ $(\alpha+1)(2\alpha+1)$	$24\beta^2(\beta-\alpha)$ $(\beta+\alpha\beta-\alpha)$ $(\alpha+1)$
31		$4\beta^3(3\beta+1)$ $(\alpha+1)^2$	$8\beta^2(\beta-\alpha)$ $(\alpha+1)(2\beta+1)$
2^2			$4\beta^2(\beta+1)$ $(2\beta+1)$ $(\alpha+1)(\alpha+2)$
21^2			
1^4			

	21^2	1^4
4	$24\beta^2(\beta-\alpha)$ $(\beta-2\alpha)(\alpha+1)$	$24\beta(\beta-\alpha)$ $(\beta-2\alpha)(\beta-3\alpha)$
31	$4\beta^2(\beta-\alpha)$ $(5\beta+3\alpha\beta+\alpha+3)$	$24\beta(\beta+1)$ $(\beta-\alpha)(\beta-2\alpha)$
2^2	$8\beta^2(\beta-\alpha)$ $(\beta+1)(\alpha+2)$	$24\beta(\beta-\alpha)$ $(\beta+1)(\beta-\alpha+1)$
21^2	$4\beta^2(\beta+1)^2$ $(\alpha+3)$	$24\beta(\beta-\alpha)$ $(\beta+1)(\beta+2)$
1^4		$24\beta(\beta+1)$ $(\beta+2)(\beta+3)$

	5	41	32
5	$120\beta^5(\alpha+1)$ $(2\alpha+1)(3\alpha+1)$ $(4\alpha+1)$	$120\beta^4(\alpha+1)$ $(2\alpha+1)(3\alpha+1)$ $(\beta-\alpha)$	$120\beta^3(\alpha+1)$ $(2\alpha+1)(\beta-\alpha)$ $(\beta+\alpha\beta-\alpha)$
41		$6\beta^4(4\beta+1)$ $(\alpha+1)(2\alpha+1)$ $(3\alpha+2)$	$12\beta^3(\beta-\alpha)$ $(\alpha+1)(3\alpha+2)$ $(3\beta+1)$
32			$4\beta^3(\alpha+2)$ $(\alpha+1)^2$ $(2\beta+1)(3\beta+1)$
31^2			
2^21			
21^3			
1^5			

	31^2	2^21	21^3
5	$120\beta^3(\alpha+1)$ $(2\alpha+1)(\beta-\alpha)$ $(\beta-2\alpha)$	$120\beta^2(\alpha+1)$ $(\beta-\alpha)(\beta-2\alpha)$ $(\beta+\alpha\beta-\alpha)$	$120\beta^2(\alpha+1)$ $(\beta-\alpha)(\beta-2\alpha)$ $(\beta-3\alpha)$
41	$12\beta^3(\beta-\alpha)$ $(\alpha+1)$ $(7\beta+8\alpha\beta+2\alpha+3)$	$12\beta^2(\alpha+1)$ $(\beta-\alpha)(2\beta+1)$ $(4\beta+\alpha\beta-5\alpha)$	$18\beta^2(\beta-\alpha)$ $(\beta-2\alpha)$ $(6\beta+4\alpha\beta+\alpha+4)$
32	$8\beta^3(\beta-\alpha)$ $(\alpha+1)(\alpha+2)$ $(3\beta+2)$	$4\beta^2(\beta-\alpha)$ $(\alpha+2)(2\beta+1)$ $(5\beta+3\alpha\beta+\alpha+3)$	$12\beta^2(\beta-\alpha)$ $(\alpha+2)(\beta+1)$ $(4\beta-5\alpha+1)$
31^2	$4\beta^3(\beta+1)$ $(\alpha+1)(2\alpha+3)$ $(3\beta+2)$	$8\beta^2(\beta-\alpha)$ $(\beta+1)(2\beta+1)$ $(2\alpha+3)$	$12\beta^2(\beta-\alpha)$ $(\beta+1)$ $(7\beta+3\alpha\beta+2\alpha+8)$
2^21		$4\beta^2(2\beta+1)$ $(\beta+1)^2$ $(\alpha+2)(\alpha+3)$	$12\beta^2(\beta-\alpha)$ $(\alpha+3)$ $(\beta+1)(2\beta+3)$
21^3			$6\beta^2(2\beta+3)$ $(\beta+1)(\beta+2)$ $(\alpha+4)$
1^5			

$$1^5$$

5	$120\beta(\beta - \alpha)$ $(\beta - 2\alpha)(\beta - 3\alpha)$ $(\beta - 4\alpha)$
41	$120\beta(\beta + 1)$ $(\beta - \alpha)(\beta - 2\alpha)$ $(\beta - 3\alpha)$
32	$120\beta(\beta + 1)$ $(\beta - \alpha)(\beta - 2\alpha)$ $(\beta - \alpha + 1)$
31^2	$120\beta(\beta + 1)$ $(\beta + 2)(\beta - \alpha)$ $(\beta - 2\alpha)$
$2^2 1$	$120\beta(\beta + 1)$ $(\beta + 2)(\beta - \alpha)$ $(\beta - \alpha + 1)$
21^3	$120\beta(\beta + 1)$ $(\beta + 2)(\beta + 3)$ $(\beta - \alpha)$
1^5	$120\beta(\beta + 1)$ $(\beta + 2)(\beta + 3)$ $(\beta + 4)$

REFERENCES

1. L. K. Hua, *Harmonic analysis of functions of several complex variables in the classical domains*, Transl. Math. Monographs, Amer. Math. Soc., Providence, RI, 1963.

2. M. Lassalle, C.R. Acad. Sci. Paris, Série I, **309** (1989), 941-944.

3. _____, C.R. Acad. Sci. Paris, Série I, **310** (1990), 253-256 et 257-260.

4. _____, *Quelques propriétés des polynômes de Jack*, en préparation.

5. I. G. Macdonald, *Symmetric functions and Hall polynomials*, Oxford University Press, London, 1979.

6. _____, in: Algebraic groups, Springer Lecture Notes in Math. **1271** (Utrecht 1986), 189-200.

7. _____, in: Actes 20ème Séminaire Lotharingien, Publications I.R.M.A. Strasbourg (1988), 131-171.

8. R. J. Muirhead, *Aspects of multivariate statistical theory*, Wiley, New York, 1982.

9. R. P. Stanley, Advances in Math **77** (1989), 76-115.

UPR 14 DU CNRS ECOLE POLYTECHNIQUE, 91128 PALAISEAU, CEDEX FRANCE

Contemporary Mathematics
Volume **138**, 1992

Generalized Hypergeometric Functions and Laguerre Polynomials in Two Variables

ZHIMIN YAN

ABSTRACT. We study a class of generalized hypergeometric functions in two variables. Some integral formulas for the Jack polynomials in two variables are obtained. Integral representations for generalized hypergeometric functions are given, generalizing the classical Euler integrals. A generalized Laplace transform is introduced and exploited to establish generalizations of many classical results for the classical Laguerre polynomials.

0. Introduction

Let $\kappa = (k_1, \ldots, k_r) \in \mathbb{Z}_+^r$ with $k_1 \geq \ldots \geq k_r \geq 0$ and $k = |\kappa| := k_1 + \cdots + k_r$. For $d > 0$, we define

$$C_\kappa^{(d)}(x_1, \ldots, x_r) = c_\kappa J_\kappa(x_1, \ldots, x_r; 2/d)$$

where $J_\kappa(x_1, \ldots, x_r; 2/d)$ is the Jack polynomial of index κ and parameter $2/d$ (cf. [M], [S]), and c_κ is chosen so that

$$(x_1 + \cdots + x_r)^k = \sum_{|\kappa|=k} C_\kappa^{(d)}(x_1, \ldots, x_r).$$

For $m = 0, 1, 2, \ldots$, let $(a)_m = a(a+1)\cdots(a+m-1)$. Then for any partition κ, define

$$(a)_\kappa = \prod_{i=1}^{r}(a - (i-1)d/2)_{k_i}.$$

1991 *Mathematics Subject Classification*. Primary 33C45, 33C80, Secondary 17B20.
This paper is in final form and no version of it will be submitted for publication elsewhere .

For $a_1, \ldots, a_p, b_1, \ldots, b_q \in \mathbb{C}$, such that $(b_j)_\kappa \neq 0$ for all κ, j, we define the hypergeometric function associated with the parameter $d > 0$ by

(1) $_pF_q^{(d)}(a_1, \ldots, a_p; b_1, \ldots, b_q; x_1, \ldots, x_r)$

$$= \sum_\kappa \frac{(a_1)_\kappa \cdots (a_p)_\kappa}{(b_1)_\kappa \cdots (b_q)_\kappa} \frac{C_\kappa^{(d)}(x_1, \ldots, x_r)}{k!}.$$

Generalized hypergeometric functions and Laguerre polynomials were first introduced by Herz [H]. Additional properties and applications in statistics were given by Constantine [C], James [J] and Muirhead [Mu]. In the case of positive definite Hermitian or quaternionic matrices, the generalized hypergeometric functions were also studied by Gross and Richards [GR1, GR2]. Generalized hypergeometric functions associated with arbitrary symmetric cones were considered by Faraut and Korányi [FK]. The more general version (1) was introduced and studied independently by Korányi [K], Macdonald [M2] and Kaneko [KK]. Generalized Laguerre polynomials associated with arbitrary symmetric cones were studied by Dib [D]. Many results known in the symmetric cone case are difficult to extend to the general case because we cannot use the machinery of Lie groups. However when $r = 2$ and d is a positive integer d, it turns out that $C_\kappa^{(d)}(x_1, x_2)$ can be expressed in terms of the classical Gegenbauer polynomials. We shall exploit this fact and obtain a number of analytic results in the case $r = 2$ for any positive d.

Throughout this paper, we shall take $r = 2$. In §2 we give an integral representation of $C_\kappa^{(d)}(x_1, x_2)$ and some other properties of $C_\kappa^{(d)}(x_1, x_2)$ which are useful for our purposes and also of independent interest. In §3 we obtain generalized Euler integral representations for $_1F_1^{(d)}$ and $_2F_1^{(d)}$. In §4 we introduce a generalized Laplace transform, which has many applications, and prove its injectivity. We also show that the Laplace transform of $_pF_q$ is a $_{p+1}F_q$ function, as in the classical case. In the last section, we introduce the generalized Laguerre polynomials L_κ^γ, and derive integral representations, orthogonality relations and a generating function for these polynomials. Furthermore, we define a generalized Hankel transform and establish some relations between Hankel transform and Laguerre polynomials as in the classical case.

ACKNOWLEDGEMENTS. This work is based on my doctoral thesis [Y3], and some of the results presented here were announced in [Y1]. I would like to express my gratitude and respect to Professor Adam Korányi for suggesting these problems, and for valuable advice and encouragement. I also wish to thank the referee for insightful comments and suggestions.

1. Notation and definitions

A *partition* is any finite or infinite sequence

(1) $\kappa = (k_1, k_2, \ldots, k_r, \ldots)$

of nonnegative integers in nonincreasing order, $k_1 \geq k_2 \geq \ldots \geq k_r \geq \ldots$, and containing only finitely many non-zero terms. The non-zero k_i in (1) are called the *parts* of κ. The number of parts is called the *length* of κ, denoted by $l(\kappa)$; and the sum of the parts is the *weight* of κ, denoted by $|\kappa| = k_1 + k_2 + \ldots + k_{l(\kappa)}$. When $l(\kappa) \leq r$, we simply write κ as $\kappa = (k_1, \ldots, k_r)$. We say that κ is a partition of k if $|\kappa| = k$. For a partition κ, hereafter, we use k to denote $|\kappa|$.

The partitions of k are ordered lexicographically; that is, if $\kappa = (k_1, k_2, \ldots)$ and $\lambda = (l_1, l_2, \ldots,)$, we write $\kappa > \lambda$ if $k_i > l_i$ for the first index i such that $k_i \neq l_i$. Suppose $y_1, \ldots, y_r$ are r variables, $\kappa > \lambda$, $l(\kappa) \leq r$ and $l(\lambda) \leq r$, then we say that the monomial $y_1^{k_1} \ldots y_r^{k_r}$ *is of higher weight* than the monomial $y_1^{l_1} \ldots y_r^{l_r}$.

We follow Stanley's introduction [S] to the Jack polynomials. For a parameter α, $\mathbb{Q}(\alpha)$ denotes the field of all rational functions of α with rational coefficients. For a partition κ, m_κ and p_κ denote the symmetric functions in infinitely many variables in the ring of symmetric functions as defined in [M3]. Then, the set $\{p_\kappa : \kappa \text{ is a partition}\}$ forms a $\mathbb{Q}$-basis of $\bigwedge \otimes \mathbb{Q}$, the ring of symmetric functions with the coefficients in $\mathbb{Q}$. We define a scalar product on the vector space $\bigwedge \otimes \mathbb{Q}(\alpha)$ of all symmetric functions of bounded degree over the field $\mathbb{Q}(\alpha)$ by

$$< p_\kappa, p_\lambda >_\alpha = \delta_{\kappa\lambda} z_\kappa \alpha^{l(\kappa)}$$

where $z_\kappa = (1^{m_1} 2^{m_2} \ldots) m_1! m_2! \ldots$, m_j is the number of k_i which are equal to j, and $\delta_{\kappa\lambda}$ denotes Kronecker's delta.

THEOREM. (Stanley [S]) *There are unique symmetric functions* $J_\kappa = J_\kappa(x; \alpha)$ $\in \bigwedge \otimes \mathbb{Q}(\alpha)$, *where* κ *ranges over all partitions, such that*

(1) $< J_\kappa, J_\lambda >_\alpha = 0$ *if* $\kappa \neq \lambda$.

(2) If we write

$$J_\kappa = \sum_\lambda v_{\kappa\lambda}(\alpha) m_\lambda,$$

then $v_{\kappa\lambda}(\alpha) = 0$ *unless* $\lambda \subseteq \kappa$ *(where* $\lambda \subseteq \kappa$ *means* $|\lambda| = |\kappa|$ *and* $\sum_{i=1}^s l_i \leq \sum_{i=1}^s k_i$ *for all* s*).*

(3) If $|\kappa| = k$, *then the coefficient of* $x_1 \cdots x_k$ *in* J_κ *equals* $k!$.

Setting $x_{r+1} = x_{r+2} = \cdots = 0$ in J_κ, we obtain a symmetric polynomial $J_\kappa(x_1, \ldots, x_r; \alpha)$, homogeneous of degree $|\kappa|$. These symmetric polynomials are called the *Jack polynomials*. The following results about Jack polynomials will be required; we refer to Stanley [S] for further details.

(i) $(y_1 + \ldots + y_r)^k = \sum_{|\kappa|=k} \alpha^k k! J_\kappa(y_1, \ldots, y_r; \alpha) j_\kappa^{-1}$ where the constant $j_\kappa = < J_\kappa, J_\kappa >_{2/d}$ is determined in [S].

(ii) The term of highest weight in $J_\kappa(y_1, \ldots, y_r; \alpha)$ is $v_{\kappa\kappa}(\alpha) y_1^{k_1} \ldots y_r^{k_r}$.

(iii) If $l(\kappa) \leq r$, $J_\kappa(y_1, \ldots, y_r; \alpha)$ is an eigenfunction of the differential operator

$$(2) \qquad \Delta_r = \sum_{i=1}^r y_i^2 \frac{\partial^2}{\partial y_i^2} + \frac{2}{\alpha} \sum_{i=1}^r \sum_{j=1, \; j \neq i}^r \frac{y_i^2}{y_i - y_j} \frac{\partial}{\partial y_i}$$

with eigenvalue $\mu_\kappa = \rho_\kappa + k(2r - \alpha)/\alpha$ where $\rho_\kappa = \sum_{i=1}^{r} k_i(k_i - 2i/\alpha)$.

For a partition κ and a positive number d, we define

$$(3) \qquad C_\kappa^{(d)}(x_1, \ldots, x_r) = k! \, (2/d)^k \, j_\kappa^{-1} \, J_\kappa(x_1, \ldots, x_r; 2/d).$$

For $a_1, \ldots, a_p, b_1, \ldots, b_q \in \mathbb{C}$ such that $(b_j)_\kappa \neq 0$ for all κ, j, we define hypergeometric functions $_pF_q^{(d)}$ and $_p\mathcal{F}_q^{(d)}$, respectively, by

$$(4) \qquad _pF_q^{(d)}(a_1, \ldots, a_p; b_1, \ldots, b_q; x_1, \ldots, x_r)$$

$$= \sum_\kappa \frac{(a_1)_\kappa \cdots (a_p)_\kappa}{(b_1)_\kappa \cdots (b_q)_\kappa} \frac{C_\kappa^{(d)}(x_1, \ldots, x_r)}{k!}.$$

and

$$(5) \qquad _p\mathcal{F}_q^{(d)}(a_1, \ldots, a_p; b_1, \ldots, b_q; x_1, \ldots, x_r | y_1, \ldots, y_r)$$

$$= \sum_\kappa \frac{(a_1)_\kappa \cdots (a_p)_\kappa}{(b_1)_\kappa \cdots (b_q)_\kappa} \frac{C_\kappa^{(d)}(x_1, \ldots, x_r)}{k!} \frac{C_\kappa^{(d)}(y_1, \ldots, y_r)}{C_\kappa^{(d)}(1, \ldots, 1)},$$

where the summations are over all partitions κ.

When $r = 1$, both $_pF_q^{(d)}$ and $_p\mathcal{F}_q^{(d)}$ become the classical generalized hypergeometric function, denoted here by $_pf_q$. In particular, $_0\mathcal{F}_0^{(d)}(x|y) = e^{xy}$; and $_1\mathcal{F}_0^{(d)}(a; x|y) = (1 - xy)^{-a}$, $|xy| < 1$.

2. The polynomials $C_\kappa^{(d)}(y_1, y_2)$

In this section, we shall establish some properties of $C_\kappa^{(d)}(y_1, y_2)$.

LEMMA 2.1. *If $P(y_1, y_2)$ is a symmetric and homogeneous polynomial with $c_\kappa y_1^{k_1} y_r^{k_2}$, c_κ constant, as its term of highest weight and an eigenfunction of Δ_2, then the corresponding eigenvalue is*

$$\mu_\kappa = \sum_{i=1}^{2} k_i(k_i - di) + k(2d - 1).$$

The lemma follows from a direct calculation following the arguments of, e.g., Muirhead [Mu].

COROLLARY 2.2. *If $P(y_1, y_2)$ is as in the lemma, then it is uniquely determined, up to a constant multiple, by its term of highest weight.*

Let $P_j^{(\alpha, \beta)}(x)$ denote the classical Jacobi polynomial, and for $\gamma = (d - 1)/2$, define

$$R_n^{(\gamma)}(x) = \frac{P_n^{(\gamma, \gamma)}(x)}{P_n^{(\gamma, \gamma)}(1)}.$$

Then we have the following result, also proved (using alternative arguments) by Kadell [KA] and Richards [R].

PROPOSITION 2.3. *For any partition* $\kappa = (k_1, k_2)$,

$$C_\kappa^{(d)}(y_1, y_2) = C_\kappa^{(d)}(1,1)(y_1 y_2)^{k/2} R_{k_1-k_2}^{(\gamma)} \left(\frac{1}{2}(y_1 + y_2)(y_1 y_2)^{-1/2}\right).$$

PROOF. Recall that the differential operator $L = \Delta_2 - \mu_\kappa$ annihilates the polynomial $C_\kappa^{(d)}(y_1, y_2)$. Substituting $u = (y_1 + y_2)(y_1 y_2)^{-1/2}$ and $v = (y_1 y_2)^{1/2}$, then the differential operator L is transformed to

$$(1) \qquad -\frac{1}{2}\left[(1-u^2)\frac{\partial^2}{\partial u^2} - v^2\frac{\partial^2}{\partial v^2}\right] - u\frac{\partial}{\partial u} + v\frac{\partial}{\partial v} - d(u\frac{\partial}{\partial u} + v\frac{\partial}{\partial v}) + 2\mu_\kappa.$$

Now it is not hard to verify that the polynomial $v^k P_{k_1-k_2}^{(\gamma,\gamma)}(u)$ is annihilated by the differential operator (1). Therefore

$$(2) \qquad (y_1 y_2)^{k/2} P_{k_1-k_2}^{\gamma,\gamma}\left(\frac{1}{2}(y_1 + y_2)(y_1 y_2)^{-1/2}\right)$$

is annihilated by L. Since the polynomial in (2) is symmetric with $y_1^{k_1} y_2^{k_2}$ as its term of highest weight then, by Corollary 2.2, we $C_\kappa^{(d)}(y_1, y_2)$ is a constant multiple of the polynomial (2). Setting $y_1 = y_2 = 1$ and substituting the definition of $R_n^{(\gamma)}$ completes the proof. $\square$

REMARK. The value of $C_\kappa^{(d)}(1,1)$ is given in [S]. The $C_\kappa^{(d)}$ are also closely related to the James-type zonal polynomials $Z_{n,k}^\gamma$ as defined by Koorwinder and Sprinkhuizen-Kuyper [KS]. In fact,

$$Z_{n,k}^\gamma(\xi, \eta) = \frac{Z_{n,k}^\gamma(2,1)}{C_\kappa^{(d)}(1,1)} C_\kappa^{(d)}(y_1, y_2)$$

where $\xi = y_1 + y_2$, $\eta = y_1 y_2$, $\kappa = (n, k)$ and the value of $Z_{n,k}^\gamma(2,1)$ can be obtained using [KS,(4.1)].

Using Proposition 2.3 and the integral formula (4.21) in [A], a direct computation establishes the following result.

PROPOSITION 2.4. *For any partition* $\kappa = (k_1, k_2)$,

$$C_\kappa^{(d)}(y_1, y_2) = C_\kappa^{(d)}(1,1)\frac{2^{k_2-k_1}\Gamma((d+1)/2)}{\sqrt{\pi}\Gamma(d/2)}(y_1 y_2)^{k_2}$$

$$\cdot \int_{-1}^{1}[y_1 + y_2 + (y_1 - y_2)s]^{k_1-k_2}(1-s^2)^{d/2-1}ds.$$

For $X = (x_1, x_2)$, $Y = (y_1, y_2) \in \mathbb{R}^2$, and $s \in (-1, 1)$, we define $\alpha : \mathbb{R}_+^2 \times \mathbb{R}^2 \to \mathbb{R}$ and $\beta_s : \mathbb{R}_+^2 \times \mathbb{R}^2 \to \mathbb{R}$ by

$$\alpha(X, Y) = (y_1 - y_2)2\sqrt{x_1 x_2}$$
$$\beta_s(X, Y) = (y_1 + y_2)(x_1 - x_2) + (y_1 - y_2)(\sqrt{x_1} - \sqrt{x_2})^2 s.$$

Further, we define $\gamma_s : \mathbb{R}^2 \times \mathbb{R}^2 \to \mathbb{R}$ and $l_s : \mathbb{R}_+^2 \times \mathbb{R}^2 \to \mathbb{R}$ by

$$\gamma_s(X,Y) = (y_1 + y_2)(x_1 + x_2) + (y_1 - y_2)(x_1 - x_2)s,$$
$$l_s(X,Y) = \sqrt{\alpha(X,Y)^2 + \beta_s(X,Y)^2 + 2\alpha(X,Y)\beta_s(X,Y)s}.$$

As a consequence of Proposition 2.3 and the product formula for Gegenbauer polynomials, cf. [A, (4.10)], we obtain

PROPOSITION 2.5. *For all $d > 0$, $X \in \mathbb{R}_+^2$, and $Y \in \mathbb{R}^2$,*

$$\frac{\Gamma(\frac{d+1}{2})}{\sqrt{\pi}\Gamma(d/2)} \int_{-1}^{1} C_\kappa^{(d)}(\frac{1}{4}[\gamma_s(X,Y) + l_s(X,Y)], \frac{1}{4}[\gamma_s(X,Y) - l_s(X,Y)])$$

$$\cdot (1 - s^2)^{(d-2)/2}ds = \frac{C_\kappa^{(d)}(X)C_\kappa^{(d)}(Y)}{C_\kappa^{(d)}(1,1)}.$$

3. Integral representations for generalized hypergeometric functions

The purpose of this section is to establish integral representations for the generalized hypergeometric functions. For every $\mathbf{s} = (s_1, \ldots, s_r)$ with $\Re s_i > \frac{1}{2}(i-1)d - 1$, $i = 1, \ldots, r$, we define the generalized gamma function

(1) $$\Gamma_d(\mathbf{s}) = (2\pi)^{r(r-1)d/4} \prod_{i=1}^{r} \Gamma(s_i - (i-1)\frac{d}{2}).$$

For $\mathbf{s} = (s, \ldots, s)$, we write $\Gamma_d(s)$ instead of $\Gamma((s, \ldots, s))$. We also define

(2) $$c_0 = (2\pi)^{r(r-1)d/4} \prod_{i=1}^{r} \frac{\Gamma((d+2)/2)}{\Gamma(id/2)}$$

and

(3) $$q = 1 + \frac{d}{2}(r-1).$$

When $r = 2$, will also denote the constant $\Gamma((d+1)/2)/\sqrt{\pi}\Gamma(d/2)$ by c_1.

In the classical case, there are well-known Euler integrals for $_1f_1$, the confluent hypergeometric function, and $_2f_1$, Gauss' hypergeometric function:

$$_1f_1(a;b;y) = \frac{\Gamma(b)}{\Gamma(a)\Gamma(b-a)} \int_0^1 e^{xy} x^{a-1}(1-x)^{b-a-1}dx$$

for $\Re a > 0$, $\Re(b-a) > 0$, and

$$_2f_1(a,b;c;y) = \frac{\Gamma(c)}{\Gamma(a)\Gamma(c-a)} \int_0^1 (1-xy)^{-b} x^{a-1}(1-x)^{c-a-1}dx$$

for $\Re a > 0$, $\Re(c-a) > 0$.

Proposition 3.4 in [Y2] gives the following generalizations of the classical Euler integrals: Let

$$dV(X,d,r) = \prod_{1 \le i < j \le r} |x_i - x_j|^d \, dx_1 \ldots dx_r,$$

then

$$(4) \qquad {}_1F_1^{(d)}(a;b;y_1,\ldots,y_r) = c_0 \frac{\Gamma_d(b)}{\Gamma_d(a)\Gamma_d(b-a)} \int_0^1 \cdots \int_0^1 {}_0\mathcal{F}_0^{(d)}(X|Y)$$

$$\cdot \prod_{i=1}^r x_i^{a-q}(1-x_i)^{b-a-q} \, dV(X,d,r)$$

if $\Re a > d(r-1)/2$, $\Re(b-a) > d(r-1)/2$, and

$$(5)$$

$$ {}_2F_1^{(d)}(a,b;c;y_1,\ldots,y_r) = c_0 \frac{\Gamma_d(c)}{\Gamma_d(a)\Gamma_d(c-a)} \int_0^1 \cdots \int_0^1 {}_1\mathcal{F}_0^{(d)}(b;X|Y)$$

$$\cdot \prod_{i=1}^r x_i^{a-q}(1-x_i)^{c-a-q} \, dV(X,d,r)$$

if $\Re a > d(r-1)/2$, $\Re(c-a) > d(r-1)/2$.

The proofs of (4) and (5) are based on Kadell's extension [KA] of Selberg's integral.

Now when $r = 2$, we express ${}_0\mathcal{F}_0^{(d)}(X|Y)$ and ${}_1\mathcal{F}_0^{(d)}(a;X|Y)$ in terms of the classical hypergeometric functions ${}_1f_1$ and ${}_2f_1$. First, an easy calculation yields

LEMMA 3.1. *If $0 < x_1, x_2 < 1, 0 < y_1, y_2 < 1$, then*

$$\frac{1}{4}|\gamma_s(X,Y) + l_s(X,Y)| < 1, \qquad \frac{1}{4}|\gamma_s(X,Y) - l_s(X,Y)| < 1.$$

It is known from [M2, Y1, Y2] that

$$(6) \qquad\qquad {}_0F_0^{(d)}(x_1,\ldots,x_r) = e^{x_1+\cdots+x_r}$$

and for $|x_i| < 1$, $i = 1,\ldots,r$,

$$(7) \qquad\qquad {}_1F_0^{(d)}(a;x_1,\ldots,x_r) = \prod_{i=1}^r (1-x_i)^{-a}.$$

In fact, (6) follows from immediately from the definition of ${}_0F_0^{(d)}$ in (1.4). The proof of (7) in [Y2] uses the result that the ${}_2F_1^{(d)}$ is the unique solution of a system of partial differential equations. Also see [M2] for an elementary proof of (7).

PROPOSITION 3.2. *For $X, Y \in \mathbb{R}^2$,*

$$(8) \qquad {}_0\mathcal{F}_0^{(d)}(X|Y) = e^{x_1y_1 + x_2y_2}\, {}_1f_1(d/2;d;-(x_1-x_2)(y_1-y_2)).$$

PROOF. Denote $c_1(1-s^2)^{d/2-1}ds$, $s \in (-1,1)$, by $dm(s)$. Applying Proposition 2.5 and (6), we have

$$
{}_0\mathcal{F}_0^{(d)}(X|Y)
$$

$$
= \sum_\kappa \frac{C_\kappa^{(d)}(X)C_\kappa^{(d)}(Y)}{k!\,C_\kappa^{(d)}(I)}
$$

$$
= \sum_\kappa \frac{1}{k!} \int_{-1}^1 C_\kappa^{(d)}(\frac{1}{4}[\gamma_s(X,Y)+l_s(X,Y)], \frac{1}{4}[\gamma_s(X,Y)-l_s(X,Y)])\, dm(s)
$$

$$
= \int_{-1}^1 e^{\frac{1}{4}[\gamma_s(X,Y)+l_s(X,Y)]+\frac{1}{4}[\gamma_s(X,Y)-l_s(X,Y)]}\, dm(s)
$$

$$
= \int_{-1}^1 e^{\frac{1}{2}\gamma_s(X,Y)}\, dm(s).
$$

Let $s = 1 - 2t$, then this last expression becomes

$$
2^{d-1}c_1 e^{x_1y_1+x_2y_2} \int_0^1 e^{-(x_1-x_2)(y_1-y_2)t}\, t^{d/2-1}(1-t)^{d/2-1}dt
$$

$$
= e^{x_1y_1+x_2y_2}\,{}_1f_1(d/2; d; -(x_1-x_2)(y_1-y_2)),
$$

where the last equality follows from the Euler integral representation for the classical confluent hypergeometric function, ${}_1f_1$, and the duplication formula for the gamma function, $\Gamma(d) = 2^{d/2-1}\pi^{-1/2}\Gamma((d+1)/2)\Gamma(d/2)$. $\square$

The following two lemmas follow from direct computations.

LEMMA 3.3. *For all $X, Y \in \mathbb{R}^2$,*

$$
[1 - \frac{1}{4}(\gamma_s(X,Y)+l_s(X,Y))][1 - \frac{1}{4}(\gamma_s(X,Y)-l_s(X,Y))/4]
$$

$$
= 1 - \frac{1}{2}(x_1+x_2)(y_1+y_2) + x_1x_2y_1y_2 - \frac{1}{2}(x_1-x_2)(y_1-y_2)s.
$$

LEMMA 3.4. *If $0 \le x_1, x_2 \le 1$, $-\infty < y_1, y_2 < 1$, then*

$$
-\frac{(x_1-x_2)(y_1-y_2)}{(1-x_1x_2)(1-y_1y_2)} < 1.
$$

PROPOSITION 3.5. *For $b \in \mathbb{C}$,*

$$
{}_1\mathcal{F}_0^{(d)}(b; X|Y) = \prod_{i=1}^2 (1-x_iy_i)^{-b}\,{}_2f_1(d/2, b; d; -\frac{(x_1-x_2)(y_1-y_2)}{\prod_{i=1}^2(1-x_iy_i)}).
$$

PROOF. By Proposition 2.5, Lemma 3.3 and (7),

$$_1\mathcal{F}_0^{(d)}(b;X|Y)$$

$$=\sum_\kappa (b)_\kappa \frac{C_\kappa^{(d)}(X)C_\kappa^{(d)}(Y)}{k!\,C_\kappa^{(d)}(I)}$$

$$=\sum_\kappa \frac{(b)_\kappa}{k!}C_\kappa^{(d)}(\frac{1}{4}(\gamma_s(X,Y)+l_s(X,Y)),\frac{1}{4}(\gamma_s(X,Y)-l_s(X,Y))dm(s)$$

$$=\int_{-1}^1 {}_1F_0^{(d)}(b;\frac{1}{4}(\gamma_s(X,Y)+l_s(X,Y)),\frac{1}{4}(\gamma_s(X,Y)-l_s(X,Y))dm(s)$$

$$=\int_{-1}^1 [1-\frac{1}{4}(\gamma_s(X,Y)+l_s(X,Y))]^{-b}[1-\frac{1}{4}(\gamma_s(X,Y)-l_s(X,Y))]^{-b}dm(s)$$

$$=\int_{-1}^1 [1-\frac{1}{2}(x_1+x_2)(y_1+y_2)+x_1x_2y_1y_2-\frac{1}{2}(x_1-x_2)(y_1-y_2)s]^{-b}dm(s).$$

The proof is completed by substituting $s=1-2t$ in this last integral, and applying the Euler integral formula for the classical Gauss hypergeometric function, $_2f_1$. □

Using similar arguments, based on (4), (5), and Propositions 3.2 and 3.5, we obtain the following result.

PROPOSITION 3.6. *If $\Re a > d/2$, $\Re(b-a) > d/2$, then*

$$(10)\qquad _1F_1^{(d)}(a;c;y_1,y_2)=c_0\frac{\Gamma_d(c)}{\Gamma_d(a)\Gamma_d(c-a)}\int_0^1\int_0^1 e^{x_1y_1+x_2y_2}$$

$$\cdot\prod_{i=1}^2 x_i^{a-d/2-1)}(1-x_i)^{c-a-d/2-1}\,|x_1-x_2|^d$$

$$\cdot {}_1f_1(d/2;d;-(x_1-x_2)(y_1-y_2))\,dx_1dx_2.$$

If $\Re a > d/2$, $\Re(c-a) > d/2$, then

(11)
$$_2F_1^{(d)}(a,b;c;y_1,y_2)$$

$$=c_0\frac{\Gamma_d(c)}{\Gamma_d(a)\Gamma_d(c-a)}\int_0^1\int_0^1\prod_{i=1}^2(1-x_iy_i)^{-b}x_i^{a-d/2-1}(1-x_i)^{c-a-d/2-1}$$

$$\cdot|x_1-x_2|^d {}_2f_1(d/2,b;d;-\frac{(x_1-x_2)(y_1-y_2)}{\prod_{i=1}^2(1-x_iy_i)})\,dx_1dx_2.$$

4. The generalized Laplace transform

Suppose that $f:\mathbb{R}_+^2\to\mathbb{C}$ is a Lebesgue measurable function. The generalized Laplace transform of f is defined by

$$(1)\;\;\mathcal{L}(f)(y_1,y_2)=c_0\int_0^\infty\int_0^\infty {}_0\mathcal{F}_0^{(d)}(-x_1,-x_2|y_1,y_2)f(x_1,x_2)|x_1-x_2|^ddx_1dx_2,$$

provided that the integral (1) is absolutely convergent. By Proposition 3.2, we also have

$$\mathcal{L}(f)(y_1, y_2) = c_0 \int_0^\infty \int_0^\infty e^{-(x_1 y_1 + x_2 y_2)} |x_1 - x_2|^d$$
$$\cdot {}_1 f_1(d/2; d; (x_1 - x_2)(y_1 - y_2)) f(x_1, x_2) \, dx_1 dx_2.$$

Let S_L denote the space of all Lebesgue measurable functions defined on $\mathbb{R}_+^2$ such that $f(x_1, x_2) = f(x_2, x_1)$ and $\mathcal{L}(|f|)(y_1, y_2) < \infty$. As in the classical case, we have

PROPOSITION 4.1. *If $f \in S_L$, and $\mathcal{L}(f) \equiv 0$, then $f \equiv 0$, a.e.*

To prove the proposition, we need some lemmas. We continue to denote $c_1(1 - s^2)^{d/2-1} ds$ by $dm(s)$, $s \in (-1, 1)$. As before, we write X for (x_1, x_2), and we will also write $I + X$ for $(1 + x_1, 1 + x_2)$.

LEMMA 4.2. *For $X, Y \in \mathbb{R}^2$,*

$$(2) \qquad {}_0\mathcal{F}_0^{(d)}(I + X|Y) = e^{y_1 + y_2} {}_0\mathcal{F}_0^{(d)}(X|Y).$$

PROOF. This follows from the first identity of Proposition 3.2. □

In analogy with Muirhead [Mu], one defines generalized binomial coefficients, $\binom{\kappa}{\sigma}$, by

$$\frac{C_\kappa^{(d)}(I + Y)}{C_\kappa^{(d)}(I)} = \sum_{s=0}^k \sum_{|\sigma|=s} \binom{\kappa}{\sigma} \frac{C_\sigma^{(d)}(Y)}{C_\sigma^{(d)}(I)}, \qquad k = |\kappa|.$$

LEMMA 4.3. *For any partition κ,*

$$(4) \qquad C_\kappa^{(d)}(Y) e^{y_1 + y_2} = k! \sum_{s=k}^\infty \sum_{|\sigma|=s} \binom{\kappa}{\sigma} \frac{C_\sigma^{(d)}(Y)}{s!}.$$

PROOF. By Lemma 4.2, we have

$$\sum_\kappa \left(e^{y_1 + y_2} C_\kappa^{(d)}(Y) \right) \frac{C_\kappa^{(d)}(X)}{k! \, C_\kappa^{(d)}(I)} = e^{y_1 + y_2} {}_0\mathcal{F}_0^{(d)}(X|Y)$$

$$= {}_0\mathcal{F}_0^{(d)}(I + X|Y)$$

$$= \sum_\sigma \frac{C_\sigma^{(d)}(I + X) C_\sigma^{(d)}(Y)}{|\sigma|! \, C_\sigma^{(d)}(I)}$$

$$= \sum_\sigma \sum_{k=0}^{|\sigma|} \sum_{|\kappa|=k} \binom{\kappa}{\sigma} \frac{C_\kappa^{(d)}(X) C_\sigma^{(d)}(Y)}{|\sigma|! \, C_\kappa^{(d)}(I)}$$

$$= \sum_\kappa \left[k! \sum_{s=k}^\infty \sum_{|\sigma|=s} \binom{\kappa}{\sigma} \frac{C_\sigma^{(d)}(Y)}{s!} \right] \frac{C_\kappa^{(d)}(X)}{k! \, C_\kappa^{(d)}(I)}.$$

Equating the coefficients of $C_\kappa^{(d)}(X)$, we obtain the result. □

Let $\mathcal{P}$ be the vector space consisting of all symmetric polynomials in y_1, y_2, and let $\mathcal{P}_s = \{P \in \mathcal{P} | P \text{ is of degree } s\}$. Denote $(\partial/\partial y_1, \partial/\partial y_2)$ by $\partial/\partial Y$, and set

$$\Phi_\sigma^d(y_1, \dots, y_r) = \frac{C_\sigma^{(d)}(Y)}{C_\sigma^{(d)}(I)}.$$

LEMMA 4.4. *For each partition σ, there exists a polynomial $P_\sigma \in \mathcal{P}_{|\sigma|}$ such that*

$$P_\sigma\left(\frac{\partial}{\partial Y}\right)\Phi_\kappa^d(Y)\big|_{Y=0} = \delta_{\kappa\sigma}.$$

PROOF. For $P, Q \in \mathcal{P}$, $< P, Q >= P(\partial/\partial Y)Q(Y)\,|_{Y=0}$ is an inner product on $\mathcal{P}$. Let $s = |\sigma|$ and $\mathcal{P}_s^-$ be the subspace of $\mathcal{P}_s$ spanned by Φ_κ^d, $\kappa \neq \sigma$, $|\kappa| = s$. Then $\dim \mathcal{P}_s^- = \dim \mathcal{P}_s - 1$ and $\Phi_\sigma^d \notin \mathcal{P}_s^-$. So there exists a $P_\sigma \perp \mathcal{P}_s^-$ under $< \cdot, \cdot >$ and $< P_\sigma, \Phi_\sigma^d >= 1$. Clearly, P_σ is the polynomial we are seeking. $\square$

LEMMA 4.5. *Let P_σ be the polynomial given by Lemma 4.4. Then*

$$P_\sigma\left(\frac{\partial}{\partial Y}\right)\Phi_\kappa^d(Y)\big|_{Y=I} = \binom{\kappa}{\sigma}$$

for all κ with $|\kappa| \geq |\sigma|$.

PROOF.

$$P_\sigma\left(\frac{\partial}{\partial Y}\right)\Phi_\kappa^d(Y)\big|_{Y=I} = P_\sigma\left(\frac{\partial}{\partial X}\right)\Phi_\kappa^d(I+X)\big|_{X=0}$$

$$= P_\sigma\left(\frac{\partial}{\partial X}\right)\left[\sum_{s=0}^k \sum_{|\alpha|=s} \binom{\kappa}{\alpha}\Phi_\alpha^d(X)\right]$$

$$= \sum_{s=0}^k \sum_{|\alpha|=s} P_\sigma\left(\frac{\partial}{\partial X}\right)\Phi_\alpha^d(X)\big|_{X=0}$$

$$= \binom{\kappa}{\sigma}.$$

$\square$

LEMMA 4.6. *If P_σ is as before, then*

(6) $$P_\sigma\left(\frac{\partial}{\partial Y}\right)_0\mathcal{F}_0^d(-X|Y)\big|_{Y=I} = (-1)^{|\sigma|}\frac{C_\sigma^{(d)}(X)}{|\sigma|!}e^{-(x_1+x_2)}$$

PROOF. By Lemma 4.3 and Lemma 4.5,

$$P_\sigma\left(\frac{\partial}{\partial Y}\right)_0\mathcal{F}_0^d(-X|Y)\big|_{Y=I} = \sum_{|\kappa|\geq|\sigma|} \binom{\kappa}{\sigma}\frac{C_\kappa^{(d)}(-X)}{k!} = (-1)^{|\sigma|}\frac{C_\sigma^{(d)}(X)}{|\sigma|!}e^{-(x_1+x_2)}.$$

$\square$

LEMMA 4.7. *If* $f(x_1, x_2) = f(x_2, x_1)$, *and for all partitions* κ

$$\int_0^\infty \int_0^\infty e^{-(x_1+x_2)} C_\kappa^{(d)}(x_1, x_2) |x_1 - x_2|^d f(x_1, x_2) dx_1 dx_2 = 0$$

then $f(x_1, x_2) \equiv 0$ *a.e.*

PROOF. Let

$$F(z_1, z_2) = \int_0^\infty \int_0^\infty e^{-(x_1 z_1 + x_2 z_2)} f(x_1, x_2) |x_1 - x_2|^d dx_1 dx_2$$

then, for $\Re z_1 > 0$, $\Re z_2 > 0$, $F(z_1, z_2)$ is a symmetric analytic function of z_1, z_2. Moreover for all κ,

$$C_\kappa^{(d)}\left(\frac{\partial}{\partial z_1}, \frac{\partial}{\partial z_2}\right) F(z_1, z_2)\Big|_{z_1 = z_2 = 1}$$

$$= \int_0^\infty \int_0^\infty e^{-(x_1+x_2)} C_\kappa^{(d)}(x_1, x_2) |x_1 - x_2|^d f(x_1, x_2) dx_1 dx_2 = 0.$$

So $F(z_1, z_2) \equiv 0$. Since the classical Laplace transform is one-to-one, we see that $f(x_1, x_2)|x_1 - x_2|^d \equiv 0$. Therefore, $f(x_1, x_2) \equiv 0$, a.e. $\square$

PROOF OF PROPOSITION 4.1. For each partition σ, let P_σ be the polynomial given by Lemma 4.4. Then

$$0 = P_\sigma\left(\frac{\partial}{\partial Y}\right)\mathcal{L}(f)(y_1, y_2)\Big|_{y_1 = y_2 = 1}$$

$$= c_0 \int_0^\infty \int_0^\infty P_\sigma\left(\frac{\partial}{\partial Y}\right){}_0\mathcal{F}_0^{(d)}(-X|Y) f(x_1, x_2) |x_1 - x_2|^d dx_1 dx_2$$

$$= c_0 \int_0^\infty \int_0^\infty \frac{(-1)^{|\sigma|}}{|\sigma|!} e^{-(x_1+x_2)} C_\sigma^{(d)}(x_1, x_2) f(x_1, x_2) |x_1 - x_2|^d dx_1 dx_2.$$

Now the proposition follows from Lemma 4.7. $\square$

As for the Laplace transform of $C_\kappa^{(d)}(x_1, x_2)$, the following result is proved in [Y3].

PROPOSITION 4.8. *For all* $d > 0$, $y_1, y_2 > 0$, $\Re a > d/2$,

(7) $\mathcal{L}((x_1 x_2)^{a-q} C_\kappa^{(d)}(x_1, x_2))(y_1, y_2) = \Gamma_d(a + \kappa) C_\kappa^{(d)}(y_1^{-1}, y_2^{-1})(y_1 y_2)^{-a}$

where $q = 1 + d/2$.

The proof of Proposition 4.8 in [Y3] is based on the facts that the result is known for all integers d by the work of Faraut and Koranyi [FK], and on a theorem of Carlson about analytic continuation.

The following proposition is proved by expanding the ${}_p\mathcal{F}_q^{(d)}$ functions in series of $C_\kappa^{(d)}$, integrating term by term, and applying Proposition 4.8.

PROPOSITION 4.9. *If $y_1, y_2 > 0$, then,*

$$\mathcal{L}((x_1 x_2)^{a-q}{}_p\mathcal{F}_q^{(d)}(a_1, \ldots, a_p; b_1, , b_q; X|Z))(y_1, y_2)$$

$$= \Gamma_d(a)(y_1 y_2)^{-a}{}_{p+1}\mathcal{F}_q^{(d)}(a_1, \ldots, a_p, a; b_1, \ldots, b_q; y_1^{-1}, y_2^{-1}|Z)$$

for $p < q$, $\Re a > d/2$; or $p = q$, $Re(a) > d/2$, $y_1 > 1$, $y_2 > 1$.

To complete this section, we note that Lemma 4.2 gives immediately the following *shifting property* of $\mathcal{L}$.

PROPOSITION 4.10. *For any $f \in S_L$,*

$$\mathcal{L}(e^{-(x_1+x_2)}f(x_1, x_2))(z_1, z_2) = \mathcal{L}(f(x_1, x_2))(z_1 + 1, z_2 + 1).$$

5. Generalized Laguerre polynomials and Hankel transformations

The results of this section are motivated by Herz [H], Dib [D] and Muirhead [Mu].

The classical Laguerre polynomials are given by

$$L_k^\gamma(x) = (\gamma + 1)_k \sum_{s=0}^{k} \binom{k}{s} \frac{(-x)^s}{(\gamma + 1)_s}$$

for $\gamma > -1$. It is well-known that they have the following properties:

(i) (Generating function)

(1) $$(1 - z)^{-\gamma - 1} e^{xz/(z-1)} = \sum_{k=0}^{\infty} \frac{L_k^\gamma(x) z^k}{k!}, \qquad |z| < 1.$$

(ii) (Orthogonality property)

(2) $$\int_0^\infty e^{-x} x^\gamma L_k^\gamma(x) L_j^\gamma(x) dx = \delta_{jk} k! \Gamma(\gamma + 1 + k).$$

(iii) (Integral representation)

(3) $$e^{-x} L_k^\gamma(x) = \frac{1}{\Gamma(\gamma + 1)} \int_0^\infty e^{-y} y^{\gamma+k} {}_0f_1(\gamma + 1; -xy) dy.$$

In the first part of this section, we shall generalize these results to the case of two variables.

We recall that when $r = 2$, $q = d/2 + 1$.

DEFINITION. The generalized Laguerre polynomial $L_\kappa^\gamma(x_1, x_2; d)$ corresponding to the parameter d and the partition κ of k is defined, for $\gamma > -1$, by

$$L_\kappa^\gamma(x_1, x_2; d) = (\gamma + q)_\kappa C_\kappa^{(d)}(1, 1) \sum_{s=0}^{k} \sum_{\sigma} \binom{\kappa}{\sigma} \frac{C_\sigma^{(d)}(-x_1, -x_2)}{(\gamma + q)_\sigma C_\sigma^{(d)}(1, 1)}.$$

In the following, we simply write $L_\kappa^\gamma(X)$ for $L_\kappa^\gamma(x_1, x_2; d)$; $C_\kappa(X)$ for $C_\kappa^{(d)}(x_1, x_2)$; X^{-1} for $(x_1^{-1}, x_2^{-1}$; and $X(I - X)^{-1}$ for $(x_1/(1 - x_1), x_2/(1 - x_2))$.

PROPOSITION 5.1. *If* $y_1, y_2 > 0$, *then*

(5) $\quad c_0 \displaystyle\int_0^\infty \int_0^\infty {}_0\mathcal{F}_0^{(d)}(-X|Y)(x_1 x_2)^\gamma L_\kappa^\gamma(X)|x_1 - x_2|^d \, dx_1 dx_2$

$$= \Gamma_d(\gamma + q + \kappa)(y_1 y_2)^{-(\gamma + q)} C_\kappa(I - Y^{-1}).$$

PROOF. By (4) and Proposition 4.8, the left-hand side of (5) equals

$$(\gamma + q)_\kappa C_\kappa(1,1) \sum_{s=0}^k \sum_\sigma \binom{\kappa}{\sigma} \frac{(-1)^s}{(\sigma + q)_\sigma C_\sigma(1,1)}$$

$$\cdot c_0 \int_0^\infty \int_0^\infty {}_0\mathcal{F}_0^{(d)}(-X|Y)(x_1 x_2)^\gamma C_\sigma(X) |x_1 - x_2|^d \, dx_1' dx_2$$

$$= (\gamma + q)_\kappa C_\kappa(1,1) \sum_{s=0}^k \sum_\sigma \binom{\kappa}{\sigma} \frac{C_\sigma(-Y^{-1})}{C_\sigma(1,1)} \Gamma_d(\gamma + q)(y_1 y_2)^{-(\gamma + q)}$$

$$= \Gamma_d(\gamma + q + \kappa)(y_1 y_2)^{-(\gamma + q)} C_\kappa(1 - 1/y_1, 1 - 1/y_2),$$

and this completes the proof. $\square$

The formula (1) has the following generalization.

PROPOSITION 5.2. *If* $x_1, x_2 > 0$, *then*

(6) $\quad \displaystyle\prod_{i=1}^2 (1 - z_i)^{-(\gamma + q)} \, {}_0\mathcal{F}_0^{(d)}(-X; Z/(I - Z)) = \sum_{k=0}^\infty \sum_\kappa \frac{L_\kappa^\gamma(X) C_\kappa(Z)}{k! \, C_\kappa(1,1)}$

for $|z_i| < 1$, $i = 1, 2$.

PROOF. The left-hand side of (6) is an analytic function of x_1, x_2, z_1, z_2 in the domain $D = \{(x_1, x_2, z_1, z_2)| \text{ all } x_1, x_2, |z_1| < 1, |z_2| < 1\}$, and may be represented by an absolutely convergent series in D. Applying the explicit formula in Proposition 3.2 for ${}_0\mathcal{F}_0^{(d)}$, we find that the left-hand side of (6) can be written in the form

(7) $\qquad\qquad \displaystyle\sum_{k=0}^\infty \sum_\kappa \frac{\tilde{L}_\kappa^\gamma(X)}{k!} \frac{C_\kappa(Z)}{C_\kappa(1,1)}$

with $\tilde{L}_\kappa^\gamma(x_1, x_2) = \tilde{L}_\kappa^\gamma(x_2, x_1)$. Now it suffices to show that

(8) $\qquad \mathcal{L}((x_1 x_2)^\gamma \tilde{L}_\kappa^\gamma(x_1, x_2))(y_1, y_2) = \mathcal{L}((x_1 x_2)^\gamma L_\kappa^\gamma(x_1, x_2))(y_1, y_2).$

We observe that for $|z_i| < 1, i = 1, 2,$

(9) $\quad \left| \displaystyle\sum_{k=0}^N \sum_\kappa \frac{\tilde{L}_\kappa^\gamma(X)}{k!} \frac{C_\kappa(Z)}{C_\kappa(1,1)} \right| \leq \prod_{i=1}^2 (1 - |z_i|)^{-(\gamma + q)} {}_0\mathcal{F}_0^{(d)}(X|\frac{|z_1|}{1 - |z_1|}, \frac{|z_2|}{1 - |z_2|}).$

For any $y_1, y_2 > 0$, there exists a $\delta > 0$ such that if $|z_i| < \delta$, $i = 1, 2$, then

(10) $\qquad\qquad \mathcal{L}_X({}_0\mathcal{F}_0^{(d)}(X|\frac{|z_1|}{1 - |z_1|}, \frac{|z_2|}{1 - |z_2|}))(y_1, y_2) < \infty.$

The formulas (9) and (10) imply that we can integrate term-by-term in (7); hence if $|z_i| < \delta, i = 1, 2$, then

$$\sum_{k=0}^{\infty} \sum_{\kappa} \frac{\mathcal{L}(\tilde{L}_{\kappa}^{\gamma}(X)(x_1 x_2)^{\gamma})(y_1, y_2)}{k!} \frac{C_{\kappa}(Z)}{C_{\kappa}(1,1)}$$

$$= \mathcal{L}_X \left(\prod_{i=1}^{2} (1 - z_i)^{-(\gamma+q)} {}_0\mathcal{F}_0^{(d)}(-X|Z/(I - Z)) \right)(Y)$$

$$= \prod_{i=1}^{2} (1 - z_i)^{-(\gamma+q)} \Gamma_d(\gamma + q)(y_1 y_2)^{-(\gamma+q)} {}_1\mathcal{F}_0^{(d)}(\gamma + q; Y^{-1}|Z/(I - Z)).$$

Since $(1 - z_i)(1 - z_i/y_i(z_i - 1)) = 1 - (1 - 1/y_i)z_i$, it follows (cf. the proof of Proposition 3.5) that

$$\prod_{i=1}^{2} (1 - z_i)^{-(\gamma+q)}(y_1 y_2)^{-(\gamma+q)} {}_1\mathcal{F}_0^{(d)}(\gamma + q; Y^{-1}| - Z/(I - Z))$$

$$= {}_1\mathcal{F}_0^{(d)}(\gamma + q; I - Y^{-1}|Z).$$

Now using the expansion of ${}_1\mathcal{F}_0^{(d)}(\gamma + q; I - Y^{-1}|Z)$, we obtain

$$\mathcal{L}(\tilde{L}_{\kappa}^{\gamma}(X)(x_1 x_2)^{\gamma})(Y) = \Gamma_d(\gamma + q + \kappa)(y_1 y_2)^{-(\gamma+q)} C_{\kappa}(I - Y^{-1})$$
$$= \mathcal{L}(L_{\kappa}^{\gamma}(X)(x_1 x_2)^{\gamma})(Y),$$

where the last equality follows from Proposition 5.1. $\square$

Next, we have the following generalization of the classical orthogonality relation (2).

PROPOSITION 5.3. *For any partitions κ, σ,*

(11) $$c_0 \int_0^{\infty} \int_0^{\infty} e^{-(x_1 + x_2)} (x_1 x_2)^{\gamma} L_{\kappa}^{\gamma}(X) L_{\sigma}^{\gamma}(X) |x_1 - x_2|^d \, dx_1 dx_2$$

$$= \delta_{\kappa\sigma} k! C_{\kappa}(1,1) \Gamma_d(\gamma + q + \kappa).$$

PROOF. By (4.3),

(12) $${}_0\mathcal{F}_0^{(d)}(-X|Z/(I - Z)) \, e^{-(x_1 + x_2)} = {}_0\mathcal{F}_0^{(d)}(-X|(I - Z)^{-1}).$$

Hence, by (6) and (12), we have

$$\sum_{k=0}^{\infty}\sum_{\kappa}C_{\kappa}(Z)\int_0^{\infty}\int_0^{\infty}\frac{L_{\sigma}^{\gamma}(X)L_{\kappa}^{\gamma}(X)}{k!C_{\kappa}(1,1)}(x_1 x_2)^{\gamma}e^{-(x_1+x_2)}|x_1-x_2|^d \, dx_1 dx_2$$

$$=\prod_{i=1}^{2}(1-z_i)^{-(\gamma+q)}c_0\int_0^{\infty}\int_0^{\infty}{}_0\mathcal{F}_0^{(d)}(-X|Z/(I-Z))$$
$$\cdot L_{\sigma}^{\gamma}(X)(x_1 x_2)^{\gamma}e^{-(x_1+x_2)}|x_1-x_2|^d \, dx_1 dx_2$$

$$=\prod_{i=1}^{2}(1-z_i)^{-(\gamma+q)}c_0\int_0^{\infty}\int_0^{\infty}{}_0\mathcal{F}_0^{(d)}(-X|(I-Z)^{-1})$$
$$\cdot L_{\sigma}^{\gamma}(X)(x_1 x_2)^{\gamma}|x_1-x_2|^d \, dx_1 dx_2.$$

By Proposition 5.1, this is equal to

$$\prod_{i=1}^{2}(1-z_i)^{-(\gamma+q)}\Gamma_d(\gamma+q+\kappa)\prod_{i=1}^{2}(1-z_i)^{(\gamma+q)}C_{\kappa}(Z)=\Gamma_d(\gamma+q+\kappa)C_{\kappa}(Z),$$

and the proof is complete. $\square$

The following result gives an integral representation for L_{κ}^{γ}.

PROPOSITION 5.4.

(13) $$e^{-(x_1+x_2)}L_{\kappa}^{\gamma}(X)=\frac{c_0}{\Gamma_d(\gamma+q)}\int_0^{\infty}\int_0^{\infty}e^{-(y_1+y_2)}(y_1 y_2)^{\gamma}C_{\kappa}(Y)$$
$$\cdot {}_0\mathcal{F}_1^{(d)}(\gamma+q;-X|Y)|y_1-y_2|^d \, dy_1 dy_2.$$

PROOF. Denote the left-hand and right-hand sides of (13) by $\phi_l(X)$ and $\phi_r(X)$, respectively. We shall prove (13) by showing that after being multiplied by $(x_1 x_2)^{\gamma}$, both $\phi_l(X)$ and $\phi_r(X)$ have the same Laplace transform.

By Propositions 4.10 and 5.1, we have

$$\mathcal{L}((x_1 x_2)^{\gamma}\phi_l(X))(z_1,z_2)=\Gamma_d(\gamma+q+\kappa)\prod_{i=1}^{2}(1+z_i)^{-(\gamma+q)}C_{\kappa}(Z/(I+Z)).$$

On the other hand, for $x_1, x_2 \geq 0$,

$$|{}_0\mathcal{F}_1^{(d)}(\gamma+q;-X|Y)| \leq {}_0\mathcal{F}_1^{(d)}(\gamma+q;X|Y).$$

It is also easy to see that for $z_1, z_2 > 0$

(14)

$$c_0^2\int_0^{\infty}\int_0^{\infty}\int_0^{\infty}\int_0^{\infty}{}_0\mathcal{F}_0^{(d)}(-X|Z)e^{-(x_1+x_2)}(x_1 x_2)^{\gamma}|x_1-x_2|^d$$
$$\cdot C_{\kappa}(Y)e^{-(y_1+y_2)}(y_1 y_2)^{\gamma}{}_0\mathcal{F}_1^{(d)}(\gamma+q;X|Y)|y_1-y_2|^d \, dx_1 dx_2 < \infty.$$

As a consequence of Proposition 4.9 with $a = \gamma + q$,

(15)
$$\mathcal{L}_X((x_1 x_2)^\gamma {}_0\mathcal{F}_1^{(d)}(\gamma + q; -X|Y))(Z)$$
$$= \Gamma_d(\gamma + q)(z_1 z_2)^{-(\gamma+q)} {}_1\mathcal{F}_1^{(d)}(\gamma + q; \gamma + q; Z^{-1}| - Y)$$
$$= \Gamma_d(\gamma + q)(z_1 z_2)^{-(\gamma+q)} {}_0\mathcal{F}_0^{(d)}(Z^{-1}| - Y).$$

Hence, by Fubini's theorem, we have

$$\mathcal{L}((x_1 x_2)^\gamma \phi_r(X))(Z)$$
$$= c_0(z_1 z_2)^{-(\gamma+q)} \int_0^\infty \int_0^\infty {}_0\mathcal{F}_0(Z^{-1}| - Y)e^{-(y_1+y_2)}$$
$$\cdot (y_1 y_2)^\gamma C_\kappa(Y)|y_1 - y_2|^d \, dy_1 dy_2$$
$$= (z_1 z_2)^{-(\gamma+q)} \mathcal{L}(e^{-(y_1+y_2)}(y_1 y_2)^\gamma C_\kappa(Y))(Z^{-1}).$$

By Propositions 4.8 and 4.10, this last expression equals

$$\Gamma_d(\gamma + q + \kappa)(z_1 z_2)^{-(\gamma+q)} C_\kappa(Z/(I + Z)) \prod_{i=1}^{2} (\frac{z_i + 1}{z_i})^{-(\gamma+q)}$$
$$= \mathcal{L}((x_1 x_2)^\gamma \phi_l(X))(Z),$$

and this completes the proof. $\square$

For $\gamma > -1$, let

$$L_\gamma^2(\mathbb{R}_+^2) := \{f : \mathbb{R}_+^2 \to \mathbb{C} | f(x_1, x_2) = f(x_2, x_1) \text{ and}$$
$$\int_0^\infty \int_0^\infty |f(x_1, x_2)|^2 (x_1 x_2)^\gamma |x_1 - x_2|^d \, dx_1 dx_2 \; < \; \infty\}.$$

For $f \in L_\gamma^2(\mathbb{R}_+^2)$, we set

$$\|f\|_\gamma = \left[c_0 \int_0^\infty \int_0^\infty |f(x_1, x_2)|^2 (x_1 x_2)^\gamma |x_1 - x_2|^d \, dx_1 dx_2 \right]^{1/2}.$$

We also introduce the notation

$$\mathcal{L}_\kappa^\gamma(x_1, x_2) := e^{-(x_1+x_2)} L_\kappa^\gamma(2x_1, 2x_2),$$

and for any partition κ,

$$d_\kappa := \frac{2(k_1 - k_2) + d}{d} \cdot \frac{B(k_1 - k_2, 1)}{B(k_1 - k_2, d)}$$

where $B(x, y)$ is the classical Beta function. Note that d_κ is well-defined (and is identically 1) if $\kappa_1 = \kappa_2$.

As a consequence of [Y2, (45)] and (5.11), we have

(16)
$$\| [\frac{2^{2\gamma+d+2}(q)_\kappa}{d_\kappa \Gamma_d(\gamma + q + \kappa)}]^{1/2} \frac{1}{k!} \mathcal{L}_\kappa^\gamma(X) \|_{L_\gamma^2} = 1.$$

Moreover, we have

PROPOSITION 5.5. *The set*

$$\{[\frac{2^{2\gamma+d+2}(q)_\kappa}{d_\kappa\Gamma_d(\gamma+q+\kappa)}]^{1/2}\frac{1}{k!}\mathcal{L}_\kappa^\gamma(X) : \kappa \text{ is a partition}\}$$

is an orthonormal basis in L_γ^2.

PROOF. Suppose $f \in L_\gamma^2$ such that for all κ

(17) $$\int_0^\infty \int_0^\infty f(x_1,x_2)\mathcal{L}_\kappa^\gamma(x_1,x_2)(x_1x_2)^\gamma|x_1-x_2|^d\,dx_1dx_2 = 0,$$

then (17) implies

$$\int_0^\infty \int_0^\infty f(x_1,x_2)C_\kappa^{(d)}(x_1,x_2)e^{-(x_1+x_2)}(x_1x_2)^\gamma|x_1-x_2|^d dx_1dx_2 = 0$$

for all κ. By Lemma 4.7, $f(x_1,x_2) \equiv 0$ a.e. Together with (16), this completes the proof. □

DEFINITION. If $f \in L_\gamma^2$ with compact support, we define the Hankel transform of f by

(18)
$$\mathcal{H}_\gamma f(y_1,y_2)$$
$$= c_0 \int_0^\infty \int_0^\infty f(x_1,x_2){}_0\mathcal{F}_1^{(d)}(\gamma+q;-X|Y)(x_1x_2)^\gamma|x_1-x_2|^d\,dx_1dx_2.$$

We shall extend $\mathcal{H}_\gamma$ to the whole space L_γ^2 (cf. [D]).
We define, for all $Z = (z_1,z_2) \in \mathbb{R}_+^2$,

$$e_Z(x_1,x_2) := {}_0\mathcal{F}_0^{(d)}(-Z|X).$$

PROPOSITION 5.6. *The closed linear space spanned by all* $e_Z(\cdot)$ *is* L_γ^2.

PROOF. Firstly, as a consequence of (3.9), for $z_1, z_2 > 0$, we have

$$\|e_Z\|_\gamma^2 \le c_0 \int_0^\infty \int_0^\infty {}_0\mathcal{F}_0^{(d)}(-Z|X)(x_1x_2)^\gamma|x_1-x_2|^d\,dx_1dx_2 < \infty.$$

Thus $e_Z \in L_\gamma^2$.
Secondly, if $f \in L_\gamma^2$ and orthogonal to all e_Z, then, by the injectivity of the generalized Laplace transformation, $f \equiv 0$, a.e. □

PROPOSITION 5.7. *We have*

(19) $$\mathcal{H}_\gamma e_Z = \Gamma_d(\gamma+q)(z_1z_2)^{-(\gamma+q)}e_{Z^{-1}},$$

and

(20) $$\|\mathcal{H}_\gamma e_Z\|_\gamma = \Gamma_d(\gamma+q)\|e_Z\|_\gamma.$$

PROOF. Firstly, since

$$|{}_0\mathcal{F}_1^{(d)}(\gamma+q;-X|Y)| \le {}_0\mathcal{F}_1^{(d)}(\gamma+q;X|Y)$$

and

$$c_0 \int_0^\infty \int_0^\infty {}_0\mathcal{F}_0(-Z|X){}_0\mathcal{F}_1^{(d)}(\gamma+q;X|Y)(x_1x_2)^\gamma |x_1-x_2|^d \, dx_1 dx_2 < \infty,$$

we see that $\mathcal{H}_\gamma e_Z$ is defined for all $z_1, z_2 > 0$. Moreover, from Proposition 4.9 it follows precisely as in the proof of Proposition 5.4 that

$$\mathcal{H}_\gamma e_Z(Y) = \mathcal{L}_X({}_0\mathcal{F}_1^{(d)}(\gamma+q;X|-Y)(Z)$$
$$= \Gamma_d(\gamma+q)(z_1 z_2)^{-(\gamma+q)} {}_0\mathcal{F}_0^{(d)}(-Z^{-1}|Y).$$

Secondly, in order to show (20), it is enough (by Proposition 5.5) to show that

(21) $$(\mathcal{H}_\gamma e_Z, \mathcal{L}_\kappa^\gamma)_\gamma^2 = [\Gamma_d(\gamma+q)]^2 (e_Z, \mathcal{L}_\kappa^\gamma)_\gamma^2$$

for all κ. On the one hand, by Propositions 4.10 and 5.1,

$$(e_Z, \mathcal{L}_\kappa^\gamma)_\gamma$$
$$= c_0 \int_0^\infty \int_0^\infty {}_0\mathcal{F}_0^{(d)}(-Z|X)e^{-(x_1+x_2)}L_\kappa^\gamma(2X)(x_1x_2)^\gamma |x_1-x_2|^d \, dx_1 dx_2$$
$$= c_0 \int_0^\infty \int_0^\infty {}_0\mathcal{F}_0^{(d)}(-(Z+I)|X)L_\kappa^\gamma(2x_1, 2x_2)(x_1x_2)^\gamma |x_1-x_2|^d \, dx_1 dx_2$$
$$= \Gamma_d(\gamma+q+\kappa). \prod_{i=1}^2 (1+z_i)^{-(\gamma+q)} C_\kappa^{(d)}\left(\frac{z_1-1}{z_1+1}, \frac{z_2-1}{z_2+1}\right).$$

On the other hand, by (19)

(22) $$(\mathcal{H}_\gamma e_Z, \mathcal{L}_\kappa^\gamma)_\gamma = \Gamma_d(\gamma+q)\Gamma_d(\gamma+q+\kappa)(z_1 z_2)^{-(\gamma+q)}$$
$$\cdot \prod_{i=1}^2 \left(\frac{1+z_i}{z_i}\right)^{-(\gamma+q)} C_\kappa^{(d)}\left(\frac{1-z_1}{1+z_1}, \frac{1-z_2}{1+z_2}\right)$$
$$= \Gamma_d(\gamma+q)(-1)^k (e_Z, \mathcal{L}_\kappa^\gamma)_\gamma.$$

We have proved (21). □

THEOREM 5.8. (Generalized Tricomi Theorem) (cf. [D],[H]) *The operator* $(1/\Gamma_d(\gamma+q))\mathcal{H}_\gamma$ *is an involutive isometry of* L_γ^2. *Moreover, if*

$$F = \mathcal{L}(f(x_1, x_2)(x_1 x_2)^\gamma)$$

and

$$G = \mathcal{L}(g(x_1, x_2)(x_1 x_2)^\gamma)$$

then $g = \mathcal{H}_\gamma f$ *iff*

$$G(z_1, z_2) = \Gamma_d(\gamma+q)(z_1 z_2)^{-(\gamma+q)} F(z_1^{-1}, z_2^{-1}).$$

PROOF. The first part of the theorem follows immediately from Proposition 5.7. Next, put $g_1 = \mathcal{H}_\gamma f$ and $G_1 = \mathcal{L}(g_1(x_1, x_2)(x_1 x_2)^\gamma)$. Since $(1/\Gamma_d(\gamma+q))H_\gamma$ is an involutive isometry then, by (19), we have

(23) $$G_1(z_1, z_2) = (e_Z, \mathcal{H}_\gamma f) = \Gamma_d(\gamma+q)(z_1 z_2)^{-(\gamma+q)} F(z_1^{-1}, z_2^{-1}).$$

If $g = \mathcal{H}_\gamma f$, then (23) implies

$$G(z_1, z_2) = \Gamma_d(\gamma + q)(z_1 z_2)^{-(\gamma+q)} F(z_1^{-1}, z_2^{-1}).$$

Conversely, the injectivity of the Laplace transform and (23) imply $g = g_1 = \mathcal{H}_\gamma f$, a.e. $\square$

Finally we have

PROPOSITION 5.9. $\mathcal{H}_\gamma(\mathcal{L}_\kappa^\gamma) = (-1)^k \Gamma_d(\gamma + q)\mathcal{L}_\kappa^\gamma.$

PROOF. Let $f = \mathcal{L}_\kappa^\gamma$ and $g = (-1)^k \Gamma_d(\gamma + q)\mathcal{L}_\kappa^\gamma$. Then (22) and (19) imply that $(e_Z, g)_\gamma = (\mathcal{H}_\gamma e_Z, f)_\gamma = \Gamma_d(\gamma + q)(z_1 z_2)^{-(\gamma+q)}(e_{Z^{-1}}, f)_\gamma$. Then the conclusion follows from Theorem 5.8. $\square$

REMARK. The generalized Laguerre polynomials have been studied independently by M. Lassalle [L], and the r-variable case of Proposition 5.5 has been obtained in [L].

REMARK 2. The results of this section on generalized Laguerre polynomials and the Hankel transform have been derived for the r variable case by I. G. Macdonald [M2] as follows:

a. Proposition 4.8 holds for the r variable case;
b. $e_Z \in L_\gamma^2$ for $Z > 0$ (=Proposition 5.6 if $r=2$);
c. $\mathcal{L}$ is injective (=Proposition 4.1 if $r=2$).

REFERENCES

[A] R. Askey, *Orthogonal Polynomials and Special Functions*, Regional Conference Series in Applied Mathematics, vol. 21, Soc. Indus. Appl. Math., Philadelphia, PA, 1975.

[C] A.G. Constantine, *Some noncentral distribution problems in multivariate analysis*, Ann. Math. Statist. **34** (1963), 1270-1285.

[D] H. Dib, *Thesis*, Université Louis Pasteur.

[F] J. Faraut, *Analyse harmonique et fonctions spéciales*, Ecole d'été d'analyse harmonique de Tunis.

[FK] J. Faraut & A. Korányi, *Fonctions hypergéométriques associées aux cônes symétriques*, C.R. Acad. Sci. Paris **307** (1988), 555-558.

[GR1] K.I. Gross & D.St.P. Richards, *Special functions of matrix argument. I: Algebraic induction, zonal polynomials, and hypergeometric functions*, Trans. Amer. Math. Soc. **301** (1987), 781-811.

[GR2] _____, *Hypergeometric functions on complex matrix space*, Bull. (N. S.) Amer. Math. Soc. **24** (1991), 349–355.

[H] C.S. Herz, *Bessel functions of matrix argument*, Ann. of Math. **61** (1955), 474-523.

[J] A.T. James, *Distributions of matrix variates and latent roots derived from normal samples*, Ann. Math. Statist. **35** (1964), 475-501.

[KA] K.W. Kadell, *The Selberg-Jack polynomials*, preprint.

[K] A. Korányi, *Hua-type integrals, hypergeometric functions and symmetric polynomials*, preprint.

[KK] J. Kaneko, *Selberg integrals and hypergeometric functions associated with Jack polynomials*, preprint, Kyushu Univ..

[KS] T.H. Koorwinder & I.G. Sprinkhuizen-Kuyper, *Generalized power series expansions for a class of orthogonal polynomials in two variables*, SIAM J. Math. Anal. **9** (1978), 457-483.

[M1] I.G. Macdonald, *Commuting differential operators and zonal spherical functions*, Algebraic Groups, Utrecht 1986, Lecture Notes in Mathematics **1271** (1987), 189-200.

[M2] _____ , *Hypergeometric functions*, unpublished manuscript.

[M3] _____ , *Symmetric functions and Hall polynomials,*, Oxford Univ. Press, Oxford, 1979.

[Mu] R.J. Muirhead, *Aspects of multivariate statistical theory,*, Wiley, New York, 1982.

[L] M. Lassalle, *Polynomes de Laguerre généralisés*, C.R. Acad. Sci. Paris Sér. I **312** (1991), 725-729.

[R] D.St.P. Richards, *Analogs and extensions of Selberg's integrals*, In: *q-Series and Combinatorics* (D. Stanton, ed.), Springer, New York, IMA Volumes in Math. and its Applications **18** (1989), 109-137.

[S] R.P. Stanley, *Some combinatorial properties of Jack symmetric functions*, Advances in Math. **77** (1989), 76-115.

[Y1] Z. Yan, *Generalized hypergeometric functions*, Série I, C.R.Acad. Paris **310** (1990), 349-354.

[Y2] _____ , *A class of generalized hypergeometric functions in several variables*, to appear, Canad. J. Math..

[Y3] _____ , *Generalized hypergeometric functions in several variables*, Thesis, Graduate School, C.U.N.Y.

DEPARTMENT OF MATHEMATICS, GRADUATE SCHOOL OF CITY UNIVERSITY OF NEW YORK, 33 W. 42 STREET, NEW YORK, NY 10036

Current address: DEPARTMENT OF MATHEMATICS, UNIVERSITY OF CALIFORNIA, BERKELEY, CA 94720

Recent Titles in This Series

(Continued from the front of this publication)

(See the AMS catalog for earlier titles)